Passive
Network Synthesis
Advances with Inerter

Passive
Network Synthesis
Advances with Inerter

Michael Z Q Chen
Nanjing University of Science and Technology, China

Kai Wang
Jiangnan University, China

Guanrong Chen
City University of Hong Kong, China

World Scientific

NEW JERSEY · LONDON · SINGAPORE · BEIJING · SHANGHAI · HONG KONG · TAIPEI · CHENNAI · TOKYO

Published by

World Scientific Publishing Co. Pte. Ltd.

5 Toh Tuck Link, Singapore 596224

USA office: 27 Warren Street, Suite 401-402, Hackensack, NJ 07601

UK office: 57 Shelton Street, Covent Garden, London WC2H 9HE

British Library Cataloguing-in-Publication Data
A catalogue record for this book is available from the British Library.

PASSIVE NETWORK SYNTHESIS
Advances with Inerter

ISBN 978-981-121-087-7

For any available supplementary material, please visit
https://www.worldscientific.com/worldscibooks/10.1142/11567#t=suppl

To the memory of Professor Rudolf E. Kalman (1930–2016)

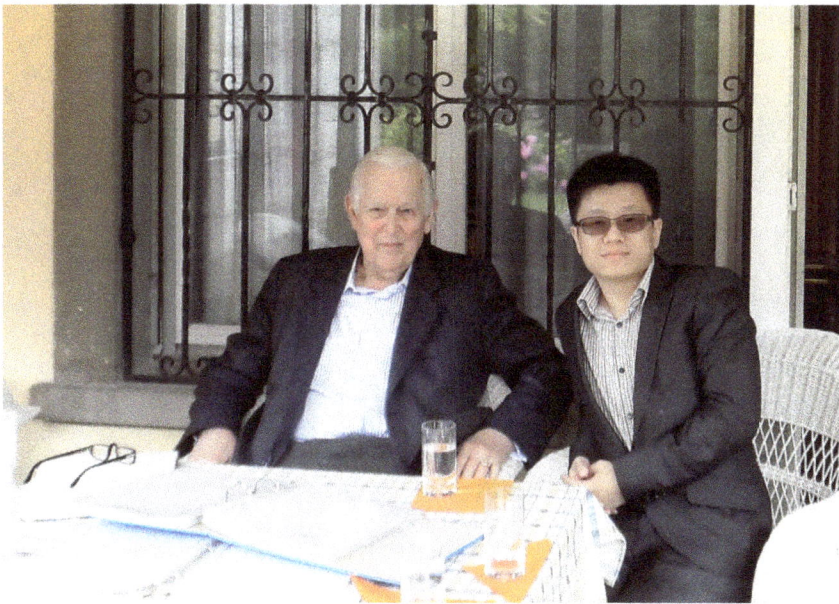

Professor Rudolf E. Kalman and Dr. Michael Z. Q. Chen
at Professor Kalman's residence, Zurich, Switzerland, on July 12, 2014

Preface

This book is concerned with the synthesis of passive electrical or mechanical networks, which is motivated by the vibration control based on a new type of mechanical elements named *inerter*. The inerter was proposed by Professor Malcolm Smith from the University of Cambridge in 2002, which is a two-terminal passive element whose force applied at the terminals are proportional to the relative acceleration. Since any n-port passive electrical network can be constructed with resistors, inductors, capacitors, and transformers, where transformers can be avoided for the one-port case, the "birth" of inerters completes the analogy between the passive mechanical networks and electrical networks under the "force-current" framework, where dampers, springs, inerters, and levers are analogous to resistors, inductors, capacitors, and transformers, respectively. As a result, the analysis and synthesis of electrical networks can be transplanted into those of mechanical networks. For instance, any positive-real impedance (or admittance) can be realized as a one-port mechanical network consisting of dampers, springs, and inerters (damper-spring-inerter network), by properly following a network synthesis procedure, such as the Bott-Duffin synthesis.

To date, many investigations have focused on applying damper-spring-inerter networks to a series of passive or semi-active vibration control systems, such as vehicle suspensions, train suspensions, building vibration systems, landing gears, wind turbines, *etc.* The passive mechanical networks are actually positive-real controllers of the vibration control systems, and system performances can certainly be enhanced by introducing inerters. For the design of inerter-based vibration control systems, the first step is to determine the transfer function of the controller according to the performance requirements, which is usually the mechanical impedance (or admittance); the second step is to realize the resulting function as a damper-spring-inerter network based on the theory of passive network synthesis. Although the realization as a one-port damper-spring-inerter network is always available, the classical synthesis methods and results always generate many redundant elements, and the synthesis problem of damper-spring-inerter networks using the least number of elements is far from being completely solved. Unlike electrical systems, the number of elements is a critical index to be considered for mechanical systems due to the limitation of space, weight, cost, *etc.*

As a consequence, the significance of investigating passive network synthesis has become appealing again, and many investigations have focused on this topic motivated by the inerter-based vibration control. Notably, Professor Rudolf E. Kalman has made an independent call for the renewed investigation on passive network synthesis. In addition, this topic can provide important impacts on electronic engineering, biometric image processing, control theory, *etc.*

In this book, some important fundamental concepts of passive network synthesis are introduced, and some recent results by the authors on this topic are presented. These results are mainly concerned with the economical realizations of low-degree functions as RLC networks (damper-spring-inerter networks), the synthesis of n-port resistive networks, and the synthesis of low-complexity mechanical networks. Many of these results can be directly applied to the optimization and design of various inerter-based vibration control systems.

In Chapter 1, the development of passive network synthesis and its application to inerter-based vibration control are introduced. In addition, an outline of this book is presented. In Chapter 2, some important fundamental results of passive network synthesis are introduced, including the properties of positive-realness, some classical synthesis procedures, and graph theory for passive networks. Some concepts and results will be utilized in the following chapters. In Chapter 3, the realization problem of biquadratic impedances as RLC networks is discussed. Since the biquadratic synthesis is a classical problem in network synthesis, electrical circuits are utilized to describe the networks. In Chapter 4, the synthesis of n-port resistive networks is discussed, including a review of the investigations on this problem, and some recent results obtained by the authors. In Chapter 5, the synthesis problems of low-complexity mechanical networks are investigated, where the number of dampers and inerters is also an important index to be considered in addition to the total number of elements. Finally, Chapter 6 presents the summary of this book.

This book is readable by graduate students and researchers in related fields. Some basic knowledge of mathematics, circuit theory, and control theory is needed to follow the content in this book.

The authors are very grateful to those who have contributed to the development of this book, and to the editorial office of the publishing company.

Michael Z. Q. Chen, Kai Wang, and Guanrong Chen

Acknowledgments

This work is supported by the National Natural Science Foundation of China under grants 61873129 and 61703184, and by the Hong Kong Research Grants Council under several GRF grants in the 1990s.

Contents

Chapter 1

Introduction

1.1 Synthesis of Passive Networks

Passive network synthesis is to physically realize a given function describing the port behavior of a passive network using only passive elements, under the assumption that the network and each element must be linear, time-invariant, and lumped. This assumption will be valid throughout this book. In contrast, network analysis is to determine the external or internal behavior of a given network. Therefore, it can be noted that network synthesis is the inverse process of network analysis.

As an important branch of system theory, passive network synthesis was widely investigated from the 1930s to the 1970s. Brune [Brune (1931)] first solved the passive realizability problem of one-port passive networks by establishing a systematic realization procedure, which is named the *Brune synthesis*. It is shown in [Brune (1931)] that the impedance (resp., admittance) of a one-port passive network is positive-real, and any positive-real impedance (resp., admittance) is realizable as a one-port passive network consisting of a finite number of resistors, inductors, capacitors, and transformers. Darlington [Darlington (1939)] proposed an alternative synthesis procedure, named the *Darlington synthesis*, through which any positive-real impedance (resp., admittance) can be realized as the cascade connection of a two-port lossless network (containing inductors, capacitors, and transformers) and one resistor. However, it should be noted that transformers are not preferred in practice. Bott and Duffin [Bott and Duffin (1949)] first showed that only resistors, inductors, and capacitors are needed to construct a one-port passive network through establishing a transformerless synthesis procedure, which is named the *Bott-Duffin synthesis*. On the other hand, it should be noted that transformers are avoided at the cost of increasing the number of redundant elements, although lately some modified approaches were further proposed [Pantell (1954); Reza (1954)]. To solve the synthesis problem of one-port RLC networks using the least number of elements, many investigations focused on the minimal realization problems of low-degree impedances (resp., admittances), especially biquadratic impedances [Ladenheim (1964); Seshu (1959); Vasiliu (1970)], which have not been completely solved so far. In addition, the synthesis problems of multi-port passive networks have also been widely investigated, which can be referred to [Anderson and Vongpanitlerd (1973); Newcomb (1966)]. Nevertheless, it should be noted that transformers cannot always be avoided in the multi-port case, even for the simple multi-port resistive networks (see Chapter 4). Notably, from the 1970s to the 1990s, the research interest in passive network synthesis declined in spite of some new developments on this topic.

After the invention of a new mechanical element called *"inerter"* in 2002 [Smith (2002)], research interest in passive network synthesis has been revived and this field has again become active and practically essential, motivated by the design of inerter-based mechanical control (see the next section for more details). Chen and Smith [Chen (2007); Chen and Smith (2009b)] first investigated the low-complexity mechanical network synthesis utilizing dampers, springs, and inerters. Then, there have been a series of new results in this field during the past decade (see [Chen *et al.* (2013a, 2015c); Hughes and Smith (2014); Jiang and Smith (2011)], for instance). Moreover, Kalman [Kalman (2010, 2014); Lin *et al.* (2011)][1] has made an independent call for the renewed investigation on this topic.

The remaining part of this section will introduce some basic concepts and results.

Definition 1.1. [Anderson and Vongpanitlerd (1973), pg. 21] Assuming that there is no energy stored at t_0, an n-port network is defined to be *passive*, if

$$\varepsilon(T) = \int_{t_0}^{T} v^T(t)i(t)dt \geq 0,$$

for any t_0, T, and port voltage vector $v(\cdot) \in \mathbb{R}^n$ and current vector $i(\cdot) \in \mathbb{R}^n$ satisfying the constraints of the network.

---

[1]Reference [Lin *et al.* (2011)] is a document summarizing the main content of Professor Rudolf E. Kalman's lecture for the Berkeley Algebraic Statistics Seminar on October 26, 2011.

Moreover, the passivity of elements can be defined based on Definition 1.1, since an element can be regarded as a specific network. It can be checked that resistors, inductors, capacitors, and transformers with constant element values are linear, time-invariant, lumped, and passive elements. An important result that can be proved by *Tellegen's Theorem* [Anderson and Vongpanitlerd (1973), pg. 22] is stated as follows.

Theorem 1.1. *[Anderson and Vongpanitlerd (1973), pg. 22] Any network consisting of a finite number of passive elements must be a passive network.*

It should be noted that the converse of Theorem 1.1 is not always true, which means that a passive network can sometimes contain active elements. However, the task of passive network synthesis requires that the passive network to be realized must only contain passive elements.

By taking the Laplace transforms, an n-port linear, time-invariant, and lumped network can be described by a transfer function matrix, whose entries are real-rational functions. For instance, the *impedance matrix* $Z(s) \in \mathbb{R}^{n \times n}(s)$ of an n-port network satisfies

$$\hat{V}(s) = Z(s)\hat{I}(s),$$

where the n-dimensional vectors $\hat{V}(s)$ and $\hat{I}(s)$ are the Laplace transforms of voltage vector $v(t)$ and current vector $i(t)$, respectively. Similarly, the *admittance matrix* $Y(s) \in \mathbb{R}^{n \times n}(s)$ of an n-port network satisfies

$$\hat{I}(s) = Y(s)\hat{V}(s).$$

Moreover, some other transfer function matrices, such as scattering matrix and hybrid matrix, can also be utilized to describe the port behavior of a linear, time-invariant, and lumped network.

Remark 1.1. Although impedance and admittance matrices are transfer function matrices that are most commonly used, the impedance or admittance matrices may not exist for some special networks. If the impedance and admittance matrices of a network simultaneously exist, then $Z(s) = Y^{-1}(s)$. Specifically, for the one-port case, these two matrices are scalars and can be called the *impedance* and *admittance* for brevity.

Definition 1.2. [Anderson and Vongpanitlerd (1973), pg. 51] A real-rational function matrix $H(s)$ is *positive-real* if $H(s)$ is analytic and $H(s) + H^*(s) \succeq 0$ for all s with $\Re(s) > 0$, that is, in the open right-half plane. Here, $H^*(s)$ denotes the complex conjugate transpose of $H(s)$, and $\succeq$ means non-negative definite.

Theorem 1.2. *[Anderson and Vongpanitlerd (1973), Sections 2.7 and 2.8] The impedance matrix $Z(s)$ (resp., admittance matrix $Y(s)$) of an n-port passive network is positive-real. Moreover, if the network is reciprocal (see [Newcomb (1966)]), then $Z(s)$ (resp., $Y(s)$) is symmetric, that is, $Z(s) = Z^T(s)$ (resp., $Y(s) = Y^T(s)$.)*

Theorem 1.3. *[Anderson and Vongpanitlerd (1973), Chapters 9 and 10] Any positive-real impedance matrix $Z(s) \in R^{n \times n}(s)$ (resp., admittance matrix $Y(s) \in R^{n \times n}(s)$) is realizable as an n-port passive network consisting of resistors, inductors, capacitors, transformers, and gyrators. Moreover, if the positive-real impedance matrix $Z(s)$ (resp., admittance matrix $Y(s)$) is symmetric, then it is realizable as an n-port passive network consisting of resistors, inductors, capacitors, and transformers.*

1.2 New Research Motivation: Inerter-Based Mechanical Control

Since 2002, the invention of a new kind of mechanical elements named *"inerter"* and its successful applications has renewed the research interest in passive network synthesis [Smith (2002); Chen *et al.* (2009)].

The inerter is a two-terminal mechanical element with its terminal dynamics satisfying $F = b(\dot{v}_1 - \dot{v}_2)$, where b is called the *inertance*, F is the force applied to its two terminals, and $\dot{v}_1$ and $\dot{v}_2$ are the accelerations of the two terminals. The inerter was first proposed and constructed by Professor Malcolm C. Smith from the University of Cambridge. The mechanical model of a rack-and-pinion inerter is shown in Fig. 1.1. Moreover, there are some other methods of constructions for inerters, such as hydraulics and screw mechanisms [Smith (2008)].

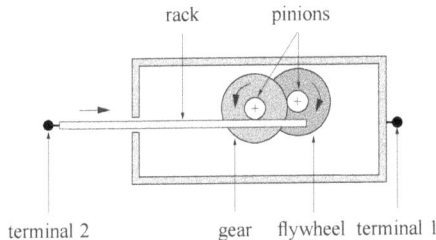

Fig. 1.1   Schematic of the mechanical model of a rack-and-pinion inerter [Smith (2002)].

Based on the conventional force-current analogy framework, which means that force is analogous to current and velocity is analogous to voltage, the passivity of a mechanical network can be defined according to Definition 1.1, that is, $\int_{t_0}^{T} v^T(t)F(t)dt \geq 0$ for any t_0, T, and external velocity vector $v(\cdot) \in \mathbb{R}^n$ and force vector $F(\cdot) \in \mathbb{R}^n$ satisfying the constraints of the network. In practice, passive mechanical networks are widely employed in many control systems, such as vehicle suspensions, train suspensions, machine vibration systems, etc. Compared with the active control approach, such a passive control method can have a lower cost and higher reliability. For instance, no power is needed and some serious practical problems such as measurement errors and actuator failures can be avoided. In order to physically realize the passive mechanical network utilizing the conventional passive elements such as springs, dampers, etc., engineers traditionally use the "trial and error" approach for design, which obviously lacks a theoretical foundation.

Based on the force-current analogy, if one can find passive mechanical elements that are analogous to the basic passive electrical elements: resistors, inductors, capacitors, and transformers (not necessarily in the one-port case), then the theory of passive electrical network synthesis can be completely transplanted into the design of passive mechanical networks. Conventionally, the damper, spring, and lever in the mechanical system can be analogous to the resistor, inductor, and transformer in the electrical system, respectively. However, for a long period of time, people used a mass to make a partial analogy to a capacitor, which is actually analogous to a grounded capacitor. A summary of the conventional incomplete analogy is presented in Table 1.1.

Recalling that the force applied to its two terminals is proportional to the relative acceleration between them, the inerter is acctaully the "missing" mechanical element that is analogous to a capacitor (see Fig. 1.2). Therefore, the "birth" of such an element completes the analogy between passive, linear, time-invariant, lumped, reciprocal mechanical networks and the electrical ones. As a result, the physical design of passive mechanical networks become much more convenient and systematic by using the theory of passive network synthesis. Specifically, based on the Bott-Duffin synthesis, any one-port mechanical network can be constructed by using at most three types of elements: dampers, springs, and inerters.

To date, the passive mechanical networks with inerters have been applied to a series of passive or semi-active vibration control systems [Chen *et al.* (2012, 2015a); Hu *et al.* (2014); Hu and Chen (2015); Papageorgiou and Smith (2006); Smith and Wang (2004); Wang *et al.* (2009, 2012)],

Table 1.1   The conventional incomplete force-current analogy between the mechanical and electrical systems, which can be referred to [Smith (2002)].

Mechanical system	Electrical system
Force	Current
Velocity	Voltage
Zero velocity point	Zero potential point
Damper	Resistor
Spring	Inductor
Mass	Grounded capacitor
Lever	Transformer
Mechanical interconnection	Electrical interconnection

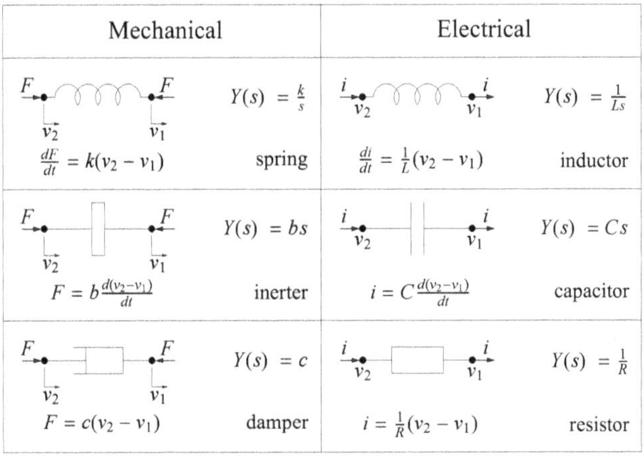

Mechanical		Electrical	
$\frac{dF}{dt} = k(v_2 - v_1)$	$Y(s) = \frac{k}{s}$ spring	$\frac{di}{dt} = \frac{1}{L}(v_2 - v_1)$	$Y(s) = \frac{1}{Ls}$ inductor
$F = b\frac{d(v_2 - v_1)}{dt}$	$Y(s) = bs$ inerter	$i = C\frac{d(v_2 - v_1)}{dt}$	$Y(s) = Cs$ capacitor
$F = c(v_2 - v_1)$	$Y(s) = c$ damper	$i = \frac{1}{R}(v_2 - v_1)$	$Y(s) = \frac{1}{R}$ resistor

Fig. 1.2   The analogy between springs, inerters, and dampers in mechanical networks and inductors, capacitors, and resistors in electrical networks [Smith (2002)].

where the mechanical networks are actually the passive control devices. The results show that introducing inerters can indeed enhance system performances. In [Hu *et al.* (2014)], a direct comparison idea is proposed to study the influence of adding one element at a specific position for vehicle suspensions, where the performance index for a complex configuration is decoupled as two parts: the part corresponding to the original configuration and the part corresponding to the added element. In [Chen *et al.* (2014a)], a fundamental property is presented that an inerter can reduce the natural frequencies of vibration systems based on a general multi-degree-of-freedom system. In [Hu and Chen (2015)], the inerter-based dynamic vibration absorber (IDVA, also known as inerter-based tuned mass damper) is proposed, and the H_∞ and H_2 performances for IDVA are evaluated. In [Hu *et al.* (2015)], inerter-based isolators are proposed, and analytic solutions for the H_∞ and H_2 performances of several inerter-based isolators are derived. In [Chen *et al.* (2015a)], the idea of decoupling the inerter-based semi-active suspensions as a passive part and a semi-active part is proposed, and the semi-active suspensions with the passive part as several given inerter-based networks are evaluated. In [Chen *et al.* (2014c)], the idea of semi-active inerter is proposed. The performance of semi-active inerter for vehicle suspensions is studied in [Chen *et al.* (2014c, 2016a)]. In [Hu *et al.* (2017a)], the physical embodiment of semi-active inerter is proposed by using a controllable-inertia flywheel. In [Hu *et al.* (2017b)], the skyhook inerter idea is proposed, and the semi-active realization of the skyhook inerter idea by using semi-active inerters is studied. In [Hu *et al.* (2018a)], a fundamental fact is revealed that mass-chain systems with inerters may have multiple natural frequencies, and a necessary and sufficient condition for natural frequency assignment problem of inerter-based mass-chain systems is derived. In [Dong *et al.* (2015)], the effect of introducing inerters to suppress the shimmy vibration of aircraft landing gear structures is investigated. In [Liu *et al.* (2015)], some nonlinearities in the landing gear model with inerters are analyzed. In [Hu and Chen (2017); Hu *et al.* (2018b)], the inerter is applied in offshore wind turbines for the first time, and its performance is evaluated by using the FAST code developed by the National Renewable Energy Laboratory. Some of the recent results on inerter-based mechanical control are presented in the book [Chen and Hu (2019)].

An illustrative example is shown in Fig. 1.3, which is the suspension control system based on a quarter-car model. In this model, m_s denotes the sprung mass, m_u denotes the unsprung mass, k_t denotes the spring stiffness of the tyre, and $K(s)$ denotes the mechanical admittance of a one-port passive network containing inerters. The control diagram of this model is shown in Fig. 1.4, where w is the external input, z is the output to be controlled, and the admittance $K(s)$ is a positive-real controller to be determined such that the control system can meet certain requirements. The complete process of designing an inerter-based vibration control system is summarized as follows.

- Given a vibration control system, determine a suitable positive-real admittance or impedance (matrix) $K(s)$ of a mechanical network, which is the system controller or part of the controller. The design process should consider the passivity of the mechanical network or some further constraints, such as the total number of elements, the structure requirement, etc.

- Using the theory of passive network synthesis, realize $K(s)$ as a passive mechanical network consisting of dampers, springs, inerters, levers (if necessary), etc.

As can be seen, the approaches and results of passive network synthesis are essential in the above design.

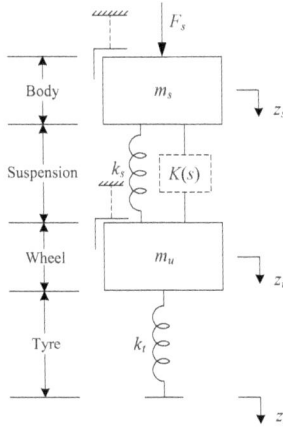

Fig. 1.3   A quarter-car vehicle suspension system model, where m_s denotes the sprung mass, m_u denotes the unsprung mass, k_t denotes the spring stiffness of the tyre, and $K(s)$ denotes the mechanical admittance of a passive network containing dampers, springs, inerters, and possibly levers.

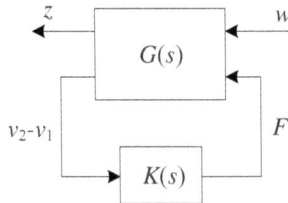

Fig. 1.4   Control diagram for the quarter-car vehicle suspension system model, where the mechanical admittance $K(s)$ is a positive-real controller, w is the external input, and z is the output to be controlled.

Since it is difficult to implement levers with unrestricted ratios in practice, no lever (transformer) is preferred. Moreover, low complexity is often required for mechanical systems due to the limitation of cost, space and weight. When no transformer (lever) is contained, the synthesis of one-port RLC (damper-spring-inerter) networks following a classical synthesis procedure yields a large number of redundant elements in many cases, and its minimal realizability problem is unsolved; the synthesis of multi-port RLC (damper-spring-inerter) networks is not solved, even for the multi-port resistive networks. As a result, the research interest in passive network synthesis has been renewed, and a series of new results have appeared since the invention of inerters [Chen and Smith (2009b); Chen *et al.* (2013a); Hughes and Smith (2014); Jiang and Smith (2011)]. In addition to the mechanical control applications, passive network synthesis can also be applied to the field of electronic engineering [Lavaei *et al.* (2011); Mukhtar *et al.* (2011)], and can provide long-term impacts on many other areas, such as biometric image processing [Saeed (2014)], passivity-preserving model reduction [Reis and Stykel (2011)], open and interconnected systems [Willems (2007)], etc.

1.3 Outline of the Book

This book introduces some recent advances on passive network synthesis, mainly including three important topics in this field: realizability of biquadratic impedances as one-port RLC networks, synthesis of n-port resistive networks, and synthesis of low-complexity mechanical networks.

In Chapter 2, some important preliminaries of passive network synthesis will be reviewed. For the synthesis of one-port passive networks, some basic properties of positive-real functions, and some classical one-port synthesis procedures will be presented. Moreover, some basic results of graph theory for network analysis and synthesis will be presented. Finally, the principle of duality will be explained.

In Chapter 3, some recent results on the realizability of biquadratic impedances as one-port RLC networks will be introduced, which are mainly referred to [Chen *et al.* (2016b, 2017); Wang and Chen (2012); Wang *et al.* (2014, 2018)]. Biquadratic synthesis of one-port RLC networks has been an important topic in passive network synthesis, where its minimal realizability problem is still unsolved.

In Chapter 4, some recent results on the synthesis of n-port resistive networks will be introduced, which are mainly referred to [Chen *et al.* (2015b); Wang and Chen (2015)]. The synthesis of n-port resistive networks is another important research topic.

In Chapter 5, some recent results on the low-complexity synthesis of passive mechanical networks will be introduced, which are mainly referred to [Chen and Smith (2009b); Chen *et al.* (2013a,b, 2015c)]. In addition, examples for mechanical control will be presented.

Chapter 6 will present some interesting problems to be further investigated in the field of passive network synthesis.

Chapter 2

Preliminaries of Passive Network Synthesis

2.1 Positive-Real Function and Foster Preamble

This section will give a brief introduction of positive-real functions and the properties, which are referred to [Baher (1984); Brune (1931); Chen and Smith (2009a); Guillemin (1957); Van Valkenburg (1960)].

Definition 2.1. [Baher (1984), pg. 27], [Van Valkenburg (1960), pg. 72] A real-rational function $H(s)$ is said to be *positive-real* if $H(s)$ is analytic and $\Re(H(s)) \geq 0$ for all s with $\Re(s) > 0$, that is, in the open right-half plane.

It is obvious that Definition 2.1 is a special case of Definition 1.2. Based on Definition 2.1, the following theorems can be obtained.

Theorem 2.1. *[Brune (1931)] If $H(s)$ and $W(s)$ are both positive-real functions, then $W(H(s))$ is positive-real.*

Theorem 2.2. *[Baher (1984), pg. 27] If $H_1(s)$ and $H_2(s)$ are two positive-real functions, then $\alpha H_1(s) + \beta H_2(s)$ is positive-real for any $\alpha > 0$ and $\beta > 0$.*

Since $1/s$, ks, and $s + k$ are all positive-real functions for any $k > 0$ by Definition 2.1, the following corollaries of Theorem 2.1 can be directly obtained.

Corollary 2.1. *If $H(s)$ is a positive-real function, then $H^{-1}(s)$ and $H(s^{-1})$ are both positive-real.*

Corollary 2.2. *If $H(s)$ is a positive-real function, then $\alpha H(\beta s) + \gamma$ is positive-real for any $\alpha > 0$, $\beta > 0$, and $\gamma \geq 0$.*

A classical criterion for testing positive-realness is presented as follows.

Theorem 2.3. *[Baher (1984), pg. 33] A real-rational function $H(s)$ is positive-real if and only if*

1. *$H(s)$ is analytic for any $\Re(s) > 0$;*
2. *$\Re(H(j\omega)) \geq 0$ for all $\omega \in \mathbb{R}$ with $s = j\omega$ not being a pole of $H(s)$;*
3. *any pole of $H(s)$ on $j\mathbb{R} \cup \infty$ is simple and have a positive residue.*

Definition 2.2. [Baher (1984), pp. 29–30] A real polynomial $P(s)$ is said to be a *Hurwitz polynomial* if all its zeros are in $\Re(s) \leq 0$ (closed left-half plane) with the zeros on $j\mathbb{R}$ (imaginary axis) being simple. Specifically, $P(s)$ is called a *strictly Hurwitz polynomial* if all its zeros are in $\Re(s) < 0$.

Since checking the residue conditions may be a complex task especially for higher-degree functions, the following necessary and sufficient condition for positive-realness might be easier to use.

Theorem 2.4. *[Weinberg and Slepian (1958)] A real-ration function $H(s) = p(s)/q(s)$ with $p(s)$ and $q(s)$ being coprime polynomials is positive-real, if and only if*

1. *$p(s) + q(s)$ is a Hurwitz polynomial;*
2. *$\Re(H(j\omega)) \geq 0$ for $\omega \in \mathbb{R}$ with $s = j\omega$ not being a pole of $H(s)$.*

In some cases, it is desirable to allow $p(s)$ and $q(s)$ to contain common roots on $j\mathbb{R}$, which means that the convenient test by Theorem 2.4 is no longer applicable. Therefore, the following new positive-real criterion is useful.

Theorem 2.5. *[Chen and Smith (2009a)] A real-ration function $H(s) = p(s)/q(s)$ with $p(s)$ and $q(s)$ having no common root in $\Re(s) > 0$ is positive-real, if and only if*

1. *$p(s) + q(s)$ has no root in $\Re(s) > 0$;*
2. *$\Re(H(j\omega)) \geq 0$ for $\omega \in \mathbb{R}$ with $s = j\omega$ not being a pole of $H(s)$.*

Note that any positive-real function $H(s)$ can be written in the form of

$$H(s) = \frac{h_0}{s} + \sum_{i=1}^{m} \frac{2h_i s}{s^2 + \omega_i^2} + h_\infty s + H_1(s), \qquad (2.1)$$

where $h_0 \geq 0$, $h_i \geq 0$, and $h_\infty \geq 0$ are the residues of the poles of $H(s)$ at $s = 0$, $s = \pm j\omega_i$, and $s = \infty$, respectively, in which the zero value means that there is no pole.

Theorem 2.6. *[Baher (1984), pg. 34] Consider a positive-real function $H(s)$ as expressed in (2.1), where $h_0 \geq 0$, $h_i \geq 0$, $i = 1, 2, \ldots, m$, and $h_\infty \geq 0$. Then, $H_1(s)$ in (2.1) is positive-real.*

Theorem 2.6 shows that any positive-real function maintains positive-realness after extracting all its poles on $j\mathbb{R} \cup \infty$ with the *McMillan degree* (or called *degree*)[1] of the function being reduced. It is noted that h_0/s, $h_\infty s$, and $2h_i s/(s^2 + \omega_i^2)$ are impedances (resp., admittances) of a capacitor (resp., an inductor), an inductor (resp., a capacitor), and the parallel (resp., series) connection of a capacitor and an inductor, respectively. Therefore, any positive-real impedance $Z(s)$ can be preliminarily realized through removing all its poles (resp., zeros) on $j\mathbb{R} \cup \infty$ as extracting these lossless components in series (resp., in parallel).

Theorem 2.7. *[Guillemin (1957); Van Valkenburg (1960)] If $H(s)$ is a positive-real function, then $H(s) - \chi$ is positive-real, where χ is no larger than the minimum value of $\Re(H(j\omega))$ for any $\omega \in \mathbb{R} \cup \infty$.*

Theorem 2.7 shows that any positive-real function maintains the positive-realness after subtracting a constant that is equal to the minimum value of the real part of the function on $j\mathbb{R} \cup \infty$. Therefore, any positive-real impedance $Z(s)$ can be preliminarily realized through extracting a series (resp., parallel) resistor whose resistance (resp., conductance) is equal to the minimum value of the real part of $Z(s)$ (resp., $Z^{-1}(s)$) on $j\mathbb{R} \cup \infty$.

Definition 2.3. [Van Valkenburg (1960), pg. 161] A real-rational function $H(s)$ is said to be a *minimum function* if (i) $H(s)$ is positive-real, (ii) $H(s)$ contains no pole and zero on $j\mathbb{R} \cup \infty$, and (iii) there exists a finite real value $\omega_1 \neq 0$ such that $H(j\omega_1) = jX_1$ with $X_1 \neq 0$.

As a consequence, the *Foster preamble* is defined as follows.

Definition 2.4. [Van Valkenburg (1960), pg. 161] Given a positive-real impedance $Z(s)$, the removal of the poles and zeros on $j\mathbb{R} \cup \infty$ and the

---

[1]For any real-rational function $H(s) = a(s)/b(s)$ with polynomials $a(s)$ and $b(s)$ being coprime, the McMillan degree of $H(s)$ is equal to the maximum degree of $a(s)$ and $b(s)$, which is denoted as $\delta(H(s)) = \max\{\deg(a(s)), \deg(b(s))\}$ [Anderson and Vongpanitlerd (1973), Chapter 3.6].

minimum constant of $\Re(Z(j\omega))$ or $\Re(Z^{-1}(j\omega))$ correspond to the extraction of resistors, capacitors or inductors (see Theorems 2.6 and 2.7). The *Foster preamble* is the successive removal of these poles, zeros, and minimum constant values, such that both the remaining impedance $Z_1(s)$ and admittance $Y_1(s) = Z_1^{-1}(s)$ are minimum functions with lower degrees or one of $Z_1(s)$ and $Y_1(s)$ is zero.

Remark 2.1. Since the admittance $Y(s)$ of any one-port network is equal to $Z^{-1}(s)$, which always exists provided that the impedance $Z(s)$ is nonzero, the realization of a positive-real admittance $Y(s)$ as a one-port passive network using the Foster preamble and any other synthesis approach can be converted into the realization of the impedance $Z(s)$. Therefore, one only needs to discuss the the realization problem of either the impedance or the admittance as one-port networks.

Remark 2.2. It can be verified that $H^{-1}(s)$ is not always a minimum function if $H(s)$ is a minimum function, e.g., $H(s) = (s^2 + s + 1/2)/(s + 1)^2$. Therefore, the Foster preamble can only terminate when both the resulting impedance and admittance are minimum functions. Otherwise, if the resulting impedance $Z_1(s)$ (resp., admittance $Z_1^{-1}(s)$) is a minimum function with the admittance $Z_1^{-1}(s)$ (resp., impedance $Z_1(s)$) not being one, then the Foster preamble can still continue by extracting the minimum value of $\Re(Z_1^{-1}(s))$ (resp., $\Re(Z_1(s))$), which further yields a pole (resp., zero) of the impedance at zero or infinity.

The Foster preamble can complete the realization of a given positive-real impedance $Z(s)$ if the resulting impedance or admittance is zero. Otherwise, other realization procedures need to be further utilized. An illustrative example is presented as follows.

Example 2.1. Consider a positive-real impedance

$$Z(s) = \frac{12s^3 + 6s^2 + 7s + 2}{4s^3 + 4s^2 + 3s + 2}.$$

Then, by the Foster preamble, it can be written as

$$Z(s) = 1 + \left(\frac{1}{2s} + \left(\frac{s}{2s^2 + 1} + 2\right)^{-1}\right)^{-1},$$

which is realizable as the configuration in Fig. 2.1 with $R_1 = 1\,\Omega$, $R_2 = 2\,\Omega$, $L_1 = 2$ H, $L_2 = 1$ H, and $C_1 = 2$ F.

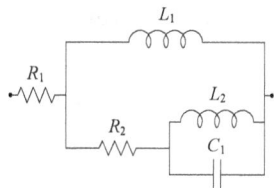

Fig. 2.1 A realization of Example 2.1.

2.2 Synthesis of One-Port Lossless Networks

Definition 2.5. [Baher (1984), pg. 48] A positive-real function $H(s)$ is called a *reactance function* (or *Foster function*) if $H(s)$ is an odd rational function, that is, $\mathrm{Ev}(H(s)) := (H(s) + H(-s))/2 = 0$ for all s.

Theorem 2.8. *[Baher (1984), pg. 51] Any reactance function $H(s)$ can be written in the form of*

$$H(s) = k \left\{ \frac{(s^2 + \omega_1^2)(s^2 + \omega_3^2) \cdots (s^2 + \omega_{2n-1}^2)}{s(s^2 + \omega_2^2)(s^2 + \omega_4^2) \cdots (s^2 + \omega_{2n-2}^2)} \right\}^{\pm 1}, \qquad (2.2)$$

where $k > 0$ and $0 \le \omega_1 < \omega_2 < \omega_3 < \omega_4 < \dots$.

A one-port *lossless network* is a special type of passive networks containing only reactive elements (inductors and capacitors) and transformers. Since no resistor is involved, the network is lossless and there is no dissipation.

Therefore, as shown in [Baher (1984), pp. 47–48], the impedance $Z(s)$ of any one-port lossless network must be a reactance function. Conversely, consider any reactance impedance function $Z(s)$. By Theorem 2.8, all the poles and zeros of $Z(s)$ must be on $j\mathbb{R} \cup \infty$ and are alternatingly interlaced with each other, in the form of (2.2). Through extracting all the poles (or zeros) of $Z(s)$ based on Theorem 2.6, which is called the *partial fraction expansion approach*, $Z(s)$ is realizable as a one-port lossless network consisting of only inductors and capacitors (LC network). Such a realization is called *Foster's form*. Any reactance impedance function $Z(s)$ is realizable as a one-port LC network, through the successive removal of poles at $s = \infty$ (or $s = 0$) from the function and the subsequently inverted remainders, which is called the *continued fraction expansion approach*. Such a realization is called *Cauer's form*. It is noted that both of these two approaches belong to special cases of the Foster preamble.

As a summary, the following result can be established.

Theorem 2.9. *[Baher (1984), Chapter 3] The impedance (resp., admittance) of a one-port lossless network must be a reactance function, and any reactance function is realizable as the impedance (resp., admittance) of a one-port lossless network consisting of only inductors and capacitors.*

Moreover, synthesis results of one-port RL and RC networks can be similarly derived, which can be referred to [Van Valkenburg (1960)] for details.

2.3 The Brune Synthesis

As discussed in the previous section, any positive-real impedance (resp., admittance) can be converted into a minimum function after the Foster preamble. Consequently, considering a minimum impedance, Brune [Brune (1931)] first established a systematic approach to realize such a function using a finite number of passive elements.

Assume that a given impedance $Z_1(s)$ is a minimum function. Then, there must exist a finite $\omega_1 > 0$ such that $\Re(Z_1(s))$ is zero at $s = \pm j\omega_1$ with $\Im(Z_1(s))$ being nonzero, that is, $Z_1(j\omega_1) = jX_1$ with $X_1 \neq 0$. It is noted that the function $Z_1(s) - sX_1/\omega_1$ must contain a zero at $s = \pm j\omega_1$.

The case of $X_1/\omega_1 < 0$ is first discussed. Letting

$$L_1 = \frac{X_1}{\omega_1}, \tag{2.3}$$

a negative inductor $L_1 < 0$ can be extracted in series based on $W_1(s) = Z_1(s) - L_1 s$. Then, $W_1(s)$ must be a positive-real function, and $W_1^{-1}(s)$ contains a pole at $s = \pm j\omega_1$ with the residue $K_1 > 0$. Therefore, it follows that

$$W_2^{-1}(s) = W_1^{-1}(s) - \frac{2K_1 s}{s^2 + \omega_1^2},$$

which implies that $W_2(s)$ is still positive-real and the extracted inductor L_2 and capacitor C_1 satisfy

$$L_2 = \frac{1}{2K_1}, \quad C_1 = \frac{2K_1}{\omega_1^2}. \tag{2.4}$$

According to the above discussion, $W_2(s)$ can be expressed as

$$W_2(s) = \frac{-L_1 s^3 + Z(s)s^2 - \omega_1^2 L_1 s + \omega_1^2 Z(s)}{(2K_1 L_1 + 1)s^2 - 2K_1 Z(s)s + \omega_1^2},$$

which implies that $W_2(s)$ must contain a pole at $s = \infty$, since $Z_1(s)$ contains no pole and no zero on $j\mathbb{R} \cup \infty$. By the extraction of a series inductor

$$L_3 = \frac{-L_1}{2K_1L_1 + 1},\tag{2.5}$$

the pole of $W_2(s)$ at infinity can be removed, yielding

$$Z_2(s) = W_2(s) - L_3s = \frac{Z_1(s)s^2 - 2\omega_1^2 K_1 L_1^2 s + \omega_1^2(2K_1L_1 + 1)Z_1(s)}{(2K_1L_1 + 1)((2K_1L_1 + 1)s^2 - 2K_1Z_1(s)s + \omega_1^2)},$$

which implies that $Z_2(s)$ is a positive-real function. It is noted that the McMillan degree of $Z_2(s)$ satisfies $\delta(Z_2(s)) = \delta(W_2(s)) - 1 = \delta(W_1(s)) - 3 = \delta(Z_1(s)) - 2$. Therefore, the above realization yields a *Brune cycle* as shown in Fig. 2.2, where the element values L_1, L_2, L_3 and C_1 satisfy (2.3)–(2.5), and the McMillan degree of $Z_2(s)$ is lower than that of $Z_1(s)$. Based on the equivalence of the "T structure" of L_1, L_2, and L_3 and a transformer, one can always obtain an equivalent Brune cycle as shown in Fig. 2.3, where

$$L_p = L_1 + L_2, \quad L_s = L_2 + L_3, \quad M = L_2.\tag{2.6}$$

It follows that

$$L_pL_s = \left(L_1 + \frac{1}{2K_1}\right)\left(\frac{1}{2K_1} - \frac{L_1}{2K_1L_1 + 1}\right) = \frac{1}{4K_1^2} > 0,\tag{2.7}$$

which implies that $L_p > 0$ due to $L_s = L_2 + L_3 > 0$. Therefore, one concludes that the Brune cycle in Fig. 2.3 contains only passive elements when $X_1/\omega_1 < 0$.

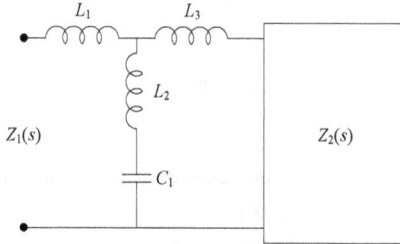

Fig. 2.2 A Brune cycle, where $L_2 > 0$, $C_1 > 0$, and $L_1L_3 < 0$ [Van Valkenburg (1960)].

The other case of $X_1/\omega_1 > 0$ can be similarly discussed, which can be referred to [Van Valkenburg (1960), pp. 170–172] for details. For this case, $Z_1(s)$ can be similarly realized as the Brune cycle in Fig. 2.3 according to (2.3)–(2.6), where $Z_2(s)$ is a positive-real function with $\delta(Z_2(s)) = \delta(Z_1(s)) - 2$. It should be noted that $W_1(s) = Z_1(s) - L_1s$ contains a zero

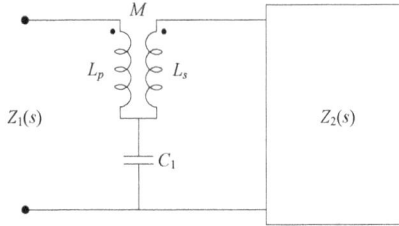

Fig. 2.3   A Brune cycle that is equivalent to Fig. 2.2, where $L_p = L_1 + L_2 > 0$, $L_s = L_2 + L_3 > 0$, and $M = L_2 > 0$ [Van Valkenburg (1960)].

at $s = \pm j\omega_1$ without preserving the positive-realness. However, after the Brune cycle, $Z_2(s)$ becomes a positive-real function again by extracting a negative inductor L_3 in series. For this case, it is clear that $L_1 > 0$, $L_2 > 0$, $L_3 < 0$, and $C_1 > 0$. Furthermore, the condition (2.7) also holds, which implies that $L_s > 0$ due to $L_p = L_1 + L_2 > 0$. Therefore, the Brune cycle in Fig. 2.3 also contains only passive elements when $X_1/\omega_1 > 0$.

As a consequence, combining the Foster preamble and Brune's work, the following theorem can be obtained.

Theorem 2.10. *[Brune (1931)] Any positive-real impedance (resp., admittance) is realizable as a one-port passive network containing a finite number of resistors, inductors, capacitors, and transformers.*

As a summary, the Brune synthesis procedure is stated as follows.

Algorithm (the Brune synthesis)

Step 1.   *Given a positive-real impedance $Z(s)$, utilize the Foster preamble to obtain a resulting impedance $Z_1(s)$. If $Z_1(s)$ or $Z_1^{-1}(s)$ is zero, then the synthesis procedure is finished. Otherwise, if $Z_1(s)$ is a minimum function, then turn to the next step.*

Step 2.   *Realize a given minimum impedance function $Z_1(s)$ as a Brune cycle shown in Fig. 2.3 according to (2.3)–(2.6).*

Step 3.   *If the resulting positive-real impedance $Z_2(s)$ is of degree zero, then the synthesis procedure is finished by realizing $Z_2(s)$ as a resistor. Otherwise, let $Z_2(s) \rightarrow Z(s)$ and return to Step 1.*

2.4 The Bott-Duffin Synthesis

Given any positive-real impedance $Z(s)$, a minimum impedance function $Z_1(s)$ can be obtained using the Foster preamble. Then, there must exist a finite $\omega_1 > 0$ such that $Z(j\omega_1) = jX_1$ with $X_1 \neq 0$. Bott and Duffin [Bott and Duffin (1949)] first established a realization procedure for any positive-real impedance as a one-port passive network without the use of transformers, which is called the *Bott-Duffin synthesis*. The derivation of such a synthesis procedure is based on *Richards's Theorem* [Richards (1947)], which is stated as follows.

Theorem 2.11 (Richards's Theorem). *[Richards (1947)] Given a positive-real function $H(s)$, the function*

$$R(s) = \frac{kH(s) - H(k)s}{kH(k) - sH(s)},$$

is also positive-real for any $k > 0$, and the McMillan degree of $R(s)$ does not exceed that of $H(s)$, that is, $\delta(R(s)) \leq \delta(H(s))$.

Applying Richards's Theorem to the minimum impedance function $Z_1(s)$, for any $k > 0$, one obtains

$$R_1(s) = \frac{kZ_1(s) - Z_1(k)s}{kZ_1(k) - sZ_1(s)}, \tag{2.8}$$

which implies that

$$Z_1(s) = \left(\frac{1}{Z_1(k)R_1(s)} + \frac{s}{kZ_1(k)} \right)^{-1} + \left(\frac{k}{Z_1(k)s} + \frac{R_1(s)}{Z_1(k)} \right)^{-1}. \tag{2.9}$$

By (2.9), $Z_1(s)$ is realizable as the configuration in Fig. 2.4, where

$$L_1 = \frac{Z_1(k)}{k}, \qquad C_1 = \frac{1}{kZ_1(k)}, \tag{2.10}$$

and the two resulting impedances $Z_1(k)R_1(s)$ and $Z_1(k)/R_1(s)$ are positive-real with $\delta(Z_1(k)R_1(s)) \leq \delta(Z(s))$ and $\delta(Z_1(k)/R_1(s)) \leq \delta(Z(s))$. It is noted that the McMillan degrees of these two impedances can be further reduced while preserving the positive-realness, provided that each of them contains a pole or a zero on $j\mathbb{R}$.

Recalling that $Z(j\omega_1) = jX_1$ with $X_1 \neq 0$, the case of $X_1 > 0$ is first discussed. Then, $X_1/\omega_1 > 0$. Since $Z_1(s)$ is a minimum function, it is clear that $Z_1(k)/k$ is a continuous function taking values from zero to infinity for $k \in (0, +\infty)$. Therefore, there always exists $k > 0$ such that

$$\frac{Z_1(k)}{k} = \frac{X_1}{\omega_1}, \tag{2.11}$$

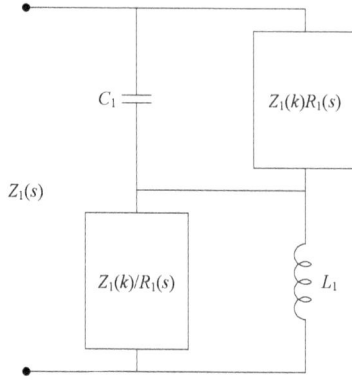

Fig. 2.4   A realization of the minimum impedance function $Z_1(s)$ based on (2.9), where $Z_1(k)R_1(s)$ and $Z_1(k)/R_1(s)$ are both positive-real impedances whose McMillan degrees do not exceed that of $Z_1(s)$.

which means that $1/R_1(s)$ has a pole at $s = \pm jw_1$, whose residue is assumed to be $\alpha > 0$, that is,

$$\frac{1}{R_1(s)} = \frac{2\alpha s}{s^2 + w_1^2} + P(s). \tag{2.12}$$

Then, letting k satisfy (2.11), one obtains

$$\frac{1}{Z_2(s)} = \frac{1}{Z_1(k)R_1(s)} - \frac{2\alpha s}{Z_1(k)(s^2 + w_1^2)} = \frac{P(s)}{Z_1(k)}, \tag{2.13}$$

$$Z_3(s) = \frac{Z_1(k)}{R_1(s)} - \frac{2\alpha Z_1(k)s}{s^2 + w_1^2} = Z_1(k)P(s), \tag{2.14}$$

where the resulting impedances $Z_2(s)$ and $Z_3(s)$ must be positive-real with $\delta(Z_2(s)) = \delta(R_1(s)) - 2 \leq \delta(Z_1(s)) - 2$ and $\delta(Z_3(s)) = \delta(R_1(s)) - 2 \leq \delta(Z_1(s)) - 2$. Therefore, the above realization yields a *Bott-Duffin cycle* as shown in Fig. 2.5, where the element values L_1 and C_1 satisfy (2.10), and

$$L_2 = \frac{Z_1(k)}{2\alpha}, \quad L_3 = \frac{2\alpha Z_1(k)}{w_1^2}, \quad C_2 = \frac{2\alpha}{w_1^2 Z_1(k)}, \quad C_3 = \frac{1}{2\alpha Z_1(k)}. \tag{2.15}$$

For the case of $X_1 < 0$, it is obvious that $w_1 X_1 < 0$. Recalling that $Z_1(s)$ is a minimum function, $kZ(k)$ is continuous and takes values from zero to infinity for $k \in (0, +\infty)$. Therefore, there always exists $k > 0$ such that

$$kZ(k) = -w_1 X_1, \tag{2.16}$$

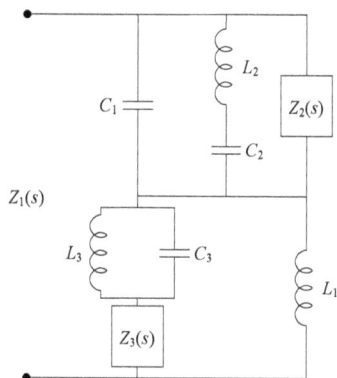

Fig. 2.5 The Bott-Duffin cycle for the case of $X_1 > 0$, where $\delta(Z_2(s)) \leq \delta(Z_1(s)) - 2$ and $\delta(Z_3(s)) \leq \delta(Z_1(s)) - 2$.

which means that $R_1(s)$ has a pole at $s = \pm j\omega_1$, whose residue is assumed to be $\beta > 0$, that is,

$$R_1(s) = \frac{2\beta s}{s^2 + \omega_1^2} + Q(s). \qquad (2.17)$$

Then, letting k satisfy (2.16), one obtains

$$Z_2(s) = Z_1(k)R_1(s) - \frac{2\beta Z_1(k)s}{s^2 + \omega_1^2} = Z_1(k)Q(s), \qquad (2.18)$$

$$\frac{1}{Z_3(s)} = \frac{R_1(s)}{Z_1(k)} - \frac{2\beta s}{Z_1(k)(s^2 + \omega_1^2)} = \frac{Q(s)}{Z_1(k)}, \qquad (2.19)$$

where the resulting impedances $Z_2(s)$ and $Z_3(s)$ are positive-real with $\delta(Z_2(s)) = \delta(R_1(s)) - 2 \leq \delta(Z_1(s)) - 2$ and $\delta(Z_3(s)) = \delta(R_1(s)) - 2 \leq \delta(Z_1(s)) - 2$. Therefore, the above realization yields a *Bott-Duffin cycle* as shown in Fig. 2.6, where the element values L_1 and C_1 satisfy (2.10), and

$$L_2 = \frac{2\beta Z_1(k)}{\omega_1^2}, \quad L_3 = \frac{Z_1(k)}{2\beta}, \quad C_2 = \frac{1}{2\beta Z_1(k)}, \quad C_3 = \frac{2\beta}{\omega_1^2 Z_1(k)}. \qquad (2.20)$$

As a consequence, by repeatedly utilizing the Foster preamble and the Bott-Duffin cycle, the following theorem can be obtained.

Theorem 2.12. *[Bott and Duffin (1949)] Any positive-real impedance (resp., admittance) is realizable as a one-port passive network containing a finite number of resistors, inductors, and capacitors, that is, a one-port RLC network.*

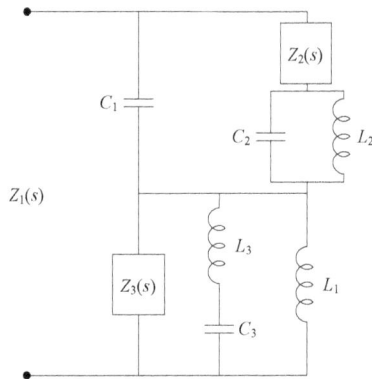

Fig. 2.6   The Bott-Duffin cycle for the case of $X_1 < 0$, where $\delta(Z_2(s)) \leq \delta(Z_1(s)) - 2$ and $\delta(Z_3(s)) \leq \delta(Z_1(s)) - 2$.

In summary, the Bott-Duffin synthesis procedure is stated as follows.

Algorithm (the Bott-Duffin synthesis)
Step 1. *Given a positive-real impedance $Z(s)$, utilize the Foster preamble to obtain a resulting impedance $Z_1(s)$. If $Z_1(s)$ or $Z_1^{-1}(s)$ is zero, then the synthesis procedure is finished. Otherwise, if $Z_1(s)$ is a minimum function, then turn to the next step.*

Step 2.   *Consider a given minimum impedance function $Z_1(s)$, where $Z_1(j\omega_1) = jX_1$ with $\omega_1 > 0$ and $X_1 \neq 0$. If $X_1 > 0$, then realize $Z_1(s)$ as the Bott-Duffin cycle in Fig. 2.5 according to (2.10)–(2.15). If $X_1 < 0$, then realize $Z_1(s)$ as the Bott-Duffin cycle in Fig. 2.6 according to (2.10) and (2.16)–(2.20).*

Step 3. *If the remaining positive-real impedances $Z_2(s)$ and $Z_3(s)$ are of degree zero, then the synthesis procedure is finished by realizing $Z_2(s)$ and $Z_3(s)$ as resistors. Otherwise, continuously repeat Steps 1 and 2 for the remaining positive-real impedances, until the McMillan degrees of them are all zero.*

It is noted that the Bott-Duffin synthesis generates many more elements than the Brune synthesis. In order to further simplify the realizations, Pantell [Pantell (1954)], Reza [Reza (1954)], *et al.* established some *modified Bott-Duffin synthesis approaches*. The realizations can be regarded as extensions of the Bott-Duffin synthesis based on the principle of balanced bridges and Y–Δ or Δ–Y transformation. Two types of Pantell's modified Bott-Duffin cycles are shown in Figs. 2.7 and 2.8 (see [Chen (2007), pp. 28–32], [Balabanian (1958), pp. 109–113] for details).

Considering the modified Bott-Duffin cycle in Fig. 2.7, when $X_1 > 0$, any minimum impedance function $Z_1(s)$ is realizable as the configuration in Fig. 2.7(a), where

$$L_5 = \frac{(C_1 + C_3)L_3}{C_1}, \quad L_6 = \frac{(C_1 + C_3)C_2 L_2}{C_1^2 + C_1 C_2 + C_2 C_3},$$
$$C_5 = \frac{C_1 C_3}{C_1 + C_3}, \quad C_6 = \frac{C_1^2 + C_1 C_2 + C_2 C_3}{C_1 + C_3}, \tag{2.21}$$

and L_1, L_2, L_3, C_1, C_2, C_3, $Z_2(s)$, and $Z_3(s)$ are determined according to (2.10)–(2.15). When $X_1 < 0$, any minimum impedance function $Z_1(s)$ is realizable as the configuration in Fig. 2.7(b), where

$$L_5 = L_1 + L_2, \quad L_6 = \frac{(L_1 + L_2)L_1 L_3}{L_1^2 + L_1 L_2 + L_2 L_3},$$
$$C_5 = \frac{L_2 C_2}{L_1 + L_2}, \quad C_6 = \frac{(L_1^2 + L_1 L_2 + L_2 L_3)C_3}{(L_1 + L_2)L_1}, \tag{2.22}$$

and L_1, L_2, L_3, C_1, C_2, C_3, $Z_2(s)$, and $Z_3(s)$ are determined according to (2.10) and (2.16)–(2.20).

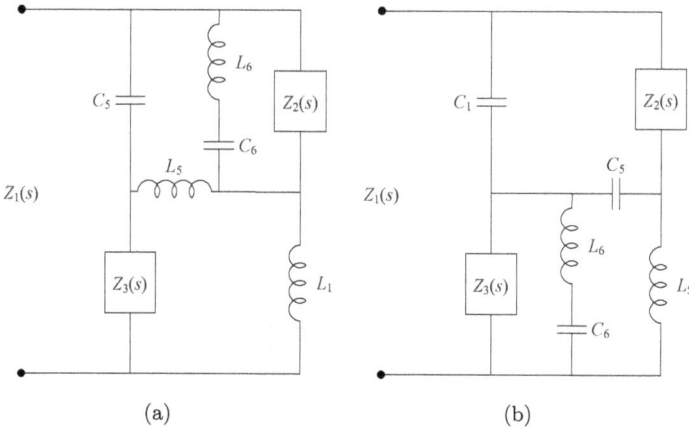

Fig. 2.7 A Pantell's modified Bott-Duffin cycle for the case of (a) $X_1 > 0$ and (b) $X_1 < 0$, where $\delta(Z_2(s)) \le \delta(Z_1(s)) - 2$ and $\delta(Z_3(s)) \le \delta(Z_1(s)) - 2$.

Consider an alternative modified Bott-Duffin cycle as shown in Fig. 2.8. When $X_1 > 0$, any minimum impedance function $Z_1(s)$ is realizable as the

configuration in Fig. 2.8(a), where

$$L_7 = \frac{L_2 C_2}{C_1 + C_2}, \quad L_8 = \frac{(C_1^2 + C_1 C_2 + C_2 C_3)L_3}{(C_1 + C_2)C_1},$$

$$C_7 = C_1 + C_2, \quad C_8 = \frac{(C_1 + C_2)C_1 C_3}{C_1^2 + C_1 C_2 + C_2 C_3}, \quad (2.23)$$

and L_1, L_2, L_3, C_1, C_2, C_3, $Z_2(s)$, and $Z_3(s)$ are determined according to (2.10)–(2.15). When $X_1 < 0$, any minimum impedance function $Z_1(s)$ is realizable as the configuration in Fig. 2.7(b), where

$$L_7 = \frac{L_1 L_3}{L_1 + L_3}, \quad C_7 = \frac{(L1 + L_3)C_3}{L_1},$$

$$L_8 = \frac{L_1^2 + L_1 L_2 + L_2 L_3}{L_1 + L_3}, \quad C_8 = \frac{(L_1 + L_3)L_2 C_2}{L_1^2 + L_1 L_2 + L_2 L_3}, \quad (2.24)$$

and L_1, L_2, L_3, C_1, C_2, C_3, $Z_2(s)$, and $Z_3(s)$ are determined according to (2.10) and (2.16)–(2.20).

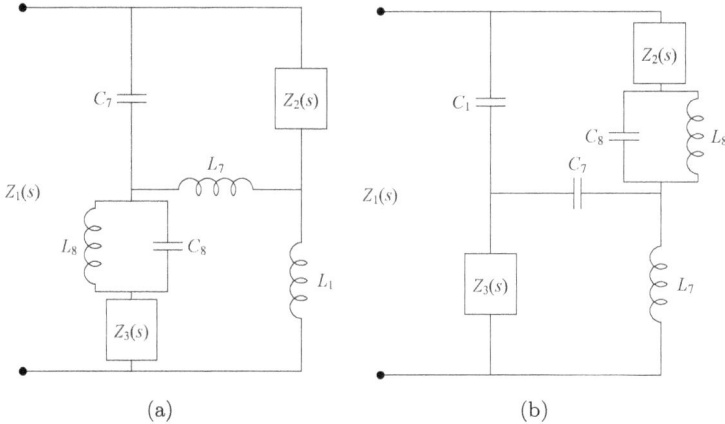

(a)   (b)

Fig. 2.8   An alternative Pantell's modified Bott-Duffin cycle for the case of (a) $X_1 > 0$ and (b) $X_1 < 0$, where $\delta(Z_2(s)) \leq \delta(Z_1(s)) - 2$ and $\delta(Z_3(s)) \leq \delta(Z_1(s)) - 2$.

Example 2.2. Consider a positive-real impedance given by

$$Z(s) = \frac{3s^2 + 2s + 3}{s^2 + s + 2}.$$

It can be checked that $\min \Re(Z(j\omega))$ occurs at $\omega_1 = 1$ such that $\Re(Z(j\omega_1)) = 1$. Therefore, through extracting a resistor $R_1 = 1\ \Omega$ in series, one can obtain a minimum impedance function as

$$Z_1(s) = Z(s) - R_1 = \frac{2s^2 + s + 1}{s^2 + s + 2}.$$

It is noted that $Z_1(j\omega_1) = jX_1 = j1$, which means that $X_1 = 1 > 0$. Following the Brune synthesis procedure, $Z(s)$ is realizable as the configuration in Fig. 2.9(a), where $R_1 = 1$ Ω, $R_2 = 1/2$ Ω, $C_1 = 1$ F, $L_p = 2$ H, $L_s = 1/2$ H, and $M = 1$ H. Following the Bott-Duffin synthesis procedure, it can be solved to obtain $k = 1$, and $Z(s)$ is realizable as the configuration in Fig. 2.9(b), where $R_1 = 1$ Ω, $R_2 = 1/2$ Ω, $R_3 = 2$ Ω, $L_1 = 1$ H, $L_2 = 1/2$ H, $L_3 = 2$ H, $C_1 = 1$ F, $C_2 = 2$ F, and $C_3 = 1/2$ F.

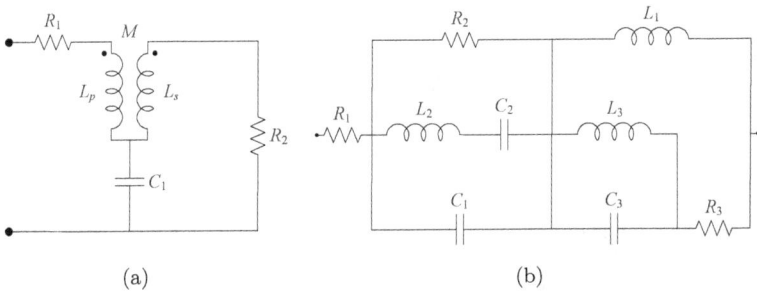

(a)   (b)

Fig. 2.9   An example for (a) the Brune synthesis and (b) the Bott-Duffin synthesis.

2.5   The Darlington Synthesis

Another well-known synthesis procedure of one-port passive networks is the *Darlington synthesis* [Darlington (1939)]. This procedure is based on the resistive extraction approach as shown in Fig. 2.10, where a resistor R is extracted such that the realization is converted into that of a two-port lossless network. As a consequence, at most one resistor is needed for realizing a positive-real impedance $Z(s)$ by following the Darlington synthesis procedure.

For the general configuration in Fig. 2.10, with $R = 1$ Ω, its impedance is obtained as

$$Z(s) = z_{11} \frac{(z_{11}z_{22} - z_{12}^2)/z_{11} + 1}{z_{22} + 1}, \tag{2.25}$$

where z_{11}, z_{22}, and z_{12} are the entries of the impedance matrix of a lossless network:

$$Z_{2 \times 2}(s) = \begin{bmatrix} z_{11} & z_{12} \\ z_{12} & z_{22} \end{bmatrix}. \tag{2.26}$$

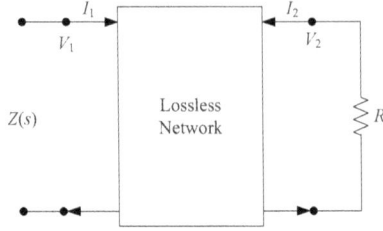

Fig. 2.10   The general realization configuration for the Darlington synthesis, which is based on the resistive extraction [Van Valkenburg (1960)].

Consider a given positive-real impedance in the form of

$$Z(s) = \frac{m_1 + n_1}{m_2 + n_2},$$

where m_1 and n_1 are the even and odd parts of the numerator, and m_2 and n_2 are the even and odd parts of the denominator. Supposing that $Z(s)$ is multiplied in numerator and denominator by a common factor $m_0 + n_0$, with m_0 and n_0 being even and odd parts, respectively, one obtains

$$\begin{aligned}
Z(s) &= \frac{m_1 + n_1}{m_2 + n_2} \frac{m_0 + n_0}{m_0 + n_0} \\
&= \frac{(m_0 m_1 + n_0 n_1) + (m_0 n_1 + n_0 m_1)}{(m_0 m_2 + n_0 n_2) + (n_0 m_2 + m_0 n_2)} =: \frac{m_1' + n_1'}{m_2' + n_2'}.
\end{aligned} \tag{2.27}$$

Comparing (2.25) with (2.27), it follows that

$$z_{11} = \frac{m_1'}{n_2'}, \quad z_{22} = \frac{m_2'}{n_2'}, \quad z_{12} = \frac{\sqrt{m_1' m_2' - n_1' n_2'}}{n_2'} \qquad \text{(Case A)}$$

or

$$z_{11} = \frac{n_1'}{m_2'}, \quad z_{22} = \frac{n_2'}{m_2'}, \quad z_{12} = \frac{\sqrt{n_1' n_2' - m_1' m_2'}}{m_2'}, \qquad \text{(Case B)}$$

where

$$m_1' m_2' - n_1' n_2' = (m_1 m_2 - n_1 n_2)(m_0^2 - n_0^2).$$

As shown in [Van Valkenburg (1960), Section 14.2], one can always determine a common factor $m_0 + n_0$ according to the zeros of $m_1 m_2 - n_1 n_2$, such that $(m_1 m_2 - n_1 n_2)(m_0^2 - n_0^2)$ is a full square polynomial. As a consequence, z_{12} is a real-rational function. Furthermore, it is shown in [Van Valkenburg (1960), Section 14.2] that z_{11}, z_{22}, and z_{12} in both Case A and Case B constitute a positive-real impedance matrix $Z_{2\times 2}(s)$ in the form of (2.26), such that $Z_{2\times 2}(s)$ can be written as

$$Z_{2\times 2}(s) = \sum_{i=1}^{N} K^{(i)} f_i(s), \qquad (2.28)$$

where $f_i(s)$ is a positive-real function in one of the three forms: $s/(s^2+\omega_i^2)$, s, and $1/s$, and

$$K^{(i)} = \begin{bmatrix} k_{11}^{(i)} & k_{12}^{(i)} \\ k_{12}^{(i)} & k_{22}^{(i)} \end{bmatrix}$$

is a non-negative definite matrix satisfying $k_{11}^{(i)} k_{22}^{(i)} - (k_{12}^{(i)})^2 = 0$. Therefore, it can be proved that $K^{(i)} f_i(s)$ is realizable as the two-port lossless configuration in Fig. 2.11, where $|n_i| = k_{22}^{(i)}/|k_{12}^{(i)}| = |k_{12}^{(i)}|/k_{11}^{(i)}$, and

$$Z^{(i)} = \frac{k_{12}^{(i)}}{n_i} f_i(s)$$

is realizable with at most two reactive elements. As a result, the given positive-real impedance $Z(s)$ is realizable as the configuration in Fig. 2.12 by following the Darlington synthesis procedure.

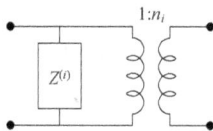

Fig. 2.11 The two-port lossless configuration realizing $K^{(i)} f_i(s)$ [Van Valkenburg (1960)].

2.6 Graph Theory for Passive Networks

For the analysis and synthesis of n-port networks, some basic concepts and results from graph theory are presented in this section.

Definition 2.6. [Seshu and Reed (1961), pg. 9] A *linear graph* is the collection of edges and vertices, where an *edge* is a line segment together with its endpoints and a *vertex* is an endpoint of an edge.

Definition 2.7. [Boesch (1966)] Consider an n-port RLC (resp., damperspring-inerter) network containing e elements and v nodes. The *augmented*

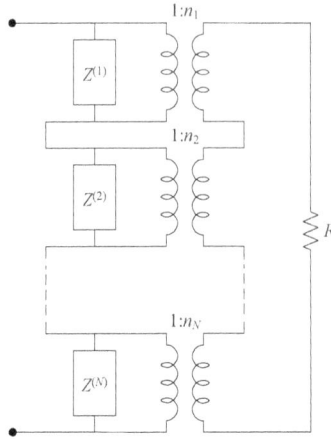

Fig. 2.12   The configuration realizing $Z(s)$ by the Darlington synthesis procedure [Van Valkenburg (1960)].

graph $\mathcal{G}$ is formulated by letting each port or element correspond to an edge and letting each node correspond to a graph vertex. The subgraph consisting of all the edges corresponding to the ports is called a *port graph* $\mathcal{G}_p$. The subgraph consisting of all the edges corresponding to the elements is called a *network graph* $\mathcal{G}_e$. Furthermore, an edge belonging to the port graph $\mathcal{G}_p$ is called a *port edge*, and an edge belonging to a network graph $\mathcal{G}_e$ is called a *network edge*.

By Definition 2.7, it is clear that $\mathcal{G}$ is the union of $\mathcal{G}_p$ and $\mathcal{G}_e$. An example to illustrate the concepts in Definition 2.7 is shown in Fig. 2.13.

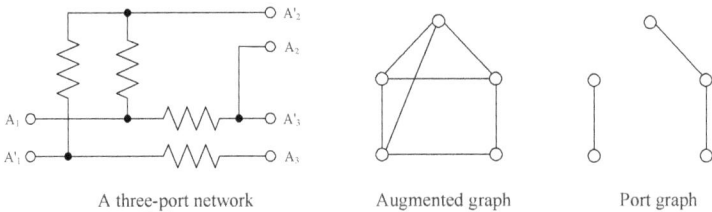

A three-port network   Augmented graph   Port graph

Fig. 2.13   An example illustrating the concepts of the port graph and network graph.

Definition 2.8. [Seshu and Reed (1961), pg. 15] A *circuit* (or called a *loop*) is a closed *edge sequence* [Seshu and Reed (1961), pg. 14] with the *degree* [Seshu and Reed (1961), pg. 14] of each vertex being two.

Definition 2.9. [Seshu and Reed (1961), pg. 24, pg. 26] For any *connected graph* [Seshu and Reed (1961), pg. 15], a subgraph that contains all the vertices without any circuit is called a *tree*. An edge of a tree is called a *branch*; an edge of the complement of a tree (co-tree) is called a *chord* (or called a *link*).

Theorem 2.13. *[Seshu and Reed (1961), pg. 24, pg. 26] A graph is a tree if and only if there is one and only one path [Seshu and Reed (1961), pg. 14] between any two vertices.*

Theorem 2.14. *[Seshu and Reed (1961), pp. 25–26] A connected graph with v vertices and $n_e \geq v-1$ edges must contain a tree, where the number of branches is $v-1$ and the number of chords is $n_e - v + 1$.*

Theorem 2.15. *[Seshu and Reed (1961), pp. 26–27] For a connected graph with v vertices, its subgraph $\mathcal{G}_s$ is made part of a tree,[2] if and only if $\mathcal{G}_s$ does not contain any circuit. Moreover, if a subgraph $\mathcal{G}_s$ contains $v-1$ edges with no circuit, then $\mathcal{G}_s$ constitutes a tree.*

Definition 2.10. [Seshu and Reed (1961), pg. 27] The *f-circuits* of a connected graph (with v vertices and $n_e \geq v-1$ edges) for a certain tree are $n_e + v - 1$ circuits, each of which is formed by a chord and its unique tree path.

Definition 2.11. [Seshu and Reed (1961), pg. 27] The *rank* of a graph with v vertices and p maximal connected subgraphs is defined to be $v - p$. Specifically, the rank of a connected graph is $v - 1$.

Definition 2.12. [Seshu and Reed (1961), pg. 28] A *cut-set* is a set of edges of a connected graph, such that the removal of these edges will reduce the rank of the graph by one and the removal of any proper subset of these edges cannot do so.[3]

Theorem 2.16. *[Seshu and Reed (1961), pg. 34] For a connected graph, its subgraph $\mathcal{G}_s$ is made part of the complement of a tree (co-tree),[4] if and only if $\mathcal{G}_s$ does not contain any cut-set.*

---

[2]The special case where $\mathcal{G}_s$ constitutes a tree is included.

[3]After removing edges, any isolated vertex is assumed to be a maximal connected subgraph for the calculation of the rank.

[4]The special case where $\mathcal{G}_s$ constitutes a co-tree is included.

Definition 2.13. [Seshu and Reed (1961), pg. 31] For a connected graph with v vertices, the *f-cut-sets* with respect to a tree are $v-1$ cut-sets, each of which is formed by one branch of the tree and some chords.

In the above definition, the chords corresponding to a certain branch can be uniquely determined based on the following theorem.

Theorem 2.17. *[Seshu and Reed (1961), pg. 31] An f-cut-set determined by a branch of a tree exactly contains the chords whose corresponding f-circuits contain the branch.*

To better analyze n-port passive networks, all the edges can be assigned with orientations to form a directed graph.

Definition 2.14. [Seshu and Reed (1961), pg. 91] For a connected graph with v vertices and $n_e \geq v - 1$ edges, the *f-circuit matrix* $B_f = [b_{ij}]$ is a $(n_e - v + 1) \times n_e$ matrix, whose rows correspond to $n_e - v + 1$ f-circuits and whose columns correspond to n_e edges. If edge j belongs to the ith f-circuit and has the same (resp., opposite) orientation as that of the f-circuit, then $b_{ij} = 1$ (resp., $b_{ij} = -1$); if edge j does not belong to the ith f-circuit, then $b_{ij} = 0$. Here, the f-circuit orientation coincides with that of the defining chord.

Definition 2.15. [Seshu and Reed (1961), pg. 97] For a connected graph with v vertices, the *f-cut-set matrix* $Q_f = [q_{ij}]$ is a $(v - 1) \times n_e$ matrix, whose rows correspond to $v - 1$ f-cut-sets and whose columns correspond to $n_e \geq v - 1$ edges. If the jth edge belongs to the ith f-cut-set and has the same (resp., opposite) orientation as that of the f-cut-set, then $q_{ij} = 1$ (resp., $q_{ij} = -1$); if the jth edge does not belong to the ith f-cut-set, then $q_{ij} = 0$. Here, the f-cut-set orientation coincides with that of the defining branch.

Consider an n-port RLC (resp., damper-spring-inerter) network containing e elements. Then, the augmented graph $\mathcal{G}$, port graph $\mathcal{G}_p$, and network graph $\mathcal{G}_e$ of the network can be formulated based on Definition 2.7. Without loss of generality, one can assume that $\mathcal{G}$ is connected with $n + e$ edges and v vertices. Otherwise, one can obtain a connected and *separable* augmented graph [Seshu and Reed (1961), pg. 35] by letting one vertex of a component be common with that of another. Therefore, a tree $\mathcal{T}$ must exist by Theorem 2.14. For each edge of the network graph $\mathcal{G}_e$, that is, network edge, its orientation is assigned with the same direction of the reference element current.[5] For each edge of the port graph $\mathcal{G}_p$, that is, port edge, its orientation is assigned with the opposite direction to the port current.[6]

Furthermore, $n+e-v+1$ f-circuits and $v-1$ f-cut-sets can be uniquely determined with respect to the tree $\mathcal{T}$, which can be denoted as

$$B_f = \begin{bmatrix} B_{f1} & B_{f2} \end{bmatrix},$$

and

$$Q_f = \begin{bmatrix} Q_{f1} & Q_{f2} \end{bmatrix},$$

respectively, where the columns of $B_{f1} \in \mathbb{R}^{(n+e-v+1)\times e}$ and $Q_{f1} \in \mathbb{R}^{(v-1)\times e}$ correspond to the edges of the network graph $\mathcal{G}_e$, and the columns of $B_{f2} \in \mathbb{R}^{(n+e-v+1)\times n}$ and $Q_{f2} \in \mathbb{R}^{(v-1)\times n}$ correspond to the edges of the port graph $\mathcal{G}_p$.

Then, Kirchhoff's laws for the network can be expressed as

$$\begin{bmatrix} B_{f1} & B_{f2} \end{bmatrix} \begin{bmatrix} \hat{U} \\ \hat{V} \end{bmatrix} = 0 \tag{2.29}$$

and

$$\begin{bmatrix} Q_{f1} & Q_{f2} \end{bmatrix} \begin{bmatrix} \hat{J} \\ -\hat{I} \end{bmatrix} = 0, \tag{2.30}$$

where $\hat{J}$ and $\hat{U}$ are the Laplace transforms of the element currents and voltages, and $\hat{I}$ and $\hat{V}$ are the Laplace transforms of the port currents and voltages.

It is known [Seshu and Reed (1961), pg. 123] that the following conditions hold:

$$\begin{bmatrix} \hat{J} \\ -\hat{I} \end{bmatrix} = B_f^T \hat{I}_m = \begin{bmatrix} B_{f1}^T \\ B_{f2}^T \end{bmatrix} \hat{I}_m, \tag{2.31}$$

and

$$\begin{bmatrix} \hat{U} \\ \hat{V} \end{bmatrix} = Q_f^T \hat{V}_n = \begin{bmatrix} Q_{f1}^T \\ Q_{f2}^T \end{bmatrix} \hat{V}_n, \tag{2.32}$$

where $\hat{I}_m$ are the Laplace transforms of the currents corresponding to the chords of the tree $\mathcal{T}$, and $\hat{V}_n$ are the Laplace transforms of the voltages corresponding to the branches of the tree $\mathcal{T}$.

---

[5]It is known that the direction of the actual current of an element may not be the same as that of the reference current, where they have the same direction when the value of the reference current is positive and they have opposite directions when the value of the reference current is negative.

[6]The directions of port currents are chosen as the actual directions, which are determined by the polarities of the ports.

By Ohm's law, it is clear that

$$\hat{J} = G_d \hat{U}, \qquad \hat{U} = D_d \hat{J}, \tag{2.33}$$

where G_d is a diagonal matrix whose diagonal entries are element admittances $(1/R_i, 1/(L_j s)$, or $C_k s)$ and D_d is a diagonal matrix whose diagonal entries are element impedances $(R_i, L_j s$, or $1/(C_k s))$.

Combining (2.29)–(2.32), one obtains

$$Q_{f1} G_d Q_{f1}^T \hat{V}_n = Q_{f2} \hat{I}, \tag{2.34}$$

and

$$B_{f1} D_d B_{f1}^T \hat{I}_m = -B_{f2} \hat{V}. \tag{2.35}$$

As discussed in [Boesch (1966)], if the impedance (resp., admittance) of an n-port RLC network exists, then any current vector $\hat{I}$ (resp., voltage vector $\hat{V}$) is permitted, which means that the port graph $\mathcal{G}_p$ cannot contain any cut-set (resp., circuit) of the augmented graph. By Theorem 2.16 (resp., Theorem 2.15), the port graph must be made part of a co-tree (resp., tree). Conversely, if the port graph $\mathcal{G}_p$ is made part of a co-tree (resp., tree), then any branch (resp., chord) must be a port edge. Therefore, Q_{f1} (resp., B_{f1}) can be written as $Q_{f1} = [I_{v-1}, Q_{f12}]$ (resp., $B_{f1} = [I_{n+e-v+1}, B_{f12}]$), where I_{v-1} (resp., $I_{n+e-v+1}$) is an identity matrix, which implies that $Q_{f1} G Q_{f1}^T$ (resp., $B_{f1} D B_{f1}^T$) must be nonsingular. Therefore, by (2.32) (resp., (2.31)), (2.34) (resp., (2.35)) is equivalent to $\hat{V} = Q_{f2}^T (Q_{f1} G_d Q_{f1}^T)^{-1} Q_{f2} \hat{I}$ (resp., $\hat{I} = B_{f2}^T (B_{f1} D_d B_{f1}^T)^{-1} B_{f2} \hat{V}$), which means that the impedance matrix (resp., admittance matrix) exists.

The above discussion can be summarized as the following two theorems.

Theorem 2.18. *[Boesch (1966)] Consider an n-port RLC (or damper-spring-inerter) network whose connected augmented graph $\mathcal{G}$ contains v vertices and $n + e$ edges. The impedance matrix of the network exists, if and only if its port graph $\mathcal{G}_p$ can be made part of a co-tree $\mathcal{G} - \mathcal{T}$ of its augmented graph $\mathcal{G}$. Moreover, the impedance matrix can be expressed as*

$$Z(s) = Q_{f2}^T (Q_{f1} G_d Q_{f1}^T)^{-1} Q_{f2}, \tag{2.36}$$

where $Q_{f1} = [I_{v-1}, Q_{f12}]$ and Q_{f2} constitute the f-cut-set matrix $Q_f = [Q_{f1}, Q_{f2}]$ of $\mathcal{G}$ with respect to the tree $\mathcal{T}$, the columns of Q_{f1} correspond to network edges, the columns of Q_{f2} correspond to port edges, and G_d is a diagonal matrix whose diagonal entries are element admittances $(a_i, b_j/s,$ or $c_k s$ for $a_i, b_j, c_k > 0)$.

The admittance matrix of the network exists, if and only if its port graph $\mathcal{G}_p$ *can be made part of a tree* $\mathcal{T}$ *of its augmented graph* $\mathcal{G}$. *Moreover, the admittance matrix can be expressed as*

$$Y(s) = B_{f2}^T (B_{f1} D_d B_{f1}^T)^{-1} B_{f2}, \tag{2.37}$$

where $B_{f1} = [I_{n+e-v+1}, B_{f12}]$ *and* B_{f2} *constitute the f-circuit matrix* $B_f = [B_{f1}, B_{f2}]$ *of* $\mathcal{G}$ *with respect to the tree* $\mathcal{T}$, *the columns of* B_{f1} *correspond to network edges, the columns of* B_{f2} *correspond to port edges, and* D_d *is a diagonal matrix whose diagonal entries are element impedances (a_i', $b_j' s$, or c_k'/s for a_i', b_j', $c_k' > 0$).*

Remark 2.3. After a proper rearrangement of rows and corresponding columns, G_d (resp., D_d) can be written as $G_d = G_{d1} + s^{-1} G_{d2} + s G_{d3}$ (resp., $D_d = D_{d1} + s D_{d2} + D_{d3}$). Specifically, G_d (resp., D_d) is a real diagonal matrix for an n-port resistive network.

Remark 2.4. If the port graph $\mathcal{G}_p$ is exactly a co-tree, that is, $n = n + e - v + 1$, then the augmented graph $\mathcal{G}$ contains n f-circuits, and by (2.36) the impedance matrix can be expressed as

$$Z(s) = L D_d L^T, \tag{2.38}$$

where $[I_n, L]$ is the f-circuit matrix of $\mathcal{G}$.

If the port graph $\mathcal{G}_p$ is a tree, that is, $n = v - 1$, then the augmented graph $\mathcal{G}$ contains n vertices, and by (2.37) the admittance matrix can be expressed as

$$Y(s) = W G_d W^T, \tag{2.39}$$

where $[I_n, W]$ is the f-cut-set matrix of $\mathcal{G}$.

Remark 2.5. By (2.36) and (2.37), it can be seen that changing the orientation of any edge in the network graph $\mathcal{G}_e$ (network edge) does not affect $Z(s)$ and $Y(s)$, and changing the orientation of an edge in the port graph $\mathcal{G}_p$ (port edge) corresponds to a *cross-sign change* [Brown and Reed (1962a)] (see Definition 4.2) of $Z(s)$ and $Y(s)$. Here, the orientation change of a port edge corresponds to switching the polarity of the port.

2.7 Principle of Duality

In addition to linear graphs, any one-port (that is, two-terminal) RLC (damper-spring-inerter) network N can be described by a *one-terminal-pair labeled graph* $\mathcal{N}$ with two distinguished *terminal vertices* (see [Seshu and

Reed (1961), pg. 14]), in which the labels designate passive circuit elements regardless of the values of the elements, namely resistors, capacitors, and inductors, which are labeled as R_i, C_i, and L_i, respectively.

Two natural maps acting on the labeled graph are defined as follows:

(1) GDu := Graph duality, which takes the one-terminal-pair graph into its dual (see [Seshu and Reed (1961), Definition 3-12]) while preserving the labeling.
(2) Inv := Inversion, which preserves the graph but interchanges the reactive elements, that is, capacitors to inductors and inductors to capacitors, with their labels C_i to L_i and L_i to C_i.

Consequently, one defines[7]

$$\text{Dual} := \text{network duality of one-terminal-pair labeled graph}$$
$$:= \text{GDu} \circ \text{Inv} = \text{Inv} \circ \text{GDu}.$$

An example to illustrate GDu, Inv, and Dual can be referred to Fig. 3.5. Denoting the one-terminal-pair labeled graphs of the configurations in Figs. 3.5(a), 3.5(b), 3.5(c), and 3.5(d) as $\mathcal{N}_{2a}$, $\mathcal{N}_{2b}$, $\mathcal{N}_{2c}$, and $\mathcal{N}_{2d}$, respectively, the following relations hold: $\mathcal{N}_{2b} = \text{GDual}(\mathcal{N}_{2a})$, $\mathcal{N}_{2c} = \text{Inv}(\mathcal{N}_{2a})$, and $\mathcal{N}_{2d} = \text{Dual}(\mathcal{N}_{2a})$.

Consider a network N whose one-terminal-pair labeled graph is $\mathcal{N}$. Let $\text{Inv}(N)$ denote the network whose one-terminal-pair labeled graph is $\text{Inv}(\mathcal{N})$, resistors are of the same values as those of N, and inductors (resp., capacitors) are replaced by capacitors (resp., inductors) with reciprocal values, which is called the *frequency inverse network* of N. Let $\text{GDu}(N)$ denote the network whose one-terminal-pair labeled graph is $\text{GDu}(\mathcal{N})$ and whose elements are of the reciprocal values to those of N, which is called the *frequency inverse dual network* of N. Let $\text{Dual}(N)$ denote the network whose one-terminal-pair labeled graph is $\text{Dual}(\mathcal{N})$, resistors are of reciprocal values to those of N, and inductors (resp., capacitors) are replaced by capacitors (resp., inductors) with same values, which is called the *dual network* of N (see [Seshu and Reed (1961), Definition 6-5]).

It can be proved that $Z(s)$ (resp., $Y(s)$) is realizable as the impedance (resp., admittance) of a network N whose one-terminal-pair labeled graph is $\mathcal{N}$, if and only if $Z(s^{-1})$ (resp., $Y(s^{-1})$) is realizable as the impedance (resp., admittance) of $\text{Inv}(N)$ whose one-terminal-pair labeled graph is

---

[7]Such an approach of defining GDu, Inv, and Dual based on one-terminal-pair labeled graphs was suggested by Professor Rudolf E. Kalman in his private communication with the first author on July 13, 2014.

Inv($\mathcal{N}$), if and only if $Z(s^{-1})$ (resp., $Y(s^{-1})$) is realizable as the admittance (resp., impedance) of GDu(N) whose one-terminal-pair labeled graph is GDu($\mathcal{N}$), and if and only if it is realizable as the admittance (resp., impedance) of Dual(N) whose one-terminal-pair labeled graph is Dual($\mathcal{N}$). Therefore, if a necessary and sufficient condition is derived for $H(s) = \sum_{i=0}^{m} a_i s^i / \sum_{j=0}^{m} b_j s^j$ to be realizable as the impedance (resp., admittance) of a one-port network whose one-terminal-pair labeled graph is $\mathcal{N}$, then the corresponding condition for Inv($\mathcal{N}$) can be obtained from that for $\mathcal{N}$ through conversions $a_k \leftrightarrow a_{m-k}$ and $b_k \leftrightarrow b_{m-k}$ for $k = 0, 1, ..., \lfloor m/2 \rfloor$ (the *principle of frequency inversion*). The corresponding condition for GDu($\mathcal{N}$) can be obtained from that for $\mathcal{N}$ through conversions $a_k \leftrightarrow b_{m-k}$ for $k = 0, 1, ..., m$ (the *principle of frequency-inverse duality*). Furthermore, the corresponding condition for Dual($\mathcal{N}$) can be obtained from that for $\mathcal{N}$ through conversions $a_k \leftrightarrow b_k$ for $k = 0, 1, ..., m$ (the *principle of duality*).

Chapter 3

Biquadratic Synthesis of One-Port RLC Networks

3.1 Introduction

The realization of biquadratic impedances as passive RLC networks has been an essential topic in passive network synthesis. Although a series of investigations have been made, this problem is still unsolved. It can be seen from the formulation of the Bott-Duffin synthesis procedure that the RLC realization of biquadratic impedances can provide important guidance on positive-real functions with higher degrees by induction. Practically, the impedances (or admittances) of many mechanical networks or electrical networks in mechatronic systems are in biquadratic forms (see [Papageorgiou and Smith (2006); Wang *et al.* (2009)], for instance). Therefore, it is important to investigate biquadratic synthesis of RLC networks, especially its minimal realization.

Through the Bott-Duffin synthesis procedure (resp. Pantell's modified Bott-Duffin synthesis procedure), nine (resp., eight) elements are needed to realize the entire class of positive-real biquadratic impedances as series-parallel (non-series-parallel) RLC networks, where the realizations contain fewer elements for special cases, such as biquadratic minimum impedances or the impedances directly realizable by the Foster preamble. However, the Bott-Duffin approach cannot guarantee the minimality of realizations, and it is necessary to discuss the realization problem of a biquadratic impedance as a k-element network for $k = 1, 2, ..., 8$. In [Ladenheim (1948)], Ladenheim first investigated the realization of biquadratic impedances by listing 108 configurations, which cover all the possible irreducible networks containing no more than two reactive elements and no more than three resistors. Furthermore, based on the method of enumeration, biquadratic synthesis of three-reactive five-element networks and six-element series-parallel networks have been investigated in [Ladenheim (1964); Vasiliu (1969)] and [Vasiliu (1970, 1971)], respectively. In addition, some other realization problems of biquadratic impedances have been investigated in [Bar-Lev (1962); Chang (1969); Eswaran and Murti (1973); Reichert (1969); Steiglitz and Zemanian (1962); Tirtoprodjo (1972)].

During recent years, by defining a new concept called *regularity* [Jiang and Smith (2011)] and investigating its properties, Jiang and Smith reconsidered the the realization problems of biquadratic impedances as five-element networks and six-element series-parallel networks in [Jiang (2010); Jiang and Smith (2011, 2012)]. As a result, the investigations are more systematic and the realization results are better combined, where it is shown [Jiang and Smith (2011)] that the regularity of a biquadratic impedance is equivalent to the realization as a two-reactive five-element series-parallel network. Following previous investigations on minimal realizations of biquadratic minimum impedances, Hughes and Smith continued to investigate such a problem in terms of the minimality of reactive elements for both series-parallel [Hughes and Smith (2014)] and non-series-parallel cases [Hughes (2017)].

This chapter presents some recent results of biquadratic synthesis in [Chen *et al.* (2016b, 2017); Wang and Chen (2012); Wang *et al.* (2014, 2018)]. In this chapter, networks are assumed to be one-port passive transformerless networks containing no more than three kinds of passive elements, which are resistors, capacitors, and inductors (RLC networks). Element values are assumed to be positive and finite if not specially mentioned.

3.2   Basic Notations and Results

The general form of a *biquadratic impedance* is

$$Z(s) = \frac{a_2 s^2 + a_1 s + a_0}{b_2 s^2 + b_1 s + b_0}, \tag{3.1}$$

where $a_i \geq 0$, $i = 0, 1, 2$, and $b_j \geq 0$, $j = 0, 1, 2$. For brevity, the following notations are introduced:[1]

$$A = a_0 b_1 - a_1 b_0, \quad B = a_0 b_2 - a_2 b_0, \quad C = a_1 b_2 - a_2 b_1,$$
$$D_a := a_1 A - a_0 B, \qquad D_b := -b_1 A + b_0 B,$$
$$E_a := a_2 B - a_1 C, \quad E_b := -b_2 B + b_1 C, \quad M := a_0 b_2 + a_2 b_0,$$
$$\Delta_a := a_1^2 - 4a_0 a_2, \quad \Delta_b := b_1^2 - 4b_0 b_2 \quad \Delta_{ab} := a_1 b_1 - 2M,$$

---

[1]These notations were suggested by Professor Rudolf E. Kalman in his private communication with the authors on July 16, 2014.

$$R := AC - B^2, \quad \Gamma_a := R + b_0 b_2 \Delta_a, \quad \Gamma_b := R + a_0 a_2 \Delta_b.$$

Define $R_0(a, b, s)$ as the *resultant* [Gantmacher (1980), Chapter XV] of $a(s) := a_2 s^2 + a_1 s + a_0$ and $b(s) := b_2 s^2 + b_1 s + b_0$ in s, that is,

$$R_0(a, b, s) = \begin{vmatrix} a_2 & a_1 & a_0 & 0 \\ 0 & a_2 & a_1 & a_0 \\ b_2 & b_1 & b_0 & 0 \\ 0 & b_2 & b_1 & b_0 \end{vmatrix}.$$

Then, it is clear that $R = -R_0(a, b, s)$. It is known from [Gantmacher (1980), Chapter XV] that there exists a common factor between $a(s)$ and $b(s)$ if and only if $R = -R_0(a, b, s) = 0$.

Lemma 3.1. *[Chen and Smith (2009a); Foster (1962)] A biquadratic function $Z(s)$ in the form of (3.1) with $a_i \geq 0$, $i = 0, 1, 2$, and $b_j \geq 0$, $j = 0, 1, 2$, is positive-real, if and only if $(\sqrt{a_2 b_0} - \sqrt{a_0 b_2})^2 \leq a_1 b_1$.*

Lemma 3.2. *[Foster (1963), pg. 527] A biquadratic function $Z(s)$ in the form of (3.1) with $a_i \geq 0$, $i = 0, 1, 2$, and $b_j \geq 0$, $j = 0, 1, 2$, is a minimum function, if and only if $a_i > 0$, $i = 0, 1, 2$, $b_j > 0$, $j = 0, 1, 2$, and $(\sqrt{a_2 b_0} - \sqrt{a_0 b_2})^2 = a_1 b_1$.*

Lemma 3.3. *[Jiang and Smith (2011), Lemma 5] A biquadratic function $Z(s)$ in the form of (3.1) with $a_i \geq 0$, $i = 0, 1, 2$, and $b_j \geq 0$, $j = 0, 1, 2$, is regular, if and only if at least one of the following four conditions holds: 1. $B \leq 0$ and $D_b \geq 0$; 2. $B \leq 0$ and $E_a \geq 0$; 3. $B \geq 0$ and $E_b \geq 0$; 4. $B \geq 0$ and $D_a \geq 0$.*

Lemma 3.4. *[Jiang (2010), Lemma 8] Any positive-real biquadratic impedance (3.1), with any of its six parameters equal to zero, can be realized by a series-parallel network with no more than two reactive elements and two resistive elements through the Foster preamble.*

Definition 3.1. Consider a one-port network N, containing no more than three kinds of elements (resistors, inductors, and capacitors). Letting each element correspond to an *edge* [Seshu and Reed (1961), pg. 9] and each voltage node correspond to a *vertex* [Seshu and Reed (1961), pg. 9] yields a linear graph, called the *network graph* of N. The *subgraph* [Seshu and Reed (1961), pg. 12] with edges corresponding to reactive elements (inductors and capacitors) is called the *reactive-element graph*, whose edges are called *reactive-element edges*. The subgraph with edges corresponding to resistors is called the *resistor graph*, whose edges are called *resistor edges*.

Let $\mathcal{P}(i,j)$ denote a path whose two end-vertices are i and j, and let $\mathcal{C}(i,j)$ denote a cut-set that separates the network graph into two connected subgraphs containing vertices i and j, respectively. Then, denote a path $\mathcal{P}(i,j)$ (resp., cut-set $\mathcal{C}(i,j)$) whose edges only correspond to resistors, inductors, or capacitors as $R\text{-}\mathcal{P}(i,j)$, $L\text{-}\mathcal{P}(i,j)$, or $C\text{-}\mathcal{P}(i,j)$ (resp., $R\text{-}\mathcal{C}(i,j)$, $L\text{-}\mathcal{C}(i,j)$, or $C\text{-}\mathcal{C}(i,j)$), respectively. Denote a path $\mathcal{P}(i,j)$ (resp., cut-set $\mathcal{C}(i,j)$) whose edges exactly correspond to two kinds of elements, which are inductors-capacitors, resistors-inductors, or resistors-capacitors, as $LC\text{-}\mathcal{P}(i,j)$, $RL\text{-}\mathcal{P}(i,j)$, or $RC\text{-}\mathcal{P}(i,j)$ (resp., $LC\text{-}\mathcal{C}(i,j)$, $RL\text{-}\mathcal{C}(i,j)$, or $RC\text{-}\mathcal{C}(i,j)$), respectively. Let $\mathcal{E}(i,j)$ denote an edge *incident with* [Seshu and Reed (1961), pg. 12] vertices i and j. Then, vertex i is said to be *adjacent to* vertex j by $\mathcal{E}(i,j)$. Moreover, let $R\text{-}\mathcal{E}(i,j)$ denote a resistor edge incident with vertices i and j.

Also, for any one-port network N whose two terminals are denoted as a and a', $\mathcal{P}(a,a')$ denotes the path whose *terminal vertices* (see [Seshu and Reed (1961), pg. 14]) are a and a', and $\mathcal{C}(a,a')$ denotes the cut-set that separates $\mathcal{N}$ into two connected subgraphs $\mathcal{N}_1$ and $\mathcal{N}_2$ containing two terminal vertices a and a', respectively.

Lemma 3.5. *The network graph of a network N with two terminals a and a' realizing a biquadratic impedance $Z(s)$ in the form of (3.1), where $a_i > 0$, $i = 0,1,2$, and $b_j > 0$, $j = 0,1,2$, can neither contain any path $\mathcal{P}(a,a')$ nor contain any cut-set $\mathcal{C}(a,a')$ whose edges correspond to only one kind of reactive elements.*

Proof. Assume that there exists such a path $\mathcal{P}(a,a')$ or cut-set $\mathcal{C}(a,a')$. Then, it is known from [Seshu (1959)] that the impedance of N must contain zeros or poles at $s = 0$ or $s = \infty$. This contradicts the assumption. $\square$

More generally, the following lemma can be obtained.

Lemma 3.6. *Any biquadratic impedance $Z(s)$ in the form of (3.1) with $a_i > 0$, $i = 0,1,2$, and $b_j > 0$, $j = 0,1,2$, is not realizable as the network shown in Fig. 3.1.*

Proof. In [Jiang (2010), Sec. 3.2], it is shown that there exist poles for the impedance of Fig. 3.1(a) and zeros for that of Fig. 3.1(b) at $s = j\omega$ or $s = \infty$, which implies that some of the impedance's coefficients must be zero. Thus, the lemma is proved. $\square$

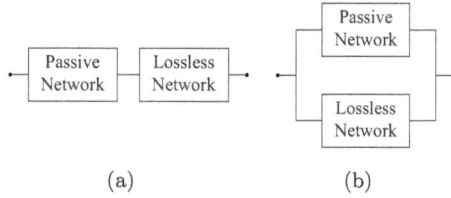

Fig. 3.1 The network structures of a lossless subnetwork and any passive subnetwork (a) in series or (b) in parallel.

The following lemma provides the equivalence of two classes of networks.

Lemma 3.7. *[Lin (1965)] Any passive network as shown in Fig. 3.2(a) is equivalent to a passive network as shown in Fig. 3.2(b), where Z_u and Z_v are positive-real impedances, $\alpha = a(a+b)/b$, $\beta = a+b$, and $\gamma = c(a+b)^2/b^2$.*

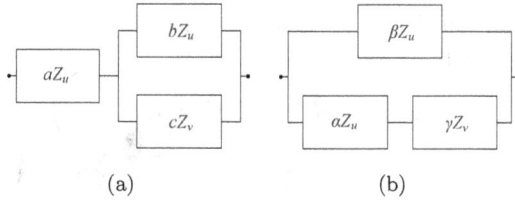

Fig. 3.2 Two equivalent networks, where Z_u and Z_v are the impedances of any two passive networks, $\alpha = a(a + b)/b$, $\beta = a + b$, and $\gamma = c(a + b)^2/b^2$ (see [Lin (1965), Fig. 3]).

3.3 A Canonical Biquadratic Impedance

A canonical form $Z_c(s)$ for biquadratic impedances (3.1), first considered in [Reichert (1969)], is expressed as

$$Z_c(s) = \frac{s^2 + 2U\sqrt{W}s + W}{s^2 + (2V/\sqrt{W})s + 1/W}, \tag{3.2}$$

where

$$W = \sqrt{\frac{a_0 b_2}{a_2 b_0}}, \quad U = \frac{a_1}{2\sqrt{a_0 a_2}}, \quad V = \frac{b_1}{2\sqrt{b_0 b_2}}. \tag{3.3}$$

Here, $Z_c(s)$ can be obtained from $Z(s)$ through $Z_c(s) = \alpha Z(\beta s)$, where $\alpha = b_2/a_2$ and $\beta = \sqrt[4]{a_0 b_0/(a_2 b_2)}$. If $Z(s)$ is realizable as a network N, then the corresponding $Z_c(s)$ must be realizable as another one N_c, with the same one-terminal-pair labeled graph by a proper transformation of the element values, and *vice versa*. Therefore, the realizability condition for $Z_c(s)$ in terms of U, V, $W > 0$, as a network whose one-terminal-pair labeled graph is $\mathcal{N}$, can be determined from that of $Z(s)$ in terms of $a_i > 0$, $i = 0, 1, 2$, and $b_j > 0$, $j = 0, 1, 2$, via the transformation

$$a_2 = 1, \ a_1 = 2U\sqrt{W}, \ a_0 = W, \ b_2 = 1, \ b_1 = 2V/\sqrt{W}, \ b_0 = 1/W. \quad (3.4)$$

Conversely, the realizability condition for $Z(s)$ as a network with one-terminal-pair labeled graph $\mathcal{N}$ in terms of $a_i > 0$, $i = 0, 1, 2$, and $b_j > 0$, $j = 0, 1, 2$, can be determined from that for $Z_c(s)$ in terms of U, V, $W > 0$, through the transformation (3.3).

Furthermore, through (3.4), it is concluded that $Z_c(s)$ is positive-real if and only if

$$\sigma_c := 4UV + 2 - (W + W^{-1}) \geq 0,$$

as stated in [Jiang and Smith (2011)]. Notations Δ_{ab}, R, Γ_a, and Γ_b, defined above are respectively converted into

$$\Delta_{ab_c} := 4UV - 2(W + W^{-1}),$$

$$R_c := -4U^2 - 4V^2 + 4UV(W + W^{-1}) - (W - W^{-1})^2,$$

$$\Gamma_{a_c} := -4V^2 + 4UV(W + W^{-1}) - (W + W^{-1})^2,$$

and

$$\Gamma_{b_c} := -4U^2 + 4UV(W + W^{-1}) - (W + W^{-1})^2.$$

Also, $MR + 2a_0a_2b_0b_2\Delta_{ab}$ is converted to $-(W+W^{-1})^3 + 4UV(W+W^{-1})^2 - 4(U^2 + V^2)(W + W^{-1}) + 8UV$. Moreover, for brevity, denote

$$\lambda_c := 4UV - 4V^2W + (W - W^{-1}).$$

With $\rho^*(U, V, W) = \rho(U, V, W^{-1})$ and $\rho^\dagger(U, V, W) = \rho(V, U, W)$ for any rational function $\rho(U, V, W)$, it can be verified that $\lambda_c^{*\dagger}W$, λ_c/W, $\lambda_c^\dagger$, and λ_c^* correspond to D_a, D_b, E_a, E_b, respectively, through (3.4). Besides, by denoting $\eta_c := 4U^2 + 4V^2 + 4UV(3W - W^{-1}) + (W - W^{-1})(9W - W^{-1})$ and $\zeta_c := -4U^2 - 4V^2 + 4UV(W + W^{-1}) - (W - W^{-1})(3W - W^{-1})$, corresponding to $-R + 4a_0b_2(a_1b_1 + 2B)$ and $R - 2a_0b_2B$, respectively, one has $\eta_c^* = \eta_c^{*\dagger}$ and $\zeta_c^* = \zeta_c^{*\dagger}$ corresponding to $-R + 4a_2b_0(a_1b_1 - 2B)$ and $R + 2a_2b_0B$, respectively.

3.4 Realizations of Biquadratic Impedances with No More than Four Elements

This section will discuss the realization problem of a biquadratic impedance $Z(s)$ in the form of (3.1) to be realizable as an RLC network containing no more than four elements. Based on Lemma 3.4, to investigate realizations with no more than four elements, it suffices to assume that $a_i > 0$, $i = 0, 1, 2$, and $b_j > 0$, $j = 0, 1, 2$.

A necessary and sufficient condition for the realization of such an impedance with no more than three elements is presented in Lemma 3.8. Furthermore, the main result of this section is shown in Theorem 3.5, which presents a necessary and sufficient condition for the realization of such an impedance with no more than four elements. Figures 3.5–3.7 are the four-element realization configurations, whose realizability conditions are summarized in Table 3.1.

3.4.1 *Realizations with No More than Three Elements*

Lemma 3.8. *A biquadratic impedance $Z(s)$ in the form of (3.1), where $a_i > 0$, $i = 0, 1, 2$, and $b_j > 0$, $j = 0, 1, 2$, is realizable with no more than three elements, if and only if $R = 0$.*

Proof. *Sufficiency.* Since $R = 0$, there exists a common factor between the numerator and denominator of $Z(s)$, which means that $Z(s)$ can be written as $Z(s) = (\alpha_1 s + \alpha_0)/(\beta_1 s + \beta_0)$, where $\alpha_i > 0$, $i = 0, 1$, and $\beta_j > 0$, $j = 0, 1$. Thus, $Z(s)$ is realizable as a configuration shown in Fig. 3.3 by the Foster preamble when $\alpha_0 \beta_1 - \alpha_1 \beta_0 \neq 0$, or as a single resistor when $\alpha_0 \beta_1 - \alpha_1 \beta_0 = 0$.

Necessity. By the principle of duality, one only needs to discuss the network graphs shown in Fig. 3.4.

For Fig. 3.4(a), the only one edge should correspond to a resistor, otherwise it will result in a path $\mathcal{P}(a, a')$ or a cut-set $\mathcal{C}(a, a')$ corresponding to one kind of reactive elements, which is impossible by Lemma 3.5. For Figs. 3.4(b), 3.4(c), and 3.4(d), the networks that can be equivalent to one containing fewer elements are not considered to avoid repetition, as the discussion is in the order of the increasing numbers of elements from Fig. 3.4(a) to Fig. 3.4(d). Furthermore, by Lemmas 3.5 and 3.6, the network graphs in Figs. 3.4(b) and 3.4(c) are directly eliminated, and Edge 1 and only one of Edge 2 or Edge 3 of the graph in Fig. 3.4(d) correspond to resistors, yielding the networks shown in Figs. 3.3(a) and 3.3(c). By the principle of duality, one obtains Figs. 3.3(b) and 3.3(d). It can be verified by calculation that impedances of these networks satisfy $R = 0$.   □

3.4.2   Realizations with Four Elements

In the remaining part of this section, it only needs to consider the case of $R \neq 0$.

Lemma 3.9. *If a biquadratic impedance $Z(s)$ in the form of (3.1), where $a_i > 0$, $i = 0, 1, 2$, $b_j > 0$, $j = 0, 1, 2$, and $R \neq 0$, is realizable as a four-element network N, then N does not contain any two elements of the same kind in series or in parallel.*

Proof. Assuming that such two elements exist, the network N can be equivalent to one containing no more than three elements, which implies $R = 0$ by Lemma 3.8. Thus, this lemma is proved.   □

Lemma 3.10. *If a biquadratic impedance $Z(s)$ in the form of (3.1), where $a_i > 0$, $i = 0, 1, 2$, $b_j > 0$, $j = 0, 1, 2$, and $R \neq 0$, is realizable with four elements, then the number of reactive elements is two or three.*

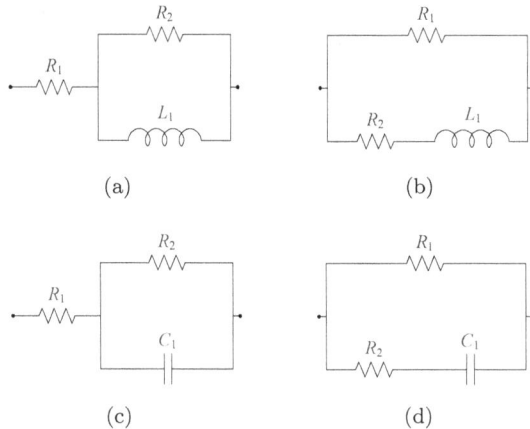

(a)   (b)

(c)   (d)

Fig. 3.3   Three-element configurations realizing $Z(s)$ in the form of (3.1), where $a_i > 0$, $i = 0, 1, 2$, and $b_j > 0$, $j = 0, 1, 2$, discussed in Lemma 3.8, whose one-terminal-pair labeled graphs are (a) $\mathcal{N}_{1a}$, (b) $\mathcal{N}_{1b}$, (c) $\mathcal{N}_{1c}$, and (d) $\mathcal{N}_{1d}$, satisfying $\mathcal{N}_{1b} = \mathrm{GDual}(\mathcal{N}_{1a})$, $\mathcal{N}_{1c} = \mathrm{Inv}(\mathcal{N}_{1a})$, and $\mathcal{N}_{1d} = \mathrm{Dual}(\mathcal{N}_{1a})$, respectively.

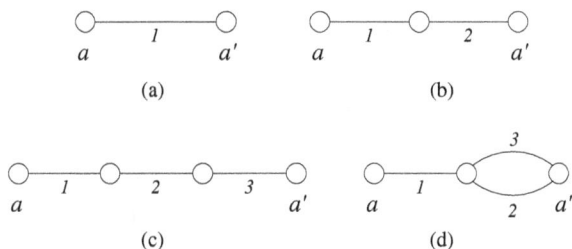

Fig. 3.4 Network graphs of the networks with at most three elements (one half).

Proof. Assume that there is no more than one reactive element. Then, it follows from [Anderson and Vongpanitlerd (1973), pg. 370] that $Z(s)$ can be expressed as a bilinear function whose McMillan degree is at most one, implying that $R = 0$. Assume that the network contains four reactive elements. Then, all the poles of $Z(s)$ must be at $s = j\omega$ or $s = \infty$ [Guillemin (1957)], which contradicts the fact that all the coefficients are positive. Thus, this lemma is proved. □

Theorem 3.1. *A biquadratic impedance $Z(s)$ in the form of (3.1), where $a_i > 0$, $i = 0, 1, 2$, $b_j > 0$, $j = 0, 1, 2$, and $R \neq 0$, is realizable with four elements, if and only if $Z(s)$ is realizable as one of the configurations shown in Figs. 3.5–3.7.*

Proof. *Necessity.* By the principle of duality, only the five network graphs shown in Fig. 3.8 need to be considered. First, the graph in Fig. 3.8(a) is directly eliminated by Lemma 3.9, since at least two elements of the same kind must be in series. For Fig. 3.8(b), Edge 1 and Edge 2 must both correspond to resistors by Lemma 3.5; otherwise, a cut-set $\mathcal{C}(a, a')$ whose edges correspond to only one kind of reactive elements must exist. Therefore, Fig. 3.8(b) is eliminated by Lemma 3.9. Similarly, Edge 1 in Figs. 3.8(c) and 3.8(d) should also correspond to resistors by Lemma 3.5. Furthermore, there are no more than two reactive elements among Edge 2, Edge 3, and Edge 4 by Lemma 3.6. Together with Lemma 3.10, it implies that the number is exactly two. Besides, for Fig. 3.8(c), the two reactive elements cannot be of the same kind. Otherwise, a cut-set $\mathcal{C}(a, a')$ whose edges correspond to only one type of reactive elements exists, contradicting the assumption by Lemma 3.5. Or, there exist two elements of the same kind in series, which is also impossible by Lemma 3.9. Therefore, the only possible networks are shown in Figs. 3.5(a), 3.5(c) and 3.7(a).

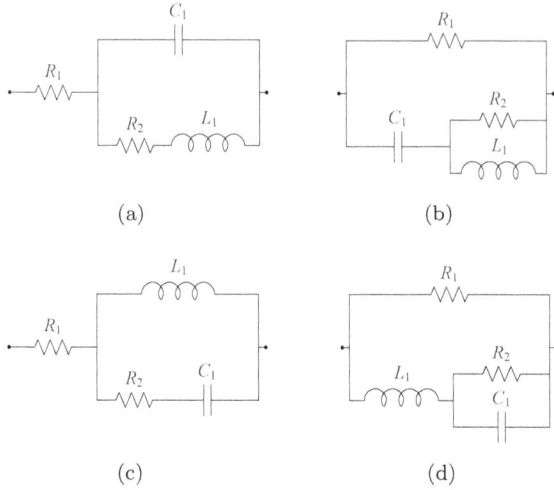

Fig. 3.5 Four-element configurations realizing $Z(s)$ in the form of (3.1), where $a_i > 0$ $i = 0, 1, 2$, and $b_j > 0$, $j = 0, 1, 2$, discussed in Theorem 3.1, whose one-terminal-pair labeled graphs are (a) $\mathcal{N}_{2a}$, (b) $\mathcal{N}_{2b}$, (c) $\mathcal{N}_{2c}$, and (d) $\mathcal{N}_{2d}$, respectively, satisfying $\mathcal{N}_{2b} = \text{GDual}(\mathcal{N}_{2a})$, $\mathcal{N}_{2c} = \text{Inv}(\mathcal{N}_{2a})$, and $\mathcal{N}_{2d} = \text{Dual}(\mathcal{N}_{2a})$.

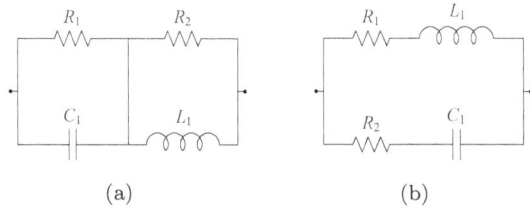

Fig. 3.6 Four-element configurations realizing $Z(s)$ in the form of (3.1), where $a_i > 0$, $i = 0, 1, 2$, and $b_j > 0$, $j = 0, 1, 2$, discussed in Theorem 3.1, whose one-terminal-pair labeled graphs are (a) $\mathcal{N}_{3a}$ and (b) $\mathcal{N}_{3b}$, respectively, satisfying $\mathcal{N}_{3b} = \text{Dual}(\mathcal{N}_{3a})$.

Then, the principle of duality yields the configurations in Figs. 3.5 and 3.7. For Fig. 3.8(d), the two reactive elements cannot be of the same kind by Lemma 3.9. The only possible network is shown in Fig. 3.9(b), which further results in the configuration in Fig. 3.9(a) by the principle of duality. Next, consider Fig. 3.8(e). There is one resistor in each of the two subnetworks in series by Lemma 3.6, and the remaining two reactive elements cannot be of the same kind by Lemma 3.5. Therefore, the only possible configuration is shown in Fig. 3.6(a), which yields Fig. 3.6 by the principle of duality. Finally, by the equivalence in Lemma 3.7 (Fig. 3.2),

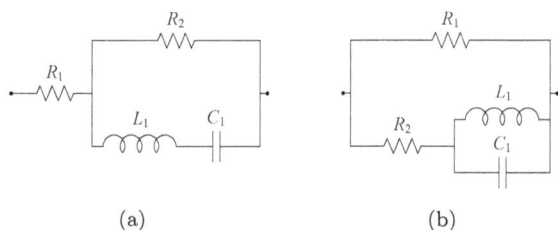

(a) (b)

Fig. 3.7 Four-element configurations realizing $Z(s)$ in the form of (3.1), where $a_i > 0$, $i = 0, 1, 2$, and $b_j > 0$, $j = 0, 1, 2$, discussed in Theorem 3.1, whose one-terminal-pair labeled graphs are (a) $\mathcal{N}_{4a}$ and (b) $\mathcal{N}_{4b}$, respectively, satisfying $\mathcal{N}_{4b} = \mathrm{Dual}(\mathcal{N}_{4a})$.

one observes that the configurations in Fig. 3.7 and Fig. 3.9 are essentially equivalent. Therefore, the configurations in Figs. 3.5–3.7 can cover all the possible cases.

Sufficiency. It is noted that all of the configurations in Figs. 3.5–3.7 contain four elements. Therefore, if a given $Z(s)$ in the form of (3.1), where $a_i > 0$, $i = 0, 1, 2$, $b_j > 0$, $j = 0, 1, 2$, and $R \neq 0$, is realizable as one of them, then $Z(s)$ is indeed realizable with four elements. $\square$

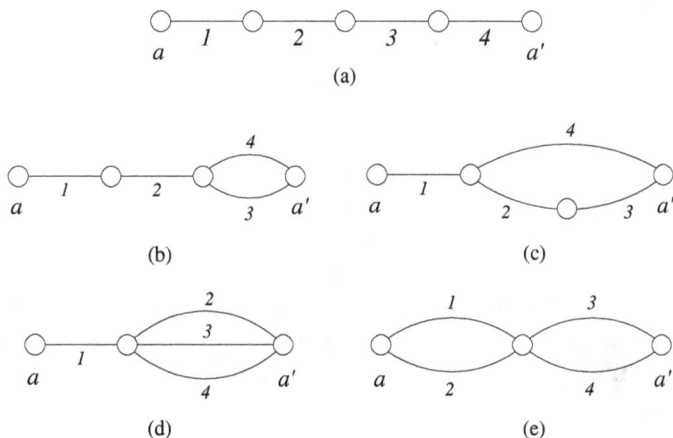

Fig. 3.8 Network graphs of the networks with no more than four elements (one half).

Next, the realizability conditions of the configurations in Figs. 3.5–3.7 are investigated in order to obtain the final conditions. Although the conditions of these configurations have been shown in [Ladenheim (1948)], the

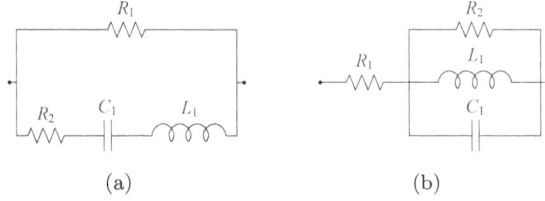

(a)   (b)

Fig. 3.9 Four-element configurations that are equivalent to the configurations in Fig. 3.7, respectively.

derivation processes were not presented there. For each figure in Figs. 3.5–3.7, one only needs to discuss one configuration, as realizability conditions of other configurations can be immediately obtained by the principle of duality or the principle of frequency inverse.

Theorem 3.2. *A biquadratic impedance $Z(s)$ in the form of (3.1), where $a_i > 0$, $i = 0, 1, 2$, $b_j > 0$, $j = 0, 1, 2$, and $R \neq 0$, is realizable as the configuration in Fig. 3.5(a), if and only if*

$$B > 0, \tag{3.5a}$$

$$E_b = 0. \tag{3.5b}$$

Furthermore, if the condition is satisfied, the values of the elements can be expressed as

$$R_1 = \frac{a_2}{b_2}, \quad R_2 = \frac{B}{b_2 b_0}, \quad L_1 = \frac{B}{b_1 b_0}, \quad C_1 = \frac{b_2 b_1}{B}. \tag{3.6}$$

Proof. *Necessity.* The impedance of the configuration in Fig. 3.5(a) is calculated, giving

$$Z(s) = \frac{R_1 C_1 L_1 s^2 + (R_1 R_2 C_1 + L_1)s + (R_1 + R_2)}{C_1 L_1 s^2 + R_2 C_1 s + 1}.$$

Suppose that the biquadratic impedance can be calculated in the form of (3.1), where $a_i > 0$, $i = 0, 1, 2$, $b_j > 0$, $j = 0, 1, 2$, and $R \neq 0$. Then, one obtains

$$R_1 C_1 L_1 = k a_2, \tag{3.7a}$$

$$R_1 R_2 C_1 + L_1 = k a_1, \tag{3.7b}$$

$$R_1 + R_2 = k a_0, \tag{3.7c}$$

$$C_1 L_1 = k b_2, \tag{3.7d}$$

$$R_2 C_1 = k b_1, \tag{3.7e}$$

$$1 = k b_0, \tag{3.7f}$$

where $k > 0$. Note that (3.7f) is equivalent to

$$k = \frac{1}{b_0}. \tag{3.8}$$

From (3.7a) and (3.7d), one can obtain the expression of R_1. Substituting (3.8) and the expression of R_1 into (3.7c), one obtains R_2, which implies Condition (3.5a). Substituting (3.8) and the expression of R_2 into (3.7e), one obtains C_1. Finally, from (3.7a) and (3.7b), one obtains the expression of L_1 and Condition (3.5b).

Sufficiency. Let R_1, R_2, L_1, and C_1 satisfy (3.6), and let k satisfy (3.8), which are all positive and finite by Condition (3.5a). Furthermore, (3.7a)–(3.7f) must hold by Condition (3.5b). Therefore, the biquadratic impedance $Z(s)$ in the form of (3.1) can be realized as the configuration in Fig. 3.5(a), with the values of the elements being expressed as (3.6). □

Theorem 3.3. *A biquadratic impedance $Z(s)$ in the form of (3.1), where $a_i > 0$, $i = 0, 1, 2$, $b_j > 0$, $j = 0, 1, 2$, and $R \neq 0$, is realizable as the configuration in Fig. 3.6(a), if and only if*

$$\Gamma_b = 0. \tag{3.9}$$

Furthermore, if the condition is satisfied, then the values of the elements can be expressed as

$$R_1 = \frac{a_0}{b_0}, \; R_2 = \frac{a_2}{b_2}, \; L_1 = \frac{a_2 a_1}{a_2 b_0 + a_0 b_2}, \; C_1 = \frac{a_2 b_0 + a_0 b_2}{a_1 a_0}. \tag{3.10}$$

Proof. *Necessity.* The impedance of the configuration in Fig. 3.6(a) is calculated, giving

$$Z(s) = \frac{R_1 R_2 L_1 C_1 s^2 + L_1 (R_1 + R_2)s + R_1 R_2}{R_1 L_1 C_1 s^2 + (L_1 + R_1 R_2 C_1)s + R_2}.$$

Then, it follows that

$$R_1 R_2 L_1 C_1 = k a_2, \; L_1(R_1 + R_2) = k a_1, \; R_1 R_2 = k a_0,$$
$$R_1 L_1 C_1 = k b_2, \; L_1 + R_1 R_2 C_1 = k b_1, \; R_2 = k b_0, \tag{3.11}$$

where $k > 0$. From the third and sixth equation of (3.11), one obtains R_1. Thus, the second equation is equivalent to $L_1 = k a_1 / (R_1 + R_2) = k a_1 b_0 / (a_0 + k b_0^2)$, and the fourth one is equivalent to $C_1 = k b_2 / (R_1 L_1) = b_2 (a_0 + k b_0^2)/(a_1 a_0)$. Substituting all these into the first equation, k is obtained as $k = a_2 / (b_2 b_0)$, resulting in the expressions of R_2, L_1, and

C_1. Finally, together with the fifth equation of (3.11), one obtains Condition (3.9).

Sufficiency. Let R_1, R_2, L_1, and C_1 satisfy (3.10), and let k satisfy $k = a_2/(b_2 b_0)$, which are all positive and finite by Condition (3.9). Furthermore, it can be verified that (3.11) must hold by Condition (3.9). Therefore, the biquadratic impedance $Z(s)$ in the form of (3.1) can be realized as the configuration in Fig. 3.6(a), with the values of the elements being expressed as (3.10). □

Theorem 3.4. *A biquadratic impedance $Z(s)$ in the form of (3.1), where $a_i > 0$, $i = 0, 1, 2$, $b_j > 0$, $j = 0, 1, 2$, and $R \neq 0$, is realizable as the configuration in Fig. 3.7(a), if and only if*

$$B = 0, \tag{3.12a}$$

$$A > 0. \tag{3.12b}$$

Furthermore, if the condition is satisfied, then the values of the elements can be expressed as

$$R_1 = \frac{a_1}{b_1}, \quad R_2 = \frac{A}{b_1 b_0}, \quad L_1 = \frac{a_2 A}{a_0 b_1^2}, \quad C_1 = \frac{b_1^2}{A}. \tag{3.13}$$

Proof. *Necessity.* The impedance of the configuration in Fig. 3.7(a) is calculated to be

$$Z(s) = \frac{L_1 C_1 (R_1 + R_2) s^2 + R_1 R_2 C_1 s + (R_1 + R_2)}{L_1 C_1 s^2 + R_2 C_1 s + 1}.$$

Then, it follows that

$$L_1 C_1 (R_1 + R_2) = k a_2, \quad R_1 R_2 C_1 = k a_1, \quad R_1 + R_2 = k a_0,$$
$$L_1 C_1 = k b_2, \quad R_2 C_1 = k b_1, \quad 1 = k b_0, \tag{3.14}$$

where $k > 0$. The sixth equation of (3.14) is equivalent to $k = 1/b_0$. By the first and third equations, one obtains $L_1 C_1 = a_2/b_2$. Together with the fourth equation (3.14), one obtains $k = a_2/(a_0 b_2)$, which implies that Condition (3.12a) holds. Furthermore, the expressions of R_1, R_2, L_1, and C_1 can be obtained, which further implies Condition (3.12b).

Sufficiency. Let R_1, R_2, L_1, and C_1 satisfy (3.13), and let $k = 1/F$, which are all positive and finite by Conditions (3.12a)–(3.12b). Furthermore, it can be verified that (3.14) must hold by Condition (3.12a). Therefore, the biquadratic impedance $Z(s)$ in the form of (3.1) is realizable as the configuration in Fig. 3.7(a), with the values of the elements satisfying (3.13). □

Table 3.1 Necessary and sufficient conditions for the biquadratic impedance in the form (3.1) to be realizable as the configurations shown in Figs. 3.5–3.7, where the second columns provide the numbers of the configurations in [Ladenheim (1948)].

Configuration	[Ladenheim (1948)]	Realizability Condition
Fig. 3.5(a)	No. 87	$B > 0$, and $E_b = 0$
Fig. 3.5(b)	No. 88	$B > 0$, and $D_a = 0$
Fig. 3.5(c)	No. 63	$B < 0$, and $D_b = 0$
Fig. 3.5(d)	No. 62	$B < 0$, and $E_a = 0$
Fig. 3.6(a)	No. 96	$\Gamma_b = 0$
Fig. 3.6(b)	No. 97	$\Gamma_a = 0$
Fig. 3.7(a)	No. 71	$B = 0$, and $A > 0$
Fig. 3.7(b)	No. 74	$B = 0$, and $A < 0$

The realizability conditions of other configurations in Figs. 3.5–3.7 can be directly derived from Figs. 3.5(a), 3.6(a), and 3.7(a) by the principle of duality ($a_2 \leftrightarrow b_2$, $a_1 \leftrightarrow b_1$, and $a_0 \leftrightarrow b_0$), the principle of inversion ($a_2 \leftrightarrow a_0$ and $b_2 \leftrightarrow b_0$), or the principle of frequency inversion ($a_2 \leftrightarrow b_0$, $a_1 \leftrightarrow b_1$, and $a_0 \leftrightarrow b_2$). The realizability conditions of these configurations are summarized in Table 3.1.

Theorem 3.5. *A biquadratic impedance $Z(s)$ in the form of (3.1), where $a_i > 0$, $i = 0, 1, 2$, $b_j > 0$, $j = 0, 1, 2$, is realizable with no more than four elements, if and only if at least one of the following conditions holds: 1) $R = 0$; 2) $B = 0$; 3) $B > 0$ and $D_a E_b = 0$; 4) $B < 0$ and $D_b E_a = 0$; 5) $\Gamma_a \Gamma_b = 0$.*

Proof. Combining Lemma 3.8, Theorem 3.1, and Table 3.1, this theorem can be proved. □

Two numerical examples of realization are provided as follows.

Example 3.1. Consider the biquadratic impedance $Z(s)$ in the form of (3.1). If $a_2 = 1$, $a_1 = 1$, $a_0 = 3$, $b_2 = 1$, $b_1 = 53/7$, and $b_0 = 4$, then calculations yield that $R = -6075/49 < 0$, and $\Gamma_b = R + a_0 a_2 \Delta_b = 0$. Therefore, $Z(s)$ is realizable as the configuration in Fig. 3.6(b), with $R_1 = 3/4$ Ω, $R_2 = 1$ Ω, $L_1 = 1/7$ H, and $C_1 = 7/3$ F.

Example 3.2. If $a_2 = 1$, $a_1 = 2$, $a_0 = 1$, $b_2 = 1$, $b_1 = 1$, and $b_0 = 2$, then it can be calculated that $R = -4 < 0$, $B = -1 < 0$, $D_b = 1 > 0$, $E_a = -3 < 0$, $\Gamma_a = -4 < 0$, and $\Gamma_b = -11 < 0$, which does not satisfy the condition of Theorem 3.5. Hence, $Z(s)$ cannot be realized with fewer than five elements. However, it can be realized as a five-element configuration in

[Jiang and Smith (2011), Fig. 8], with $R_1 = 1/2\ \Omega$, $R_2 = 1/2\ \Omega$, $R_3 = 4\ \Omega$, $L_1 = 3/4$ H, and $C_1 = 3/4$ F, which is consequently a minimal realization of $Z(s)$ in terms of the total number of elements.

Consider the quarter car vehicle model shown in Fig. 1.3, where the parameters are chosen as $m_s = 250$ kg, $m_u = 35$ kg, and $k_t = 150$ kN/m, and $K(s)$ is the biquadratic impedance in the form of (3.1) with $\alpha_i > 0$, $i = 1, 2, 3$ and $\beta_j > 0$, $j = 1, 2, 3$, of a damper-spring-inerter mechanical network. The parallel connection of the spring k_s and the network $K(s)$ constitute the suspension strut, where k_s is the static stiffness. For brevity, only consider the ride comfort performance measure J_1 defined in [Chen et al. (2012); Smith and Wang (2004)], that is,

$$J_1 = 2\pi(V\kappa)^{1/2}||sT_{\hat{z}_r \to \hat{z}_s}||_2,$$

where $T_{\hat{z}_r \to \hat{z}_s}$ denotes the transfer function from z_r to z_s, $||\cdot||_2$ denotes the standard H_2 norm, and the values of V and κ are taken as $V = 25$ m/s and $\kappa = 5 \times 10^{-7}$ m^{3}cycle^{-1} (a typical British principal road).

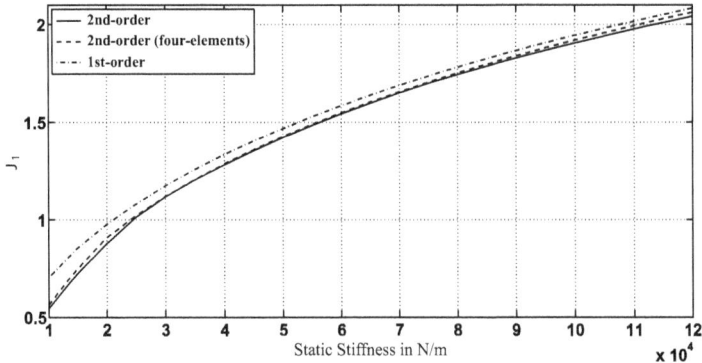

Fig. 3.10   Comparison of the J_1 performance with the results in [Papageorgiou and Smith (2006)].

An example is given based on the results in [Papageorgiou and Smith (2006)], where $K(s)$ in Fig. 1.3 is the impedance of a passive mechanical network consisting of finite interconnections of springs, dampers and inerters. In [Papageorgiou and Smith (2006)], there is no restriction on the number of elements. Here, at most four elements are permitted. The simulation results are shown in Fig. 3.10 and Fig. 3.11, where one can see that the performance J_1 is slightly degraded by restricting the maximum

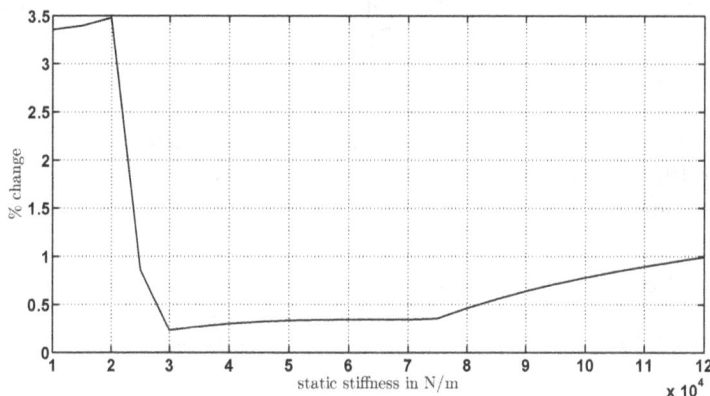

Fig. 3.11 Performance degradation compared with [Papageorgiou and Smith (2006)].

number of elements to be four. It is shown in Fig. 3.11 that at most 3.5% degradation is obtained compared with the results in [Papageorgiou and Smith (2006)].

3.5 Realization of Biquadratic Impedances as Five-Element Bridge Networks

This section is concerned with the realization of biquadratic impedances as five-element bridge networks. As discussed in Section 3.1, this problem remains unsolved. Pantell's simplification [Pantell (1954)] shows that non-series-parallel networks may often contain less redundancy. Moreover, the non-series-parallel structure sometimes has its own advantages in practice [Chen *et al.* (2012)]. It is essential to construct a five-element bridge network, the simplest non-series-parallel network, to solve the minimal realization problem of biquadratic impedances.

This section continues to discuss the biquadratic impedance $Z(s)$ in the form (3.1), that is,

$$Z(s) = \frac{a_2 s^2 + a_1 s + a_0}{b_2 s^2 + b_1 s + b_0},$$

where $a_i > 0$, $i = 0, 1, 2$, and $b_j > 0$, $j = 0, 1, 2$. In the consideration of minimal realizations, it is further assumed that the condition of Theorem 3.5 does not hold, which means that $Z(s)$ cannot be realized with fewer than five elements. For brevity, the set of all such biquadratic functions (with positive coefficients and with the condition of Theorem 3.5 not being satisfied) is denoted by $\mathcal{Z}_b$ in this section.

This section presents a necessary and sufficient condition for a biquadratic impedance $Z(s) \in \mathcal{Z}_b$ to be realizable as a two-reactive five-element bridge network (Theorem 3.12) and the corresponding result for a general five-element bridge network (Theorem 3.13). The corresponding realizations are shown in Figs. 3.12–3.16 and 3.18.

3.5.1   *Preliminary Lemmas*

Lemma 3.11. *[Hughes and Smith (2012)] If a biquadratic impedance $Z(s) \in \mathcal{Z}_b$ is realizable with two reactive elements of different types and an arbitrary number of resistors, then $R < 0$. If a biquadratic impedance $Z(s) \in \mathcal{Z}_b$ is realizable with two reactive elements of the same type and an arbitrary number of resistors, then $R > 0$.*

Lemma 3.12. *If U, V, $W > 0$ satisfy $W \neq 3$,*

$$-4U^2 - 4V^2 + 4UV(W + W^{-1}) - (W - W^{-1})^2 > 0, \tag{3.15}$$

and

$$\begin{aligned} -4U^2 - 4V^2 + 4UV(W - 3W^{-1}) \\ - (W - W^{-1})(W - 9W^{-1}) \geq 0, \end{aligned} \tag{3.16}$$

then

$$4\zeta_c^* W^{-1}(UW - V)(UW - 3V) + 8\lambda_c^*(V^2 - U^2) > 0, \tag{3.17}$$

where ζ_c^* and λ_c^* are defined in Section 3.3.

Proof. It follows from condition (3.16) and the assumption of $W \neq 3$ that $W > 3$. In order to simplify the proof of this lemma, the following coordinate transformation will be used:

$$\begin{aligned} U &= x \cos \frac{\pi}{4} - y \sin \frac{\pi}{4} = \frac{\sqrt{2}}{2}(x - y), \\ V &= x \sin \frac{\pi}{4} + y \cos \frac{\pi}{4} = \frac{\sqrt{2}}{2}(x + y), \end{aligned} \tag{3.18}$$

which does not affect the nature and proof of the lemma. Thus, conditions (3.15)–(3.17) are converted into

$$\begin{aligned} \Psi_1(x, y) := 2W^{-1}(W - 1)^2 x^2 - 2W^{-1}(W + 1)^2 y^2 \\ - W^{-2}(W^2 - 1)^2 > 0, \end{aligned} \tag{3.19}$$

$$\Psi_2(x,y) := 2W^{-1}(W+1)(W-3)x^2 - 2W^{-1}(W+3)(W-1)y^2$$
$$- W^{-2}(W^2-9)(W^2-1) \geq 0, \tag{3.20}$$

and

$$\Sigma(x,y) := -4W^{-2}(W+3)(W+1)^3y^4$$
$$+ 8W^{-2}(W+3)(W+1)(W^2-2W-1)xy^3$$
$$+ 16W^{-1}(W^2-5)x^2y^2$$
$$- 8W^{-2}(W-1)(W-3)(W^2+2W-1)x^3y$$
$$+ 4W^{-2}(W-3)(W-1)^3x^4 \tag{3.21}$$
$$- 2W^{-3}(W-1)(W+3)(W^2-3)(W+1)^2y^2$$
$$+ 4W^{-3}(W^2-9)(W^2-1)^2xy$$
$$- 2W^{-3}(W-3)(W+1)(W^2-3)(W-1)^2x^2 > 0,$$

where $x > 0$ and $W > 3$. Note that (3.66) and (3.67) are equivalent to

$$x > \frac{W+1}{W-1}\sqrt{\frac{2Wy^2+(W-1)^2}{2W}} =: x_{\Psi_1}(y), \tag{3.22}$$

$$x \geq \sqrt{\frac{(W+3)(W-1)(2Wy^2+(W+1)(W-3))}{2W(W+1)(W-3)}} =: x_{\Psi_2}(y), \tag{3.23}$$

respectively. Hence, it can be verified that condition (3.22) yields condition (3.23) when $y \in [-y_1, y_1]$, and condition (3.23) yields condition (3.22) when $y \in (-\infty, -y_1) \cup (y_1, +\infty)$, where

$$y_1 = \frac{W-1}{2W}\sqrt{\frac{(W+1)(W-3)}{2}}.$$

For the case of $y \in [-y_1, y_1]$, it suffices to show that condition (3.22) implies condition (3.21). It can be verified that

$$\Sigma_x(x,y) := \frac{\partial}{\partial x}\Sigma(x,y)$$
$$= 16W^{-2}(W-1)^3(W-3)x^3$$
$$- 24W^{-2}(W-1)(W-3)(W^2+2W-1)x^2y$$
$$+ 32W^{-1}(W^2-5)xy^2 \tag{3.24}$$
$$- 4W^{-3}(W-3)(W+1)(W^2-3)(W-1)^2x$$
$$+ 8W^{-2}(W+3)(W+1)(W^2-2W-1)y^3$$
$$+ 4W^{-3}(W^2-1)^2(W^2-9)y,$$

and

$$
\begin{aligned}
\Sigma_{xx}(x,y) &:= \frac{\partial^2}{\partial x^2}\Sigma(x,y) \\
&= 48W^{-2}(W-1)^3(W-3)x^2 \\
&\quad - 48W^{-2}(W-1)(W-3)(W^2+2W-1)xy \\
&\quad + 32W^{-1}(W^2-5)y^2 \\
&\quad - 4W^{-3}(W+1)(W-1)^2(W^2-3)(W-3).
\end{aligned} \tag{3.25}
$$

Regarding (3.25) as a quadratic function of x, its symmetric axis $x_s(y) = (W^2+2W-1)y/(2(W-1)^2)$ satisfies $x_s(y) < x_{\Psi_1}(y)$, since $4W(W-1)^4\left(((x_s(y))^2 - (x_{\Psi_1}(y))^2\right) = W(W^2+2W-1)^2y^2 - 2(W+1)^2(2Wy^2 + (W-1)^2) = -W(3W^2+2W-3)(W^2-2W-1)y^2 - 2(W+1)^2(W-1)^4 < 0$. Furthermore, one obtains $\Sigma_{xx}(x_{\Psi_1}(y),y) = \Sigma_{xx1}(x_{\Psi_1}(y),y) - \Sigma_{xx2}(x_{\Psi_1}(y),y)$, where

$$
\begin{aligned}
\Sigma_{xx1}(x_{\Psi_1}(y),y) &= 16W^{-2}(3W^4-4W^3-12W^2-4W+9)y^2 \\
&\quad + 4W^{-3}(W+1)(5W^2-3)(W-1)^2(W-3),
\end{aligned}
$$

and

$$
\begin{aligned}
\Sigma_{xx2}(x_{\Psi_1}(y),y) &= 24\sqrt{2}W^{-2}(W+1)(W-3)(W^2+2W-1)y \\
&\quad \times \sqrt{2y^2 + W^{-1}(W-1)^2}.
\end{aligned}
$$

It is obvious that $\Sigma_{xx1}(x_{\Psi_1}(y),y) > 0$. Hence, if $y \in [-y_1,0]$, then $\Sigma_{xx2}(x_{\Psi_1}(y),y) \le 0$, indicating that $\Sigma_{xx}(x_{\Psi_1}(y),y) > 0$. If $y \in (0,y_1]$, then $\Sigma_{xx2}(x_{\Psi_1}(y),y) > 0$. One further obtains $(\Sigma_{xx1}(x_{\Psi_1}(y),y))^2 - (\Sigma_{xx2}(x_{\Psi_1}(y),y))^2 = -2048W^{-3}(W^2-3W-2)(3W^4-2W^3-18W^2-8W+9)y^4 + 256W^{-4}(W+1)(W-3)(3W^5-19W^4+6W^3+86W^2+27W-39)(W-1)^2y^2 + 16W^{-6}(W-3)^2(W+1)^2(5W^2-3)^2(W-1)^4$, which is positive for $y \in (0,y_1]$. Hence, $\Sigma_{xx}(x,y) > 0$ for $y \in [-y_1,y_1]$ and $x > x_{\Psi_1}(y)$, indicating that $\Sigma_x(x,y)$ increases monotonically with x when $x > x_{\Psi_1}(y)$ for any fixed value y satisfying $y \in [-y_1,y_1]$. For (3.24), one has $\Sigma_x(x_{\Psi_1}(y),y) = \Sigma_{x2}(x_{\Psi_1}(y),y) - \Sigma_{x1}(x_{\Psi_1}(y),y)$, where

$$
\begin{aligned}
\Sigma_{x1}(x_{\Psi_1}(y),y) &= 16W^{-2}(W-1)^{-1}(W+1)(W^4-8W^2-8W+3)y^3 \\
&\quad + 8W^{-2}(W-1)(W-3)(W+2)(W+1)^2y,
\end{aligned}
$$

and

$$
\begin{aligned}
\Sigma_{x2}(x_{\Psi_1}(y),y) &= 2\sqrt{2}W^{-3}(W-1)^{-1}(W+1)\sqrt{2y^2 + W^{-1}(W-1)^2} \\
&\quad \times \big(4W(W^4-4W^2-8W+3)y^2 \\
&\quad\quad + (W-3)(W+1)(W^2+1)(W-1)^2\big).
\end{aligned}
$$

It can be verified that $\Sigma_{x2}(x_{\Psi_1}(y), y) > 0$. Hence, if $y \in [-y_1, 0]$, then $\Sigma_{x1}(x_{\Psi_1}(y), y) \leq 0$, implying that $\Sigma_x(x_{\Psi_1}(y), y) > 0$. If $y \in (0, y_1]$, then $\Sigma_{x1}(x_{\Psi_1}(y), y) > 0$. One obtains

$$\begin{aligned}
f(Y) := & (\Sigma_{x2}(x_{\Psi_1}(y), y))^2 - (\Sigma_{x1}(x_{\Psi_1}(y), y))^2 \\
= & 2048 W^{-2}(W-1)^{-2}(W+1)^2 S_1 Y^3 \\
& - 256 W^{-4}(W+1)^2 S_2 Y^2 \\
& + 16 W^{-6}(W+1)^3 (W-1)^2 (W-3) S_3 Y \\
& + 8 W^{-7}(W+1)^4 (W^2+1)^2 (W-1)^4 (W-3)^2,
\end{aligned}$$
(3.26)

where $S_1 = W^4 - 6W^2 - 8W + 3 > 0$, $S_2 = W^6 - 8W^5 - 5W^4 + 36W^3 + 63W^2 - 4W - 3$, $S_3 = W^6 - 10W^5 + 15W^4 + 44W^3 + 39W^2 - 34W + 9$, and $Y = y^2$. The *Sturm chain* [Gantmacher (1980), Sec. XV.2] for (3.26) can be obtained using the Euclidean algorithm, as $f_0(Y) = f(Y)$, $f_1(Y) = f'(Y)$, $f_2(Y) = -\text{rem}(f_0, f_1)$, and $f_3(Y) = -\text{rem}(f_1, f_2)$, where $\text{rem}(p_i, p_j)$ denotes the remainder of the polynomial long division of p_i by p_j. The sign of this chain at $Y = 0$ and $Y = Y_1 = y_1^2$ is shown in Table 3.2 (the special case when $f_3(0) = f_3(Y_1) = -\infty$ is excluded, which does not affect the result), where $T = W^{12} + 18W^{11} + 18W^{10} - 242W^9 - 685W^8 + 60W^7 + 3516W^6 + 6260W^5 + 2407W^4 - 1422W^3 + 1234W^2 - 258W - 27$ and $X = 3W^{10} + 4W^9 - 47W^8 - 104W^7 + 54W^6 + 552W^5 + 1178W^4 + 552W^3 - 1065W^2 - 1356W + 549$, and $V(Y)$ denotes the number of sign variations in the Sturm chain for (3.26).

By investigating the roots of $S_3 = 0$, $T = 0$, and $X = 0$ in W for $W > 3$, it is noted that $V(0) = V(Y_1)$. By the *Theorem of Sturm* [Gantmacher (1980), Chapter XV], the number of distinct roots of $f(Y)$ in $Y \in (0, Y_1)$ is $I_0^{Y_1}(f'(Y)/f(Y)) = V(0) - V(Y_1) = 0$. Together with $f(0) > 0$ and $f(Y_1) > 0$, it follows that $f(Y) > 0$ for $Y \in [0, Y_1]$, which further implies that $\Sigma_x(x_{\Psi_1}(y), y) > 0$ for $y \in [-y_1, y_1]$. This means that $\Sigma(x, y)$ increases monotonically with $x > x_{\Psi_1}(y)$ for any $y \in [-y_1, y_1]$. Substituting $x = x_{\Psi_1}(y)$ into $\Sigma(x, y)$ yields $\Sigma(x_{\Psi_1}(y), y) = \Sigma_1(x_{\Psi_1}(y), y) - \Sigma_2(x_{\Psi_1}(y), y)$, where

$$\begin{aligned}
\Sigma_1(x_{\Psi_1}(y), y) = & - 64 W^{-1}(W-1)^{-2}(W+1)^2 y^4 \\
& + 8 W^{-3}(W^2 - 4W - 3)(W+1)^2 y^2 \\
& + 2 W^{-4}(W+1)^3 (W-1)^2 (W-3),
\end{aligned}$$

and

$$\begin{aligned}
\Sigma_2(x_{\Psi_1}(y), y) = & 4\sqrt{2} W^{-3}(W-1)^{-2}(W+1)^2 y \\
& \times (-8W^2 y^2 + (W+1)(W-3)(W-1)^2) \\
& \times \sqrt{2y^2 + W^{-1}(W-1)^2}.
\end{aligned}$$

Table 3.2   The sign of the Sturm chain for (3.26).

sign($f_i(Y)$)	$i=0$	$i=1$	$i=2$	$i=3$	Variations
$Y=0$	$+$	sign(S_3)	sign($-T$)	$-$	$V(0)$
$Y=Y_1$	$+$	$+$	sign($-X$)	$-$	$V(Y_1)$

It is also noted that $\Sigma_1(x_{\Psi_1}(y),y) \geq 0$ for $y \in [-y_1, y_1]$. Moreover, it is observed that if $y \in (-y_1, 0)$, then $\Sigma_2(x_{\Psi_1}(y),y) \leq 0$, implying that $\Sigma(x_{\Psi_1}(y),y) \geq 0$. If $y \in [0, y_1]$, then one has $(\Sigma_1(x_{\Psi_1}(y),y))^2 - (\Sigma_2(x_{\Psi_1}(y),y))^2 = 4W^{-8}(W+1)^4(8W^2y^2-(W+1)(W-3)(W-1)^2)^2 \geq 0$. Hence, $\Sigma(x_{\Psi_1}(y),y) \geq 0$ for $y \in [-y_1, y_1]$. As a result, $\Sigma(x,y) > 0$ for $y \in [-y_1, y_1]$ and $x > x_{\Psi_1}(y)$.

Now, it remains to consider the case of $y \in (-\infty, -y_1) \cup (y_1, +\infty)$. In this case, one only needs to consider condition (3.23), that is, $x \geq x_{\Psi_2}(y)$, as it has been checked that $x_{\Psi_2}(y) > x_{\Psi_1}(y)$. It suffices to show that $\Sigma_x(x,y) > 0$ for $y \in (-\infty, -y_1) \cup (y_1, +\infty)$ and $x \geq x_{\Psi_2}(y)$, since it has been checked that $\Sigma(x_{\Psi_2}(y),y) > 0$ for $y \in (-\infty, -y_1) \cup (y_1, +\infty)$. These can be straightforwardly verified but the tedious detail is omitted for the similarity in the derivation.   □

3.5.2   *Five-Element Bridge Networks with Two Reactive Elements of the Same Type*

Lemma 3.13. *A biquadratic impedance $Z(s) \in \mathcal{Z}_b$ is realizable as a five-element bridge network containing two reactive elements of the same type if and only if $Z(s)$ is realizable as one of the configurations in Figs. 3.12 and 3.13.*

Proof. By Lemma 3.5 and the method of enumeration, this lemma can be immediately proved.   □

Theorem 3.6. *A biquadratic impedance $Z(s) \in \mathcal{Z}_b$ is realizable as the configuration in Fig. 3.12 if and only if $R - 4a_0a_2b_0b_2 \geq 0$. Furthermore, if $R - 4a_0a_2b_0b_2 \geq 0$ and $B > 0$, then $Z(s)$ is realizable as the configuration*

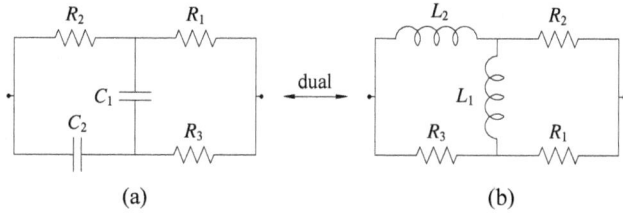

Fig. 3.12 The two-reactive five-element bridge configurations containing the same type of reactive elements, whose one-terminal-pair labeled graphs are (a) $\mathcal{N}_{5a}$ and (b) $\mathcal{N}_{5b}$, respectively, satisfying $\mathcal{N}_{5b} = \text{Dual}(\mathcal{N}_{5a})$. Here, (a) is No. 85 configuration in [Ladenheim (1948)] and (b) is No. 60 configuration in [Ladenheim (1948)].

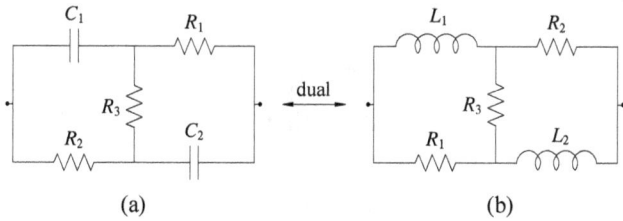

Fig. 3.13 The two-reactive five-element bridge configurations containing the same type of reactive elements, whose one-terminal-pair labeled graphs are (a) $\mathcal{N}_{6a}$ and (b) $\mathcal{N}_{6b}$, respectively, satisfying $\mathcal{N}_{6b} = \text{Dual}(\mathcal{N}_{6a})$. Here, (a) is No. 86 configuration in [Ladenheim (1948)] and (b) is No. 61 configuration in [Ladenheim (1948)].

in Fig. 3.12(a), where the element values satisfy

$$R_2 = \frac{a_0 - b_0 R_1}{b_0}, \tag{3.27a}$$

$$R_3 = \frac{a_2 R_1}{b_2 R_1 - a_2}, \tag{3.27b}$$

$$C_1 = \frac{(a_1 a_2 b_0 + a_0 C) R_1 - a_0 a_1 a_2}{(a_0 - b_0 R_1) R_1^2 M}, \tag{3.27c}$$

$$C_2 = \frac{(a_2 b_0 b_1 + b_2 A) - b_0 b_1 b_2 R_1}{(a_0 - b_0 R_1) M}, \tag{3.27d}$$

and R_1 is the positive root of the following quadratic equation:

$$b_0 b_2 \Gamma_a R_1^2 - (MR + 2a_0 a_2 b_0 b_2 \Delta_{ab}) R_1 + a_0 a_2 \Gamma_b = 0. \tag{3.28}$$

Proof. *Necessity.* The impedance of the configuration in Fig. 3.12(a) is $Z(s) = a(s)/b(s)$, where $a(s) = R_1 R_2 R_3 C_1 C_2 s^2 + ((R_1 R_2 + R_2 R_3 +$

$R_1 R_3)C_1 + (R_1 + R_2)R_3 C_2)s + (R_1 + R_2)$ and $b(s) = (R_1 + R_3)R_2 C_1 C_2 s^2 + ((R_1 + R_3)C_1 + (R_1 + R_2 + R_3)C_2)s + 1$. Since $Z(s) \in \mathcal{Z}_b$ is realizable as the configuration in Fig. 3.12(a), the following equations are satisfied:

$$R_1 R_2 R_3 C_1 C_2 = ka_2, \tag{3.29a}$$

$$(R_1 R_2 + R_3(R_1 + R_2))C_1 + (R_1 + R_2)R_3 C_2 = ka_1, \tag{3.29b}$$

$$R_1 + R_2 = ka_0, \tag{3.29c}$$

$$(R_1 + R_3)R_2 C_1 C_2 = kb_2, \tag{3.29d}$$

$$(R_1 + R_3)C_1 + (R_1 + R_2 + R_3)C_2 = kb_1, \tag{3.29e}$$

$$1 = kb_0. \tag{3.29f}$$

From (3.29f), it follows that

$$k = \frac{1}{b_0}. \tag{3.30}$$

From (3.29a) and (3.29d), it follows that $1/R_1 + 1/R_3 = b_2/a_2$. The assumption that $R_1 > 0$ and $R_3 > 0$ yields

$$b_2 R_1 - a_2 > 0. \tag{3.31}$$

Therefore, R_3 can be solved as (3.27b). Based on (3.29c) and (3.30), R_2 is solved as (3.27a), implying that

$$a_0 - b_0 R_1 > 0. \tag{3.32}$$

Substituting (3.27a), (3.27b), and (3.30) into (3.29b) and (3.29e), C_1 and C_2 are solved as (3.27c) and (3.27d), implying that

$$(a_1 a_2 b_0 + a_0 C)R_1 - a_0 a_1 a_2 > 0, \tag{3.33}$$

$$(a_2 b_0 b_1 + b_2 A) - b_0 b_1 b_2 R_1 > 0. \tag{3.34}$$

Substituting (3.27a)–(3.27d) and (3.30) into (3.29a) yields (3.28). The discriminant of (3.28) in R_1 is obtained as

$$\begin{aligned}
\delta &= (-MR - 2a_0 a_2 b_0 b_2 \Delta_{ab})^2 - 4b_0 b_2 \Gamma_a a_0 a_2 \Gamma_b \\
&= M^2 R(R - 4a_0 a_2 b_0 b_2),
\end{aligned} \tag{3.35}$$

which must be nonnegative. Together with Lemma 3.11, it follows that $R - 4a_0 a_2 b_0 b_2 \geq 0$. Moreover, based on (3.31) and (3.32), one obtains $B > 0$. Therefore, if $Z(s) \in \mathcal{Z}_b$ is realizable as the configuration in Fig. 3.12(b), then $R - 4a_0 a_2 b_0 b_2 \geq 0$ and $B < 0$, which are obtained by the principle of duality ($a_2 \leftrightarrow b_2$, $a_1 \leftrightarrow b_1$, and $a_0 \leftrightarrow b_0$).

Sufficiency. By the principle of duality, one only needs to show that if $R - 4a_0 a_2 b_0 b_2 \geq 0$ and $B > 0$, then $Z(s) \in \mathcal{Z}_b$ is realizable as the configuration in Fig. 3.12(a). Since $R - 4a_0a_2b_0b_2 \geq 0$, it follows that $\Gamma_a \geq 4a_0a_2b_0b_2 + b_0b_2\Delta_a = a_1^2b_0b_2 > 0$, $\Gamma_b \geq 4a_0a_2b_0b_2 + a_0a_2\Delta_b = b_1^2a_0a_2 > 0$, $MR + 2a_0a_2b_0b_2(a_1b_1 - 2M) \geq 4a_0a_2b_0b_2M + 2a_0a_2b_0b_2(a_1b_1 - 2M) = 2a_0a_1a_2b_0b_1b_2 > 0$, and the discriminant of (3.28) in R_1 as expressed in (3.35) is nonnegative. Hence, (3.28) in R_1 has one or two nonzero real roots, which must be positive.

Moreover, $AC > 0$ due to $R = AC - B^2 > 0$. Assume that $A < 0$, that is, $a_0b_1 < a_1b_0$. Together with $B > 0$, that is, $a_2b_0 < a_0b_2$, it is obvious that $(a_0b_1)(a_2b_0) < (a_1b_0)(a_0b_2)$, which is equivalent to $C > 0$. This contradicts the fact that $AC > 0$, which has been derived above. Therefore, it is only possible that $A > 0$ and $C > 0$. This further implies that $a_1a_2b_0 + a_0C > 0$ and $a_2b_0b_1 + b_2A > 0$. Replacing R_1 in (3.28) by a_0/b_0 and a_2/b_2 yields $a_0^2C^2 > 0$ and $a_2^2A^2 > 0$, respectively. Therefore, $a_0 - b_0R_1 \neq 0$ and $b_2R_1 - a_2 \neq 0$, provided that R_1 is the positive root of (3.28).

Let the element values of the configuration in Fig. 3.12(a) satisfy (3.27a)–(3.27d), where R_1 is the positive root of (3.28), and let k satisfy (3.30). It can be verified that (3.29a)–(3.29f) hold. Hence, $k^4R = ((R_1 + R_3)R_1C_1 + (R_1 + R_2)R_3C_2)^2R_2^2C_1C_2$. Therefore, the values of C_1 and C_2 must be simultaneously positive or negative. This implies that $(a_1a_2b_0 + a_0C)R_1 - a_0a_1a_2$ and $(a_2b_0b_1 + b_2A) - b_0b_1b_2R_1$ are simultaneously positive or negative. Assume that $(a_1a_2b_0 + a_0C)R_1 - a_0a_1a_2 < 0$ and $(a_2b_0b_1 + b_2A) - b_0b_1b_2R_1 < 0$. Then, one obtains $(a_1a_2b_0 + a_0C)(a_2b_0b_1 + b_2A) < a_0a_1a_2b_0b_1b_2$, which is equivalent to $AC < 0$. This contradicts the fact that $AC > 0$. Therefore, conditions (3.33) and (3.34) must hold, which is equivalent to $(a_0a_1a_2)/(a_1a_2b_0 + a_0C) < R_1 < (a_2b_0b_1 + b_2A)/(b_0b_1b_2)$. Since it follows from $A > 0$ and $C > 0$ that $(a_0a_1a_2)/(a_1a_2b_0 + a_0C) > a_2/b_2$ and $(a_2b_0b_1 + b_2A)/(b_0b_1b_2) < a_0/b_0$, conditions (3.32) and (3.31) must hold. Hence, the element values as expressed in (3.27a)–(3.28) must be positive and finite. Consequently, the given impedance $Z(s)$ is realizable as the specified configuration. $\qquad\square$

Lemma 3.14. *A biquadratic impedance $Z(s) \in \mathcal{Z}_b$ is realizable as the configuration in Fig. 3.13(a), where $R_1 \neq R_2$, if and only if there exists a positive root of*

$$b_0E_bR_3^2 - (R + 2a_2b_0B)R_3 + a_2D_a = 0 \tag{3.36}$$

in R_3 such that

$$b_2 R_3 - a_2 > 0, \tag{3.37a}$$

$$a_0 - b_0 R_3 > 0, \tag{3.37b}$$

$$b_0(C - 2a_2 b_1)CR_3 + a_2(a_0 a_2 b_1^2 - a_1^2 b_0 b_2) > 0. \tag{3.37c}$$

Furthermore, if the condition is satisfied under the assumption that $R_1 > R_2$, then the element values are expressed as

$$R_1 = \frac{(a_0 - b_0 R_3)(1 + \sqrt{\Lambda})}{2b_0}, \tag{3.38a}$$

$$R_2 = \frac{(a_0 - b_0 R_3)(1 - \sqrt{\Lambda})}{2b_0}, \tag{3.38b}$$

$$C_1 = \frac{a_1 - b_1 R_2}{b_0(R_2 + R_3)(R_1 - R_2)}, \tag{3.38c}$$

$$C_2 = \frac{b_1 R_1 - a_1}{b_0(R_1 + R_3)(R_1 - R_2)}, \tag{3.38d}$$

where

$$\Lambda = 1 - \frac{4a_2 b_0 R_3}{(b_2 R_3 - a_2)(a_0 - b_0 R_3)}, \tag{3.39}$$

and R_3 is the positive root of (3.36), satisfying (3.37a)–(3.37c).

Proof. *Necessity.* The impedance of the configuration in Fig. 3.13(a) is calculated as $Z(s) = a(s)/b(s)$, where $a(s) = R_1 R_2 R_3 C_1 C_2 s^2 + ((R_2 + R_3)R_1 C_1 + (R_1 + R_3)R_2 C_2)s + (R_1 + R_2 + R_3)$ and $b(s) = (R_1 R_2 + R_2 R_3 + R_3 R_1)C_1 C_2 s^2 + ((R_2 + R_3)C_1 + (R_1 + R_3)C_2)s + 1$. Then, it follows that

$$R_1 R_2 R_3 C_1 C_2 = ka_2, \tag{3.40a}$$

$$(R_2 + R_3)R_1 C_1 + (R_1 + R_3)R_2 C_2 = ka_1, \tag{3.40b}$$

$$R_1 + R_2 + R_3 = ka_0, \tag{3.40c}$$

$$(R_1 R_2 + R_2 R_3 + R_3 R_1)C_1 C_2 = kb_2, \tag{3.40d}$$

$$(R_2 + R_3)C_1 + (R_1 + R_3)C_2 = kb_1, \tag{3.40e}$$

$$1 = kb_0. \tag{3.40f}$$

Note that (3.40f) gives

$$k = \frac{1}{b_0}. \tag{3.41}$$

Together with (3.40b) and (3.40e), the expressions of C_1 and C_2 are solved as (3.38c) and (3.38d). It follows from (3.40a) and (3.40d) that

$$\frac{1}{R_1} + \frac{1}{R_2} + \frac{1}{R_3} = \frac{b_2}{a_2}. \tag{3.42}$$

Consequently, one can derive condition (3.37a). Without loss of generality, assume that $R_1 > R_2$ based on the symmetry of this configuration. Therefore, from (3.40c), (3.41), and (3.42), the expressions of R_1 and R_2 are solved as (3.38a) and (3.38b), implying condition (3.37b). Substituting the expressions of (3.38a)–(3.38d) and (3.41) into $R_1R_2R_3C_1C_2 - ka_2$ yields

$$R_1R_2R_3C_1C_2 - ka_2 = \frac{a_2}{b_0} \cdot \frac{b_0 E_b R_3^2 - (R + 2a_2b_0B)R_3 + a_2D_a}{B(b_0b_2R_3^2 + (2a_2b_0 - B)R_3 + a_0a_2)}. \quad (3.43)$$

Then, one obtains (3.36). Since Λ must be nonnegative and $b_0b_2R_3^2 + (2a_2b_0 - B)R_3 + a_0a_2$ cannot be zero, it follows that

$$b_0b_2R_3^2 + (2a_2b_0 - B)R_3 + a_0a_2 < 0. \quad (3.44)$$

Substituting the expressions of the roots of (3.36) into (3.44) yields $((R + 2a_2b_0B) + \sqrt{(R + 2a_2b_0B)^2 - 4a_2b_0D_aE_b})(C - 2a_2b_1)C + 2a_2E_b(a_0a_2b_1^2 - a_1^2b_0b_2) > 0$ or $((R + 2a_2b_0B) - \sqrt{(R + 2a_2b_0B)^2 - 4a_2b_0D_aE_b})(C - 2a_2b_1)C + 2a_2E_b(a_0a_2b_1^2 - a_1^2b_0b_2) > 0$, which is equivalent to condition (3.37c).

Sufficiency. Let R_1, R_2, C_1, and C_2 satisfy (3.38a)–(3.38d), R_3 be a positive root of (3.36) satisfying (3.37a) and (3.37b), and k satisfy (3.41). Then, it can be verified that (3.40a)–(3.40f) hold. It suffices to prove that the element values must be positive and finite. From the discussion in the necessity part, it is noted that condition (3.37c) yields $\Lambda > 0$, and conditions (3.37a) and (3.37b) imply $R_1 > 0$ and $R_2 > 0$. Moreover, one has

$$\begin{aligned} C_1C_2 &= \frac{b_1^2(a_0 - b_0R_3)^2\Lambda - (A - a_1b_0 - b_0b_1R_3)^2}{(a_0 - b_0R_3)^2((a_0 + b_0R_3)^2 - (a_0 - b_0R_3)^2\Lambda)\Lambda} \\ &= \frac{(b_2R_3 - a_2)(b_0b_1CR_3^2 - ACR_3 + a_1a_2A)}{(a_0 - b_0R_3)(b_0b_2R_3^2 + (2a_2b_0 - B)R_3 + a_0a_2)R_3^2B} \\ &= \frac{(b_2R_3 - a_2)}{(a_0 - b_0R_3)R_3^2} > 0, \end{aligned} \quad (3.45)$$

where the third equality is (3.36). Since $R_1 > R_2$, it follows from (3.38c) and (3.38d) that $C_1 > 0$ and $C_2 > 0$. $\qquad\square$

Now, the realizability condition for the configuration in Fig. 3.13(a) with $R_1 = R_2$ is investigated.

Lemma 3.15. *A biquadratic impedance $Z(s) \in \mathcal{Z}_b$ is realizable as the configuration in Fig. 3.13(a), where $R_1 = R_2$, if and only if*

$$a_1b_1^2(2A - a_1b_0) - 4b_2A^2 \geq 0, \quad (3.46a)$$

$$a_1b_2(A - a_1b_0) = a_2b_1(2A - a_1b_0) > 0. \quad (3.46b)$$

Furthermore, if the condition is satisfied, then the element values are ex-pressed as

$$R_1 = \frac{a_1}{b_1}, \tag{3.47a}$$

$$R_2 = \frac{a_1}{b_1}, \tag{3.47b}$$

$$R_3 = \frac{A - a_1 b_0}{b_0 b_1}, \tag{3.47c}$$

$$C_1 = \frac{b_1^2 b_2}{a_1(2A - a_1 b_0)C_2}, \tag{3.47d}$$

and C_2 is a positive root of

$$a_1 A(2A - a_1 b_0)C_2^2 - a_1 b_1^2(2A - a_1 b_0)C_2 + b_1^2 b_2 A = 0. \tag{3.48}$$

Proof. *Necessity.* Since $R_1 = R_2$, (3.40a)–(3.40f) become

$$R_1^2 R_3 C_1 C_2 = ka_2, \tag{3.49a}$$

$$(R_1 + R_3)R_1 C_1 + (R_1 + R_3)R_1 C_2 = ka_1, \tag{3.49b}$$

$$2R_1 + R_3 = ka_0, \tag{3.49c}$$

$$(R_1 + 2R_3)R_1 C_1 C_2 = kb_2, \tag{3.49d}$$

$$(R_1 + R_3)C_1 + (R_1 + R_3)C_2 = kb_1, \tag{3.49e}$$

$$1 = kb_0. \tag{3.49f}$$

It follows from (3.49f) that

$$k = \frac{1}{b_0}. \tag{3.50}$$

From (3.49b) and (3.49e), it follows that R_1 satisfies (3.47a), which implies that R_2 satisfies (3.47b). Then, substituting (3.47a) and (3.50) into (3.49c) yields that R_3 satisfies (3.47c). Therefore, $A - a_1 b_0 > 0$. Thus, it follows from (3.49d) and (3.49e) that C_1 satisfies (3.47d) and C_2 is a positive root of (3.48). Consequently, $2A - a_1 b_0 > 0$. Since the discriminant of (3.48) should be nonnegative, one obtains condition (3.46a). Finally, substituting (3.47a)–(3.48) and (3.50) into (3.49a) gives $a_1 b_2(A - a_1 b_0) - a_2 b_1(2A - a_1 b_0) = 0$. Since $A - a_1 b_0 > 0$, one obtains condition (3.46b).

Sufficiency. Let the values of the elements satisfy (3.47a)–(3.48), and let k satisfy (3.50). Both $A - a_1 b_0 > 0$ and condition (3.46a) guarantee that all the elements are positive and finite. Since condition (3.46b) holds, it can be verified that (3.40a)–(3.40f) hold.   □

Combining Lemmas 3.14 and 3.15, the following theorem can be derived, where the realizability condition of the configuration in Fig. 3.13(b) follows from that of Fig. 3.13(a) based on the principle of duality.

Theorem 3.7. *A biquadratic impedance $Z(s) \in \mathcal{Z}_b$ is realizable as the configuration in Fig. 3.13(a) (resp., Fig. 3.13(b)) if and only if $R > 0$ and $R - 4a_2b_0(a_1b_1 - 2B) \geq 0$ (resp., $R > 0$ and $R - 4a_0b_2(a_1b_1 + 2B) \geq 0$).*

Proof. First, it is to show the equivalence between the condition of Lemma 3.14 and

$$R > 0, \ B > 0, \ R - 4a_2b_0(a_1b_1 - 2B) \geq 0, \tag{3.51a}$$

$$2a_2b_0E_b < b_2(R + 2a_2b_0B) < 2a_0b_2E_b, \tag{3.51b}$$

$$(R + 2a_2b_0B)(C - 2a_2b_1)C + 2(a_0a_2b_1^2 - a_1^2b_0b_2)a_2E_b > 0. \tag{3.51c}$$

Suppose that the condition of Lemma 3.14 holds. The discriminant of (3.36) is calculated as $\delta = R(R - 4a_2b_0(a_1b_1 - 2B))$. Lemma 3.11 implies that $R > 0$. Together with $\delta \geq 0$, one concludes that $R - 4a_2b_0(a_1b_1 - 2B) \geq 0$. From (3.37a) and (3.37b), one obtains $B > 0$. Therefore, $R > 0$ indicates $AC > B^2 > 0$, which further implies that $R + 2a_2b_0B > 0$ and

$$E_b = \frac{-b_2B^2 + b_1BC}{B} > \frac{-b_2AC + b_1BC}{B} = \frac{b_0C^2}{B} > 0. \tag{3.52}$$

By [Gantmacher (1980), Ch.XV, Theorems 11 and 13], $R > 0$ implies that $\Delta_a > 0$ and $\Delta_b > 0$. Substituting $R_3 = a_2/b_2$ and $R_3 = a_0/b_0$ into the left-hand side of (3.36) yields, respectively,

$$b_0E_bR_3^2 - (R + 2a_2b_0B)R_3 + a_2D_a\big|_{R_3 = \frac{a_2}{b_2}} = \frac{a_2^2\Delta_bB}{b_2^2} > 0, \tag{3.53}$$

$$b_0E_bR_3^2 - (R + 2a_2b_0B)R_3 + a_2D_a\big|_{R_3 = \frac{a_0}{b_0}} = \Delta_aB > 0. \tag{3.54}$$

Therefore, condition (3.51b) holds. If $C - 2a_2b_1 = 0$, then condition (3.51c) holds because of (3.37c). Otherwise, substituting $R_3 = -a_2(a_0a_2b_1^2 - a_1^2b_0b_2)/(b_0C(C - 2a_2b_1))$ into the left-hand side of (3.36) yields

$$\frac{a_2^2E_b(3a_1a_2b_0b_1 - 2a_0a_2b_1^2 + a_0a_1b_1b_2 - 2a_1^2b_0b_2)^2}{b_0C^2(C - 2a_2b_1)^2} \geq 0. \tag{3.55}$$

Therefore, condition (3.51c) also holds. Conversely, based on the above discussion, one can also prove that (3.51a)–(3.51c) imply the condition of Lemma 3.14.

By (3.4), one can convert (3.51a)–(3.51c) into $W > 1$ as well as

$$R_c := -4U^2 - 4V^2 + 4UV(W + W^{-1}) - (W - W^{-1})^2 > 0, \tag{3.56}$$

$$- 4U^2 - 4V^2 + 4UV(W - 3W^{-1}) - (W - W^{-1})(W - 9W^{-1}) \geq 0, \quad (3.57)$$

$$\begin{aligned} -4U^2 - 4V^2 + 4UV(W - W^{-1}) \\ - (W - W^{-1})(W - 5W^{-1}) + 8V^2W^{-2} > 0, \end{aligned} \quad (3.58)$$

$$- 4V^2 + 4U^2 + 4UV(W - W^{-1}) - (W - W^{-1})(W + 3W^{-1}) > 0, \quad (3.59)$$

and

$$4W^{-1}\zeta_c^*(UW - V)(UW - 3V) + 8\lambda_c^*(V^2 - U^2) > 0. \quad (3.60)$$

It is noted that condition (3.56) yields $U > 1$ and $V > 1$, and condition (3.57) yields $W \geq 3$. If $W = 3$, then $U = V$ by (3.57), contradicting condition (3.60). Hence, $W > 3$. Thus, Lemma 3.12 shows that conditions (3.56) and (3.57) with $W \neq 3$ indicate condition (3.60). If $UV \leq (W - W^{-1})/2$, then conditions (3.58) and (3.59) hold. One has $-4U^2 - 4V^2 + 4UV(W - W^{-1}) - (W - W^{-1})(W - 5W^{-1}) + 8V^2W^{-2} > (W - W^{-1})^2 - 8UVW^{-1} - (W - W^{-1})(W - 5W^{-1}) + 8V^2W^{-2} = 4W^{-1}(W - W^{-1}) - 8UVW^{-1} + 8V^2W^{-2} \geq 8V^2W^{-2} > 0$ and $-4V^2 + 4U^2 + 4UV(W - W^{-1}) - (W - W^{-1})(W + 3W^{-1}) > (W - W^{-1})^2 + 8U^2 - 8UVW^{-1} - (W - W^{-1})(W + 3W^{-1}) = -4W^{-1}(W - W^{-1}) - 8UVW^{-1} + 8U^2 \geq -8W^{-1}(W - W^{-1}) + 8U^2 > 8(U^2 - 1) > 0$, by condition (3.56). Similarly, if $UV > (W - W^{-1})/2$, then one can also show that conditions (3.58) and (3.59) hold because of condition (3.57). Therefore, the condition of $W \neq 3$ and (3.56)–(3.57) together is equivalent to that of $W > 1$ and (3.56)–(3.60). Moreover, through (3.4), the condition of Lemma 3.15 is converted into

$$2WV - 3U > 0, \quad (3.61)$$

$$2WU^2 + 2WV^2 - UV(W^2 + 3) = 0, \quad (3.62)$$

$$U^2 + W^2V^2 + 3U^2V^2 - 2WUV^3 - 2WUV \leq 0. \quad (3.63)$$

When $W = 3$ and conditions (3.56) and (3.57) hold, one obtains $U = V > 2\sqrt{3}/3$, implying that (3.61)–(3.63) hold.

The proof is completed if one can show that conditions (3.61)–(3.63) imply conditions (3.56) and (3.57). Indeed, by the transformation (3.66), conditions (3.61)–(3.63) are further converted into $(2W - 3)x + (2W + 3)y > 0$, $(W + 1)(W + 3)y^2 - (W - 1)(W - 3)x^2 = 0$, and $(3 - 2W)x^4 - 4Wx^3y - 6x^2y^2 + 4Wxy^3 + (2W + 3)y^4 + 2((W - 1)x + (W + 1)y)^2 \leq 0$ with $x > 0$, which are in turn equivalent to

$$y = \pm\sqrt{\frac{(W - 1)(W - 3)}{(W + 1)(W + 3)}}x, \quad (3.64)$$

$$x \geq \frac{W + 1}{2W}\sqrt{\frac{(W - 1)(W + 3)}{2}}. \quad (3.65)$$

Then, conditions (3.56) and (3.57) are converted into

$$2W^{-1}(W-1)^2x^2 - 2W^{-1}(W+1)^2y^2 \\ - W^{-2}(W^2-1)^2 > 0, \tag{3.66}$$

$$2W^{-1}(W+1)(W-3)x^2 \\ - 2W^{-1}(W+3)(W-1)y^2 \\ - W^{-2}(W^2-9)(W^2-1) \geq 0. \tag{3.67}$$

Substituting (3.64) into (3.66) and (3.67) yields

$$\frac{8(W-1)x^2}{W+3} - \frac{(W+1)^2(W-1)^2}{W^2} \geq 0, \tag{3.68}$$

$$\frac{8(W-3)x^2}{W+1} - \frac{(W^2-1)(W^2-9)}{W^2} \geq 0, \tag{3.69}$$

respectively. It is obvious that conditions (3.64) and (3.65) imply conditions (3.66) and (3.67).

Therefore, $Z_c(s) \in \mathcal{Z}_{b_c}$ is realizable as the configuration in Fig. 3.13(a) if and only if conditions (3.56) and (3.57) hold. Through (3.3), the condition for $Z(s) \in \mathcal{Z}_b$ is obtained as stated in the theorem. □

Theorem 3.8. *A biquadratic impedance $Z(s) \in \mathcal{Z}_b$ is realizable as a five-element bridge network containing two reactive elements of the same type, if and only if $R > 0$ and $R \geq \min\{4a_0b_2(a_1b_1 + 2B), 4a_2b_0(a_1b_1 - 2B), 4a_0a_2b_0b_2\}$.*

Proof. This theorem can be directly proved by combining Lemma 3.13 and Theorems 3.6 and 3.7. □

3.5.3 Five-Element Bridge Networks with One Inductor and One Capacitor

Lemma 3.16. *A biquadratic impedance $Z(s) \in \mathcal{Z}_b$ is realizable as a five-element bridge network with two reactive elements of different types, if and only if $Z(s)$ is realizable as one of the configurations in Figs. 3.14–3.16.*

Proof. This lemma can be proved by the approach of enumeration. □

The realizability condition of Fig. 3.14 has already been established in [Jiang and Smith (2011)], which is stated as follows.

Lemma 3.17. *[Jiang and Smith (2011)] A biquadratic impedance $Z(s) \in \mathcal{Z}_b$ is realizable as the configuration in Fig. 3.14(a) (resp., Fig. 3.14(b))*

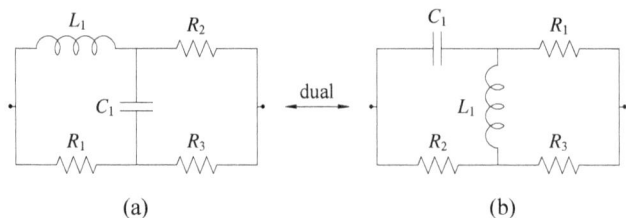

Fig. 3.14   The two-reactive five-element bridge configurations containing different types of reactive elements, whose one-terminal-pair labeled graphs are $\mathcal{N}_{7a}$ and $\mathcal{N}_{7b}$, respectively, satisfying $\mathcal{N}_{7b} = \mathrm{Dual}(\mathcal{N}_{7a})$. Here, (a) is No. 70 configuration in [Ladenheim (1948)], and (b) is No. 95 configuration in [Ladenheim (1948)].

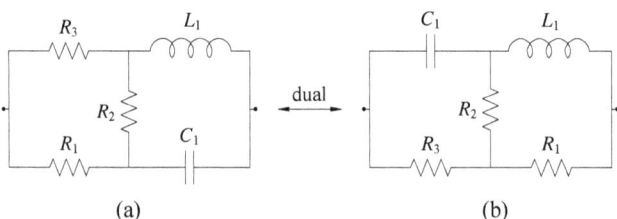

Fig. 3.15   The two-reactive five-element bridge configurations containing different types of reactive elements, whose one-terminal-pair labeled graphs are $\mathcal{N}_{8a}$ and $\mathcal{N}_{8b}$, respectively, satisfying $\mathcal{N}_{8b} = \mathrm{Dual}(\mathcal{N}_{8a})$. Here, (a) is No. 105 configuration in [Ladenheim (1948)], and (b) is No. 107 configuration in [Ladenheim (1948)].

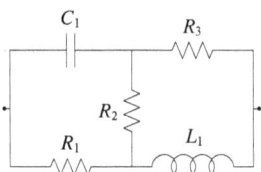

Fig. 3.16   The two-reactive five-element bridge configuration containing different types of reactive elements, whose one-terminal-pair labeled graph is $\mathcal{N}_9$, and it is No. 108 configuration in [Ladenheim (1948)].

if and only if $B < 0$, $R - 4a_0 b_2(a_1 b_1 + 2B) \leq 0$ (resp., $B > 0$, $R - 4a_2 b_0(a_1 b_1 - 2B) \leq 0$), and signs of D_b, E_a, and $(R - 2a_2 b_0 B)$ (resp., D_a, E_b, and $(R + 2a_2 b_0 B)$) are not all the same. If $R - 2a_0 b_2 B = 0$ (resp., $R + 2a_2 b_0 B = 0$), then $D_b E_a < 0$ (resp., $D_a E_b < 0$).

By the star-mesh transformation [Versfeld (1970)], it can be verified that the configuration in Fig. 3.15(a) is equivalent to the configuration in

Fig. 3.17. The element values for configurations in Figs. 3.15–3.17 have been listed in [Ladenheim (1948)], without detailed derivation.

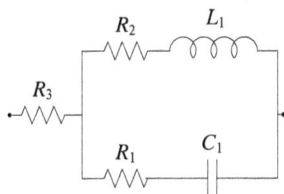

Fig. 3.17   The two-reactive five-element series-parallel configuration that is equivalent to Fig. 3.15(a), which is No. 104 configuration in [Ladenheim (1948)].

Lemma 3.18. *A biquadratic impedance $Z(s) \in \mathscr{Z}_b$ is realizable as the configuration in Fig. 3.17 if and only if $R < 0$ and one of the following two conditions is satisfied:*

1. *$\Gamma_a < 0$, either $D_b > 0$ for $B < 0$ or $E_b > 0$ for $B > 0$, and the signs of Δ_b, $-\Delta_{ab}$, and Γ_a are not all the same (when one of them is zero, the other two are nonzero and have different signs);*
2. *$\Gamma_a > 0$, $\Delta_b > 0$, $\Delta_{ab} > 0$, and either $D_b + b_0 B < 0$ for $B < 0$ or $E_b - b_2 B < 0$ for $B > 0$.*

Furthermore, if the condition is satisfied, then the element values are expressed as

$$R_1 = \frac{a_2 - b_2 R_3}{b_2}, \tag{3.70a}$$

$$R_2 = \frac{a_0 - b_0 R_3}{b_0}, \tag{3.70b}$$

$$L_1 = \frac{M - 2b_0 b_2 R_3}{b_0 b_1}, \tag{3.70c}$$

$$C_1 = \frac{b_1 b_2}{M - 2b_0 b_2 R_3}, \tag{3.70d}$$

and R_3 is a positive root of

$$b_0 b_2 \Delta_b R_3^2 - 2b_0 b_2 \Delta_{ab} R_3 + \Gamma_a = 0, \tag{3.71}$$

satisfying

$$R_3 < \min\{a_2/b_2, a_0/b_0\}. \tag{3.72}$$

Proof. *Necessity.* The impedance of the configuration in Fig. 3.17 is calculated as $Z(s) = a(s)/b(s)$, where $a(s) = (R_1 + R_3)L_1C_1s^2 + ((R_1R_3 + R_2R_3 + R_1R_2)C_1 + L_1)s + (R_2 + R_3)$ and $b(s) = L_1C_1s^2 + (R_1 + R_2)C_1s + 1$. Then, it follows that

$$(R_1 + R_3)L_1C_1 = ka_2, \tag{3.73a}$$

$$(R_1R_3 + R_2R_3 + R_1R_2)C_1 + L_1 = ka_1, \tag{3.73b}$$

$$R_2 + R_3 = ka_0, \tag{3.73c}$$

$$L_1C_1 = kb_2, \tag{3.73d}$$

$$(R_1 + R_2)C_1 = kb_1, \tag{3.73e}$$

$$1 = kb_0. \tag{3.73f}$$

From (3.73f), one obtains

$$k = \frac{1}{b_0}. \tag{3.74}$$

Based on (3.73a) and (3.73d), it is concluded that R_1 satisfies (3.70a). From (3.73c), it follows that the expression of R_2 satisfies (3.70b). Therefore, condition (3.72) holds. Substituting (3.70a), (3.70b), and (3.74) into (3.73e) gives the expression of C_1 as (3.70d). As a result, the expression of L_1 is obtained from (3.73d) as (3.70c). Finally, substituting (3.70a)–(3.70d) into (3.73b) gives (3.71). It follows that the discriminant of (3.71) in R_3 is

$$\delta = (2b_0b_2\Delta_{ab})^2 - 4b_0b_2\Delta_b\Gamma_a = -4b_0b_1^2b_2R. \tag{3.75}$$

Since (3.71) must have at least one positive root, one has $R < 0$, and at most one of Δ_b, Δ_{ab}, and Γ_a is zero. Substituting $R_3 = a_0/b_0$, $R_3 = a_2/b_2$, and $R_3 = (a_0/b_0 + a_2/b_2)/2 = M/(2b_0b_2)$ into the left-hand side of (3.71), respectively, implies that

$$b_0b_2\Delta_bR_3^2 - 2b_0b_2\Delta_{ab}R_3 + \Gamma_a\big|_{R_3 = \frac{a_0}{b_0}} = -\frac{BD_b}{b_0}, \tag{3.76}$$

$$b_0b_2\Delta_bR_3^2 - 2b_0b_2\Delta_{ab}R_3 + \Gamma_a\big|_{R_3 = \frac{a_2}{b_2}} = \frac{BE_b}{b_2}, \tag{3.77}$$

$$b_0b_2\Delta_bR_3^2 - 2b_0b_2\Delta_{ab}R_3 + \Gamma_a\big|_{R_3 = \frac{M}{2b_0b_2}} = \frac{b_1^2B^2}{4b_0b_2} > 0. \tag{3.78}$$

Since the condition of Theorem 3.5 does not hold, $\Gamma_a \neq 0$. When $\Gamma_a < 0$, it follows from (3.78) that (3.71) has only one positive root in R_3 such that (3.72) holds. Therefore, the signs of Δ_b, $-\Delta_{ab}$, and Γ_a are not all the same (when one of them is zero, the other two are nonzero and have different signs). Moreover, if $B < 0$, then $a_0/b_0 < M/(2b_0b_2) < a_2/b_2$, which implies that $D_b > 0$ to guarantee (3.76) to be positive; if $B > 0$, then $a_0/b_0 > M/(2b_0b_2) > a_2/b_2$, which implies that $E_b > 0$ to guarantee (3.77) to be positive.

When $\Gamma_a > 0$, it follows from (3.78) that $\Delta_b > 0$ and $-\Delta_{ab} < 0$. If $B < 0$, then $a_0/b_0 < M/(2b_0b_2) < a_2/b_2$. Therefore, in either the case when (3.76) is negative or the case when (3.76) is nonnegative, it implies that $\Delta_{ab}/\Delta_b < a_0/b_0$ holds. The above two cases correspond to $-b_1A < -b_0B$ and $-b_0B \leq -b_1A < -2b_0B$, respectively. Hence, combining them yields $-b_1A < -2b_0B$, which is equivalent to $D_b + b_0B < 0$. Similarly, one can prove that if $B > 0$ then $E_b - b_2B < 0$.

Sufficiency. Let the element values in Fig. 3.17 satisfy (3.70a)–(3.70d), R_3 be a positive root of (3.71) satisfying (3.72), and k satisfy (3.74). Then, $a_2 - b_2R_3 > 0$, $a_0 - b_0R_3 > 0$, and $M - 2b_0b_2R_3 > 0$. It can be verified that (3.73a)–(3.73f) hold. $R < 0$ implies that the discriminant of (3.71) as expressed in (3.75) is positive.

If condition 1 is satisfied, then as discussed in the necessity part there exists a unique positive root of (3.71) in terms of R_3 such that (3.72) holds.

If condition 2 holds, and either $-b_1A < -b_0B$ for $B < 0$ or $-b_2B < -b_1C$ for $B > 0$, then it can be proved that there exists a unique positive root for (3.71) in terms of R_3 such that (3.72) holds.

If condition 2 holds, and either $-b_0B \leq -b_1A < -2b_0B$ for $B < 0$ or $-2b_2B < -b_1C \leq -b_2B$ for $B > 0$, then there are two positive roots for (3.71) in terms of R_3 such that (3.72) holds.

Consequently, the element values must be positive and finite. The given impedance $Z(s)$ is realizable as the specified network. $\qquad\square$

The element values in Fig. 3.15(a) can be obtained from those in Fig. 3.17 through the following transformation: $R_P/R_2 \to R_1$, $R_P/R_3 \to R_2$, $R_P/R_1 \to R_3$, $C_1 \to C_1$, and $L_1 \to L_1$, where $R_P = R_1R_2 + R_2R_3 + R_3R_1$.

Since the realizability condition of the configuration in Fig. 3.15(a) is equivalent to that of Lemma 3.18, a necessary and sufficient condition for the realization as the configurations in Fig. 3.15 is obtained as follows.

Theorem 3.9. *A biquadratic impedance $Z(s) \in \mathcal{Z}_b$ is realizable as one of the configurations in Fig. 3.15 if and only if $R < 0$ and one of the following three conditions is satisfied:*

    1. $\Gamma_a < 0$, either $D_b > 0$ for $B < 0$ or $E_b > 0$ for $B > 0$, and the signs of Δ_b, $-\Delta_{ab}$, and Γ_a are not all the same (when one of them is zero, the other two are nonzero and have different signs);

2. $\Gamma_b < 0$, *either $D_a > 0$ for $B > 0$ or $E_a > 0$ for $B < 0$, and the signs of Δ_a, $-\Delta_{ab}$, and Γ_b are not all the same (when one of them is zero, the other two are nonzero and have different signs);*

3. $\Gamma_a > 0$, $\Gamma_b > 0$, *and $\Delta_{ab} > 0$.*

Proof. Conditions 1 and 2 can be obtained from Lemma 3.18 based on the principle of duality.

To obtain condition 3, it suffices to show that

$$\Gamma_{a_c} > 0, \ \Gamma_{b_c} > 0, \ \Delta_{ab_c} > 0 \qquad (3.79)$$

is equivalent to the union of the following two conditions:

a. $\Gamma_{a_c} > 0$, $V > 1$, $\Delta_{ab_c} > 0$, and either $2UV - 2WV^2 + (W - W^{-1}) < 0$ for $W < 1$ or $2UV - 2W^{-1}V^2 - (W - W^{-1}) < 0$ for $W > 1$;

b. $\Gamma_{b_c} > 0$, $U > 1$, $\Delta_{ab_c} > 0$, and either $2UV - 2WU^2 + (W - W^{-1}) < 0$ for $W < 1$ or $2UV - 2W^{-1}U^2 - (W - W^{-1}) < 0$ for $W > 1$.

First, one can show that condition a or b implies (3.79). Without loss of generality, assume that $W > 1$. From condition a, one obtains $W + W^{-1} < 2UV < (W - W^{-1}) + 2V^2W^{-1}$. Therefore, it follows that $\Gamma_{b_c} = -4U^2 + 4UV(W + W^{-1}) - (W + W^{-1})^2 = -V^{-2}(2UV - V(V - \sqrt{V^2 - 1})(W + W^{-1}))(2UV - V(V + \sqrt{V^2 - 1})(W + W^{-1})) > 0$, since $(W + W^{-1}) - V(V - \sqrt{V^2 - 1})(W + W^{-1}) = \sqrt{V^2 - 1}(V - \sqrt{V^2 - 1})(W + W^{-1}) > 0$ and $(W - W^{-1}) + 2V^2W^{-1} - V(V + \sqrt{V^2 - 1})(W + W^{-1}) = -(W - W^{-1})(V^2 - 1) - V\sqrt{V^2 - 1}(W + W^{-1}) < 0$. As a result, condition a implies condition (3.79). Similarly, condition b implies $\Gamma_{a_c} > 0$, which also yields condition (3.79). The case of $W < 1$ can be similarly proved.

Now, it remains to show that condition (3.79) implies condition a or b. Assume that $W > 1$. Since $\Gamma_{a_c} > 0$ and $\Gamma_{b_c} > 0$ yield respectively $U > 1$ and $V > 1$, it follows that if $2UV - 2W^{-1}V^2 - (W - W^{-1}) < 0$ then condition a holds. Otherwise, one obtains $U - \sqrt{U^2 - 2W^{-1}(W - W^{-1})} \leq 2VW^{-1} \leq U + \sqrt{U^2 - 2W^{-1}(W - W^{-1})}$. It can be verified that $U(W + W^{-1}) + (W - W^{-1})\sqrt{U^2 - 1} - W(U + \sqrt{U^2 - 2W^{-1}(W - W^{-1})}) > 0$. Together with $R_c < 0$, one has $V < (U(W + W^{-1}) - (W - W^{-1})\sqrt{U^2 - 1})/2$, which implies that $2UV - 2W^{-1}U^2 - (W - W^{-1}) < U^2(W + W^{-1}) - (W - W^{-1})U\sqrt{U^2 - 1} - 2W^{-1}U^2 - (W - W^{-1}) = (W - W^{-1})\sqrt{U^2 - 1}(\sqrt{U^2 - 1} - U) < 0$. Therefore, condition b is obtained. The case of $W < 1$ can be similarly proved. $\square$

Theorem 3.10. *A biquadratic impedance* $Z(s) \in \mathcal{Z}_b$ *is realizable as the configuration in Fig. 3.16 if and only if* $R < 0$ *and the signs of* Γ_a, Γ_b, *and* $(MR + 2a_0a_2b_0b_2\Delta_{ab})$ *are not all the same (when* $MR + 2a_0a_2b_0b_2\Delta_{ab} = 0$, $\Gamma_a\Gamma_b < 0$). *Furthermore, if the condition holds, then the element values are expressed as*

$$R_1 = \frac{a_0}{b_0}, \tag{3.80a}$$

$$R_3 = \frac{a_2}{b_2}, \tag{3.80b}$$

$$L_1 = \frac{(a_1a_2b_0 + a_0C)R_2 + a_0a_1a_2}{(b_0R_2 + a_0)M}, \tag{3.80c}$$

$$C_1 = \frac{b_0b_1b_2R_2 + (a_0b_1b_2 - b_0C)}{(b_0R_2 + a_0)M}, \tag{3.80d}$$

and R_2 *is a positive root of*

$$b_0b_2\Gamma_aR_2^2 + (MR + 2a_0a_2b_0b_2\Delta_{ab})R_2 + a_0a_2\Gamma_b = 0, \tag{3.81}$$

satisfying

$$b_0b_1b_2R_2 + (a_0b_1b_2 - b_0C) > 0, \tag{3.82a}$$

$$(a_1a_2b_0 + a_0C)R_2 + a_0a_1a_2 > 0. \tag{3.82b}$$

Proof. *Necessity.* The impedance of the configuration in Fig. 3.16 is calculated as $Z(s) = a(s)/b(s)$, where $a(s) = (R_1+R_2)R_3L_1C_1s^2+(R_1R_2R_3C_1+(R_1+R_2+R_3)L_1)s+(R_2+R_3)R_1$ and $b(s) = (R_1+R_2)L_1C_1s^2+((R_1R_2+R_2R_3+R_1R_3)C_1+L_1)s+R_2+R_3$. Then, it follows that

$$(R_1 + R_2)R_3L_1C_1 = ka_2, \tag{3.83a}$$

$$R_1R_2R_3C_1 + (R_1 + R_2 + R_3)L_1 = ka_1, \tag{3.83b}$$

$$(R_2 + R_3)R_1 = ka_0, \tag{3.83c}$$

$$(R_1 + R_2)L_1C_1 = kb_2, \tag{3.83d}$$

$$(R_1R_2 + R_2R_3 + R_1R_3)C_1 + L_1 = kb_1, \tag{3.83e}$$

$$R_2 + R_3 = kb_0. \tag{3.83f}$$

From (3.83a) and (3.83d), the expression of R_3 satisfies (3.80b). From (3.83c) and (3.83f), it follows that R_1 satisfies (3.80a). Substituting (3.80b) into (3.83f) yields

$$k = \frac{b_2R_2 + a_2}{b_0b_2}. \tag{3.84}$$

Therefore, L_1 and C_1 can be solved from (3.83b) and (3.83e) as (3.80c) and (3.80d). The assumption that all the element values are positive and finite implies conditions (3.82a) and (3.82b). Substituting (3.80a)–(3.80d) into (3.83d) yields (3.81). The discriminant of (3.81) in terms of R_2 is obtained as

$$\delta = M^2 R(R - 4a_0 a_2 b_0 b_2). \tag{3.85}$$

Since the discriminant must be nonnegative to guarantee the existence of real roots, together with Lemma 3.11, one obtains $R < 0$, implying that $\delta > 0$ and at most one of Γ_a, Γ_b, and $(MR + 2a_0 a_2 b_0 b_2 \Delta_{ab})$ is zero. If one of them is zero, then it is only possible that $MR + 2a_0 a_2 b_0 b_2 \Delta_{ab} = 0$ and $\Gamma_a \Gamma_b < 0$. If none of them is zero, then it follows that the signs of them cannot be the same to guarantee the existence of a positive root.

Sufficiency. Let the element values in Fig. 3.16 be (3.80a)–(3.80d), R_2 be a positive root of (3.81) satisfying conditions (3.82a) and (3.82b), and k satisfy (3.84). It can be verified that (3.83a)–(3.83f) hold.

It suffices to show that (3.81) always has a positive root, such that R_1, R_3, L_1, C_1 expressed as (3.80a)–(3.80d) are positive and finite. Since R_1 and R_3 in (3.80a) and (3.80b) are obviously positive and finite, one only needs to discuss L_1 and C_1.

It can be verified that if the signs of Γ_a, Γ_b, and $MR + 2a_0 a_2 b_0 b_2 \Delta_{ab}$ satisfy the given conditions, then (3.81) must have at least one positive root, since $R < 0$ implies that the discriminant of (3.81) shown in (3.85) is always positive. Furthermore, $-k^4 R = (R_1 + R_2)^2 (R_2 + R_3)^2 (R_1 R_3 C_1 - L_1)^2 L_1 C_1$. Together with $R < 0$, it implies that L_1 and C_1 are both positive or both negative. Hence, $\chi_1 \chi_2 > 0$, where $\chi_1 := b_0 b_1 b_2 R_2 + (a_0 b_1 b_2 - b_0 C)$ and $\chi_2 := (a_1 a_2 b_0 + a_0 C) R_2 + a_0 a_1 a_2$. Assume that $\chi_1 < 0$ and $\chi_2 < 0$. By letting $a_0 a_2 \chi_1 + b_0 b_2 \chi_2$, one obtains $a_0 a_2 \chi_1 + b_0 b_2 \chi_2 = (a_1 b_0 b_2 R_2 + a_0 a_2 b_1)M < 0$. This contradicts the assumption that all the coefficients are positive. Therefore, $\chi_1 > 0$ and $\chi_2 > 0$, implying that $L_1 > 0$ and $C_1 > 0$.   □

3.5.4   Main Results

Combining Lemma 3.17, Theorems 3.9 and 3.10, the following result is obtained.

Theorem 3.11. *A biquadratic impedance $Z(s) \in \mathcal{Z}_b$ is realizable as a five-element bridge network containing one inductor and one capacitor if and* *only if $R < 0$, and $Z(s)$ is regular² or satisfies the condition of Lemma 3.17.*

Proof. *Necessity.* Lemma 3.11 implies that $R < 0$. It is shown in [Jiang and Smith (2011)] that the biquadratic impedance realizing any configuration in Figs. 3.15 and 3.16 must be regular. By Lemma 3.16, the necessity part is proved.

Sufficiency. Based on Lemma 3.17, it suffices to consider the case when $R < 0$ and $Z(s)$ is regular, which means that the corresponding $Z_c(s)$ is regular.

Assuming that the condition of Theorem 3.5 does not hold, $Z_c(s)$ is regular if and only if (1) $\lambda_c > 0$ or $\lambda_c^\dagger > 0$ when $W < 1$; (2) $\lambda_c^* > 0$ or $\lambda_c^{*\dagger} > 0$ when $W > 1$. It suffices to show that if $R_c < 0$ and $Z_c(s) \in \mathcal{Z}_{b_c}$ is regular then $Z_c(s)$ is realizable as one of the configurations in Figs. 3.15 and 3.16.

Case 1: $\Gamma_{a_c} < 0$ and $\Gamma_{b_c} < 0$. If $U < 1$ and $V < 1$, then $\Delta_{ab_c} = 4UV - 2(W + W^{-1}) < 0$. Suppose that $U \geq 1$. Then, $\Gamma_{a_c} < 0$ implies that $V < (W + W^{-1})(U - \sqrt{U^2 - 1})/2$ or $V > (W + W^{-1})(U + \sqrt{U^2 - 1})/2$, and $\Gamma_{b_c} < 0$ implies $V < (4U^2 + (W + W^{-1})^2)/(4U(W + W^{-1}))$. Since $2U(W+W^{-1})^2(U+\sqrt{U^2 - 1})-((W+W^{-1})^2+4U^2) > 4(2U(U+\sqrt{U^2 - 1})-1)-4U^2 = 4(U^2-1)+8U\sqrt{U^2 - 1} > 0$, it implies that $V < (W+W^{-1})(U-\sqrt{U^2 - 1})/2$. Hence, $\Delta_{ab_c} = 4UV - 2(W + W^{-1}) < 2(W + W^{-1})(U^2 - U\sqrt{U^2 - 1})-2(W+W^{-1}) = 2(W+W^{-1})\sqrt{U^2 - 1}(\sqrt{U^2 - 1}-U) < 0$. Utilizing Theorem 3.9 and (3.4), $Z_c(s)$ is realizable as one of the configurations in Fig. 3.15.

Case 2: Only one of Γ_{a_c} and Γ_{b_c} is negative. By Theorem 3.10 and (3.4), $Z_c(s)$ is realizable as the configuration in Fig. 3.16.

Case 3: $\Gamma_{a_c} > 0$ and $\Gamma_{b_c} > 0$. It implies that $U > 1$ and $V > 1$. If $\Delta_{ab_c} > 0$, then $Z_c(s)$ is realizable as one of the configurations in Fig. 3.15 by Theorem 3.9 and (3.4). If $\Delta_{ab_c} \leq 0$, then $R_c = -4U^2 - 4V^2 + 4UV(W + W^{-1}) - (W - W^{-1})^2 < 0$ yields $-(W+W^{-1})^3+4UV(W+W^{-1})^2-4(U^2+V^2)(W+W^{-1})+8UV < -(W+W^{-1})^3+4UV(W+W^{-1})^2+((W-W^{-1})^2-4UV(W+W^{-1}))(W+W^{-1})+8UV = -4(W+W^{-1})+8UV = 2\Delta_{ab_c} \leq 0$. Therefore, $Z_c(s)$ is realizable as the configuration in Fig. 3.16, based on Theorem 3.10 and (3.4). □

Corollary 3.1. *A biquadratic impedance $Z(s) \in \mathcal{Z}_b$ with $R < 0$ is regular if and only if it is realizable as one of the configurations in Figs. 3.15 and 3.16.*

---

²A necessary and sufficient condition for a biquadratic impedance to be regular has been derived in [Jiang and Smith (2011), Lemma 5] (see Lemma 3.3).

Proof. This corollary can be obtained from the proof of Theorem 3.11. □

Corollary 3.2. *A biquadratic impedance, which is realizable as an irreducible[3] five-element series-parallel network containing one inductor and one capacitor, can always be realized as a five-element bridge network containing one inductor and one capacitor.*

Proof. In [Jiang and Smith (2011)], it has been shown that if the biquadratic impedance $Z(s)$ in the form of (3.1) is realizable as a five-element series-parallel network containing one inductor and one capacitor, then $Z(s)$ must be regular. Since the network is irreducible, it follows that $Z(s) \in \mathcal{Z}_b$ and $R < 0$. Hence, the conclusion directly follows from Theorem 3.11. □

Corollary 3.3. *If a biquadratic impedance $Z(s) \in \mathcal{Z}_b$ is realizable as a network containing one inductor, one capacitor, and at least three resistors, then $Z(s)$ is always realizable as a five-element bridge network containing one inductor and one capacitor.*

Proof. This corollary can be proved by Theorem 3.11 and a theorem in [Reichert (1969)]. □

Theorem 3.12. *A biquadratic impedance $Z(s) \in \mathcal{Z}_b$ is realizable as a two-reactive five-element bridge network if and only if one of the following two conditions holds:*

1. $R > 0$, and $R \geq \min\{4a_0b_2(a_1b_1 + 2B), 4a_2b_0(a_1b_1 - 2B), 4a_0a_2b_0b_2\}$;
2. $R < 0$, and $Z(s)$ is regular or satisfies the condition of Lemma 3.17.

Proof. Combining Theorems 3.8 and 3.11 directly yields the result. □

Next, a corresponding result for general five-element bridge networks is presented.

Theorem 3.13. *A biquadratic impedance $Z(s) \in \mathcal{Z}_b$ is realizable as a five-element bridge network if and only if $Z(s)$ satisfies the condition of Theorem 3.12 or is realizable as a configuration in Fig. 3.18.[4]*

---

[3]An irreducible network means that it can never become equivalent to one containing fewer elements.

[4]A necessary and sufficient condition for the realization of Fig. 3.18 is presented in [Jiang and Smith (2011), Theorem 7].

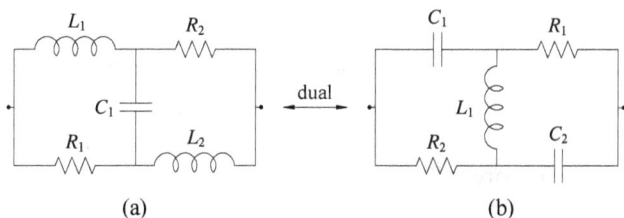

Fig. 3.18 The three-reactive five-element bridge configurations in Theorem 3.13, whose one-terminal-pair labeled graphs are $\mathcal{N}_{10a}$ and $\mathcal{N}_{10b}$, respectively, satisfying $\mathcal{N}_{10b} = \text{Dual}(\mathcal{N}_{10a})$.

Proof. *Sufficiency.* The sufficiency part is obvious.

Necessity. Since $R \neq 0$, the McMillan degree of $Z(s) \in \mathcal{Z}_b$ satisfies $\delta(Z(s)) = 2$. Since the McMillan degree is equal to the minimal number of reactive elements for realizations of $Z(s)$ [Anderson and Vongpanitlerd (1973), pg. 370], there must exist at least two reactive elements. Since $Z(s) \in \mathcal{Z}_b$ has no pole or zero on $j\mathbb{R} \cup \infty$, the number of reactive elements cannot be five. If the number of reactive elements is four (only one resistor), then $Z(0) = Z(\infty)$, which is equal to the element value of the resistor. This implies that $B = 0$, contradicting the assumption that the condition of Theorem 3.5 does not hold. Hence, there must be two or three reactive elements. If there are two reactive elements, then the condition of Theorem 3.12 holds. If there are three reactive elements, then they cannot be of the same type by Lemma 3.5. Furthermore, by the discussion in [Ladenheim (1964)], the network is equivalent to either a five-element series-parallel structure containing one inductor and one capacitor, or a configuration in Fig. 3.18. By Corollary 3.2, the necessity is proved. $\square$

Some notes are in order.

Remark 3.1. Without any detail of derivation, Ladenheim [Ladenheim (1948)] only listed element values for configurations in Figs. 3.12–3.16. In [Ladenheim (1948)], neither explicit conditions in terms of the coefficients of $Z(s)$ alone (like the condition of Theorem 3.6) nor a complete set of conditions are given. Moreover, the special case when $R_1 = R_2$ for Fig. 3.13(a) (No. 86 configuration in [Ladenheim (1948)]) is not discussed in [Ladenheim (1948)].

Remark 3.2. In [Jiang and Smith (2011)], necessary and sufficient conditions are derived only for the realizations as bridge networks that sometimes cannot realize regular biquadratic impedances. In this section, through discussing other two-reactive five-element bridge networks and by combining their conditions, a complete result is obtained as Theorem 3.12. Together with the results in [Ladenheim (1964)], the above result has been further extended to the realizability condition for the general five-element bridge network (Theorem 3.13).

Remark 3.3. Corollary 3.1 shows that a regular biquadratic impedance $Z(s) \in \mathcal{Z}_b$ with $R < 0$ is always realizable as one of the three five-element bridge configurations in Figs. 3.15 and 3.16. It has been shown in [Jiang and Smith (2011)] that four five-element series-parallel configurations are needed to realize such a function. Therefore, the number of configurations covering the case of regularity with $R < 0$ is reduced by one in the present section.

Finally, some numerical examples are presented to illustrate the theoretical results.

Example 3.3. As shown in [Wang *et al.* (2012)], the function
$$Z_{e,2}^{sy} = \frac{s^2 + 2.171 \times 10^8 s + 4.824 \times 10^9}{1.632s^2 + 1.575 \times 10^8 s + 2.838 \times 10^8}$$
is the impedance of an external circuit in a machatronic suspension system, which optimizes the settling time at a certain velocity range and is realizable as a five-element series-parallel configuration in [Wang *et al.* (2012), Fig. 18]. Since $R > 0$, $R - 4a_0a_2b_0b_2 > 0$, and $R - 4a_2b_0(a_1b_1 - 2B) > 0$, $Z_{e,2}^{sy}$ satisfies the condition of Theorem 3.12, so it is realizable as a five-element bridge network. Furthermore, $Z_{e,2}^{sy}$ is realizable as the configuration in Fig. 3.12(a) with $R_1 = 16.232 \ \Omega$, $R_2 = 0.637 \ \Omega$, $R_3 = 0.766 \ \Omega$, $C_1 = 0.0329$ F, and $C_1 = 1.411 \times 10^{-8}$ F, and $Z_{e,2}^{sy}$ is also realizable as the configuration in Fig. 3.13(a) with $R_1 = 14.425 \ \Omega$, $R_2 = 1.378 \ \Omega$, $R_3 = 1.194 \ \Omega$, $C_1 = 4.157 \times 10^{-9}$ F, and $C_2 = 0.0355$ F.

Example 3.4. As shown in [Wang *et al.* (2009)], the function
$$Z_{e,J_1}^{2nd} = \frac{1.665 \times 10^5 s^2 + 5.776 \times 10^5 s + 5.466 \times 10^7}{s^2 + 1.544 \times 10^6 s + 0.342}$$
is the impedance of an external circuit in a machatronic suspension system (LMIS3 layout), which optimizes J_1 (ride comfort) and is realizable as a five-element series-parallel configuration in [Wang *et al.* (2009), Fig. 2(c)]. It can be verified that $R < 0$ and Z_{e,J_1}^{2nd} is regular, implying that Z_{e,J_1}^{2nd} is realizable as a five-element bridge network by Theorem 3.12. Furthermore, Z_{e,J_1}^{2nd} is realizable as the configuration in Fig. 3.16 with $R_1 = 1.598 \times 10^8 \ \Omega$, $R_2 = 0.374 \ \Omega$, $R_3 = 1.665 \times 10^5 \ \Omega$, $C_1 = 0.0282$ F, and $L_1 = 0.108$ H.

3.6 Generalized Synthesis without Real-Part Minimization for Biquadratic Impedances

This section investigates the realization of biquadratic impedances as series-parallel RLC networks. Through the Bott-Duffin synthesis, any positive-real biquadratic impedance is realizable as a series-parallel RLC network containing no more than nine elements. Although the enumeration approach [Ladenheim (1948); Vasiliu (1970)] has been successfully applied to investigate the realization with no more than five and six elements, it is much more complex to apply this method to solve the realization problem with more than six elements. Hence, the investigation on some more general realization procedures are needed, such as the realization procedure for biquadratic impedances proposed in [Tirtoprodjo (1972)], where the resulting network does not contain more elements than the network by the Bott-Duffin synthesis. However, the biquadratic impedance in [Tirtoprodjo (1972)] contains *multiple* real zeros (poles) and arbitrary poles (zeros). Therefore, it is necessary to generalize this procedure to a broader class of biquadratic impedances.

3.6.1 *Preliminary Lemmas*

Lemma 3.19. *Consider a biquadratic impedance $Z(s)$ in the form of (3.1), where $a_i > 0$, $i = 0, 1, 2$, and $b_j > 0$, $j = 0, 1, 2$. If $B = 0$, then $Z(s)$ is realizable as a series-parallel network containing no more than four elements. In addition, if either a_1 or b_1 is zero, then the number of elements is no more than three; if $C = 0$, then $Z(s)$ is realizable as a single resistor.*

Proof. This lemma directly follows from Lemma 3.4 and Theorem 3.5. When $C > 0$ and $B = 0$, $Z(s)$ can be decomposed as $Z(s) = a_2/b_2 + Cs/(b_2^2 s^2 + b_2 b_1 s + b_2 b_0)$. When $C \leq 0$ and $B = 0$, $Z(s)$ can be decomposed as $Z(s) = \left(b_2/a_2 + (-C)s/(a_2^2 s^2 + a_2 a_1 s + a_2 a_0)\right)^{-1}$. Then, this lemma can be easily proved, with realization configurations shown in Fig. 3.19, where some element values may be zero or infinity. $\square$

By Lemmas 3.4 and 3.19, it is assumed that $a_i > 0$, $i = 0, 1, 2$, $b_j > 0$, $j = 0, 1, 2$, and $B \neq 0$ for the remaining part of this section.

Lemma 3.20. *[Baher (1984), Corollary 3.1] For a strictly Hurwitz polynomial $Q(s)$, let $Q(s) = m(s) + n(s)$, where $m(s)$ and $n(s)$ are even and odd parts of $Q(s)$, respectively. Then, $Z(s) = m(s)/n(s)$ must be a reactance function.*

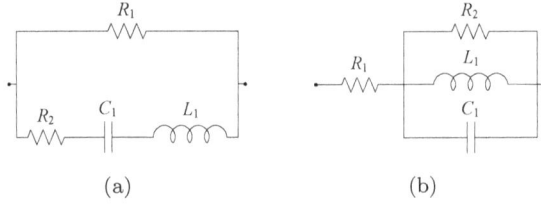

Fig. 3.19   The configurations discussed in the proof of Lemma 3.19, where some element values can be zero or infinite.

Lemma 3.21. *[Foster (1962); Tirtoprodjo (1972)] Consider a biquadratic impedance $Z(s)$ in the form of (3.1), where $a_i > 0$, $i = 0, 1, 2$, $b_j > 0$, $j = 0, 1, 2$, and $B \neq 0$. If $|B| \leq a_1 b_1$, then $Z(s)$ is realizable as a series-parallel network containing no more than seven elements.*

Proof. Supposing that $B > 0$, decompose $Z(s)$ as $Z(s) = (a_2/b_2) \cdot m_2/(m_2 + n_2) + (a_2/b_2) \cdot (m_1 + n_1 - m_2)/(m_2 + n_2) := Z_1(s) + Z_2(s)$, where $m_1 = s^2 + a_0/a_2$, $n_1 = a_1 s/a_2$, $m_2 = s^2 + b_0/b_2$, and $n_2 = b_1 s/b_2$. Based on Lemma 3.20 and by removing the pole of $Z_2^{-1}(s)$ at $s = \infty$, $Z(s)$ is realizable as a network shown in Fig. 3.20(a), where the impedance of N' is at most of degree one while N' contains no more than one reactive element and no more than two resistors. By duality, when $B < 0$, one obtains the configuration in Fig. 3.20(b), whose one-terminal-pair labeled graph is due to that in Fig. 3.20(a).   □

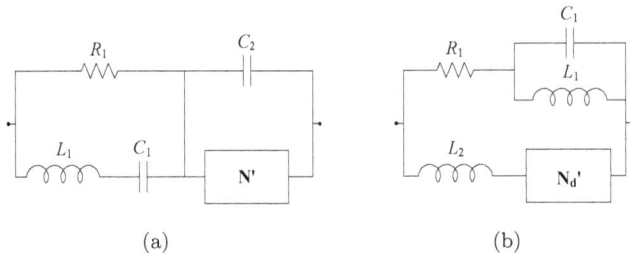

Fig. 3.20   The configurations discussed in the proof of Lemma 3.21, where N' (resp., N'_d) is a series-parallel network containing no more than one reactive element and no more than two resistors.

3.6.2 Biquadratic Impedances with Real Zeros and Arbitrary Poles

It was shown in [Tirtoprodjo (1972)] that there exists $q > 0$ such that $Z(s) = H(s + z)^2/(s^2 + \beta_1 s + \beta_0)$ or $Z(s) = H(s^2 + \alpha_1 s + \alpha_0)/(s + p)^2$ is realizable as a series-parallel network containing no more than nine elements, by simultaneous multiplication of $(s + q)$ when the condition of Lemma 3.21 does not hold. This class will first be generalized to the biquadratic impedances with real zeros (poles) and arbitrary poles (zeros). By the principle of duality, one only needs to consider

$$Z(s) = H\frac{(s + z_1)(s + z_2)}{s^2 + \beta_1 s + \beta_0}, \qquad (3.86)$$

where H, β_0, $\beta_1 > 0$, $z_2 \geq z_1 > 0$, and $z_1 z_2 \neq \beta_0$. Besides, by the principle of frequency inversion, one can assume that $z_1 z_2 > \beta_0$. Since $Z(s)$ is positive-real, with the assumption that it does not satisfy the condition of Lemma 3.21, the following condition should be further satisfied:

$$\frac{(\sqrt{z_1 z_2} - \sqrt{\beta_0})^2}{z_1 + z_2} \leq \beta_1 < \frac{z_1 z_2 - \beta_0}{z_1 + z_2}. \qquad (3.87)$$

Obviously, $Z(s)$ is a minimum function if and only if $\beta_1 = (\sqrt{z_1 z_2} - \sqrt{\beta_0})^2/(z_1 + z_2)$.

An alternative constructive procedure to synthesize a biquadratic impedance $Z(s)$ in the form of (3.86), different from the Bott-Duffin synthesis, is presented in the proof of the following theorem.

Theorem 3.14. *Consider a biquadratic impedance $Z(s)$ in the form of (3.86), where H, b_0, $b_1 > 0$, $z_2 \geq z_1 > 0$, $z_1 z_2 > b_0$, and (3.87) is satisfied. Then, through a constructive procedure, $Z(s)$ is realizable as a series-parallel network containing no more than nine elements. Specially, if $Z(s)$ is a minimum function, then the number of elements is no more than eight.*

Proof. An alternative synthesis procedure is presented here to prove the theorem.

First, multiplying the numerator and denominator with a common factor $(s + q)$, $q > 0$, gives

$$Z(s) = H\frac{s^2 + (z_1 + z_2)s + z_1 z_2}{s^2 + \beta_1 s + \beta_0} \cdot \frac{s + q}{s + q} =: H\frac{m_1 + n_1}{m_2 + n_2},$$

where $m_1 = (q + z_1 + z_2)s^2 + z_1 z_2 q$, $n_1 = s^3 + ((z_1 + z_2)q + z_1 z_2)s$, $m_2 = (q + \beta_1)s^2 + \beta_0 q$, and $n_2 = s^3 + (\beta_1 q + \beta_0)s$. Then, decompose $Z(s)$ as $Z(s) = Z_1(s) + Z_2(s)$, where $Z_1(s) = Hn_2/(m_2 + n_2) = H(s^3 + (\beta_1 q + \beta_0)s)/(s^3 +$

$(q+\beta_1)s^2+(\beta_1q+\beta_0)s+\beta_0q)$ and $Z_2(s) = H(m_1+n_1-n_2)/(m_2+n_2) = H((q+z_1+z_2)s^2+((z_1+z_2-\beta_1)q+z_1z_2-\beta_0)s+z_1z_2q)/(s^3+(q+\beta_1)s^2+(\beta_1q+\beta_0)s+\beta_0q)$. Since m_2+n_2 is a strictly Hurwitz polynomial for β_1, $\beta_0 > 0$, n_2/m_2 must be a reactance by Lemma 3.20. Furthermore, $n_2/(Hm_2)$ is also a reactance, and it is realizable as a lossless network containing two inductors and one capacitor. Therefore, $Z_1(s) = 1/(1/H + m_2/(Hn_2))$ is realizable as a network with one resistor in parallel with a lossless three-element subnetwork.

For $Z_2(s)$, removing the pole of $Z_2^{-1}(s)$ at $s = \infty$ gives $Z_2^{-1}(s) = s/(H(q + z_1 + z_2)) + Z_3^{-1}(s)$, where $Z_3(s)$ is a biquadratic function in the form of

$$Z_3(s) = \frac{\alpha_2's^2+\alpha_1's+\alpha_0'}{\beta_2's^2+\beta_1's+\beta_0'} \tag{3.88}$$

with $\alpha_2' = H(q+z_1+z_2)^2$, $\alpha_1' = H((z_1+z_2-\beta_1)q+(z_1z_2-\beta_0))(q+z_1+z_2)$, $\alpha_0' = Hz_1z_2q(q+z_1+z_2)$, $\beta_2' = q^2+2\beta_1q+(\beta_1z_1+\beta_1z_2+\beta_0-z_1z_2)$, $\beta_1' = \beta_1q^2+(\beta_1z_1+\beta_1z_2+\beta_0-z_1z_2)q+\beta_0(z_1+z_2)$, and $\beta_0' = \beta_0q(q+z_1+z_2)$. Letting $\alpha_2'\beta_0' = \alpha_0'\beta_2'$ yields the following quadratic equation in q

$$(z_1z_2-\beta_0)q^2+2(z_1z_2\beta_1-\beta_0z_1-\beta_0z_2)q$$
$$+\left(z_1z_2(z_1+z_2)\beta_1-(z_1^2z_2^2+\beta_0z_1^2+\beta_0z_2^2+\beta_0z_1z_2)\right)=0. \tag{3.89}$$

To investigate the property of its roots, denote

$$\gamma(\beta_1) := z_1z_2(z_1+z_2)\beta_1-(z_1^2z_2^2+\beta_0z_1^2+\beta_0z_2^2+\beta_0z_1z_2).$$

Then, $\gamma(\beta_1)$ can be regarded as a polynomial function of β_1. Furthermore, denote the lower and upper bounds of β_1 in (3.87) as $x_1 = (\sqrt{z_1z_2}-\sqrt{\beta_0})^2/(z_1+z_2)$ and $x_2 = (z_1z_2-\beta_0)/(z_1+z_2)$. After calculation, one obtains $\gamma(x_1) = -(\beta_0z_1^2+\beta_0z_2^2+2z_1z_2\sqrt{z_1z_2}\sqrt{\beta_0}) < 0$ and $\gamma(x_2) = -\beta_0(z_1+z_2)^2 < 0$. Since $\beta_1 \in [x_1, x_2]$, $\gamma(\beta_1)$ has no pole in $[x_1, x_2]$ by the theorem of Sturm. Consequently, it implies that $\gamma(\beta_1) < 0$. Together with the assumption that $z_1z_2 > \beta_0$, this indicates that (3.89) has a unique positive root q_0.

It is obvious that α_2', α_0', $\beta_0' > 0$ when $q = q_0$. In addition, it can be verified that $\beta_1 < (z_1z_2-\beta_0)/(z_1+z_2)$ is equivalent to $z_1+z_2-\beta_1 > (\beta_1z_1+\beta_0+z_2^2)/z_2$, which implies that $z_1+z_2-\beta_1 > 0$. Hence, $\alpha_1' > 0$ when $q = q_0$. It is noted that the equation $\beta_2' = 0$ in q has one and only one positive root, $q = q_1$, because $\beta_1z_1+\beta_1z_2+\beta_0-z_1z_2 < 0$. To show that $\beta_2' > 0$, it suffices to prove that $q_1 < q_0$. Substituting $q_1^2+2\beta_1q_1+(\beta_1z_1+\beta_1z_2+\beta_0-z_1z_2) = 0$ into the left-hand side of (3.89), that is, $W(q_1) := (z_1z_2-\beta_0)q_1^2+2(z_1z_2\beta_1-\beta_0z_1-\beta_0z_2)q_1+(z_1z_2(z_1+z_2)\beta_1-(z_1^2z_2^2+\beta_0z_1^2+\beta_0z_2^2+\beta_0z_1z_2))$, one obtains that $W(q_1) = 2\beta_0(\beta_1 - z_1 - z_2)q_1 + \beta_0(\beta_1 z_1 + \beta_1 z_2 + \beta_0 - z_1 z_2 - (z_1 + z_2)^2) < 0$, which shows that $q_1 < q_0$.

Now, to show that all the coefficients of (3.88) are nonnegative when $q = q_0$, it remains to show that $\beta_1' \geq 0$ when $q = q_0$. If regarding β_1' as a quadratic function of q, $q > 0$, then the discriminant of β_1' can be calculated as $\delta(\beta_1) = (\beta_1 z_1 + \beta_1 z_2 + \beta_0 - z_1 z_2)^2 - 4\beta_1\beta_0(z_1 + z_2) = (z_1 + z_2)^2\beta_1^2 - 2(z_1 z_2(z_1 + z_2) + \beta_0(z_1 + z_2))\beta_1 + (z_1 z_2 - \beta_0)^2$. Therefore, the two roots of $\delta(\beta_1) = 0$ are obtained as $\beta_1^{(1)} = (\sqrt{z_1 z_2} - \sqrt{\beta_0})^2/(z_1 + z_2)$, and $\beta_1^{(2)} = (\sqrt{z_1 z_2} + \sqrt{\beta_0})^2/(z_1 + z_2)$. It can be checked that $\beta_1^{(1)} = x_1$, and $\beta_1^{(2)} - x_2 = (2\sqrt{\beta_0}(\sqrt{z_1 z_2} + \sqrt{\beta_0}))/(z_1 + z_2) > 0$. Hence, $\delta(\beta_1) < 0$ when $x_1 < \beta_1 < x_2$, and $\delta(\beta_1) = 0$ when $\beta_1 = x_1$. When $\beta_1 \in (x_1, x_2)$, it follows that $\beta_1' > 0$ for $q = q_0$. Thus, by Lemma 3.19, $Z_3(s)$ is realizable as the configuration in Fig. 3.19 (some elements may be short- or open-circuited), where there are no more than four elements. When $\beta_1 = x_1$, that is, $Z(s)$ is a minimum function, the quadratic equation $\beta_1' = 0$ in q has a double positive root, which is $q^{(0)} = (z_1 z_2 - (z_1 + z_2)\beta_1 - \beta_0)/(2\beta_1) = (\sqrt{\beta_0}(z_1 + z_2))/(\sqrt{z_1 z_2} - \sqrt{\beta_0})$. It is noted that (3.89) holds when $q = q^{(0)}$ and $\beta_1 = x_1$, which indicates that $q_0 = q^{(0)}$ when $\beta_1 = x_1$. Therefore, $\beta_1' = 0$ when $q = q^{(0)}$ and $\beta_1 = x_1$. By Lemma 3.19, $Z_3(s)$ is realizable with no more than three elements.

The proof of this theorem is thus completed. $\qquad\square$

Summarizing the realization procedure in Theorem 3.14, it has the following steps.

1. Solve the positive root q_0 of Equation (3.89) in q. The positive root is proved to exist and to be unique, which guarantees the solution of this procedure to be unique.

2. Multiply the numerator and denominator of $Z(s)$ with a common factor $(s + q_0)$, and express it as $Z(s) = H(m_1 + n_1)/(m_2 + n_2)$. Since the degrees of both numerator and denominator increase by one, this step makes the realization possible at the cost of the realization complexity.

3. Decompose $Z(s)$ as $Z(s) = Z_1(s) + Z_2(s)$, where $Z_1(s) = Hn_2/(m_2 + n_2)$, and $Z_2(s) = H(m_1 + n_1 - n_2)/(m_2 + n_2)$. This means that the final realization contains two subnetworks in series, whose impedances are $Z_1(s)$ and $Z_2(s)$.

4. Since $Z_1(s) = (1/H + m_2/(Hn_2))^{-1}$, $Z_1(s)$ is realizable as a resistor in parallel with a lossless subnetwork, whose admittance is $m_2/(Hn_2)$, where $m_2/(Hn_2)$ is further realizable with two inductors and one capacitor in Foster's second form [Baher (1984), pg. 49].

5. By removing the pole of $Z_2(s)$ at $s = \infty$, $Z_2(s)$ is realizable as a capacitor in parallel with a subnetwork N, whose impedance is $Z_3(s)$. It is shown in the proof of Theorem 3.14 that $Z_3(s)$ is always realizable as the configuration in Fig. 3.19, based on the proof of Lemma 3.19.

The final network is a series connection of the realizations of $Z_1(s)$ and $Z_2(s)$, as shown in Fig. 3.21(a), where N is a series-parallel network containing no more than four elements as discussed in Lemma 3.19. The uniqueness of the realization is guaranteed by the unique solution of Equation (3.89) in q.

Since any regular biquadratic impedance $Z(s)$ is realizable with no more than five elements [Jiang (2010), Lemma 8] by the Foster preamble, one may assume that $Z(s)$ is non-regular in the next theorem.

Theorem 3.15. *Consider a biquadratic impedance $Z(s)$ in the form of (3.1), which is non-regular and positive-real with $a_i > 0$, $i = 0, 1, 2$, and $b_j > 0$, $j = 0, 1, 2$. If $\Delta_a \geq 0$ or $\Delta_b \geq 0$, then $Z(s)$ is realizable as the configuration in Fig. 3.20, through the method in the proof of Lemma 3.21, or as the configuration in Fig. 3.21, following the procedure outlined in Theorem 3.14 (or its dual, or frequency inverse). Furthermore, by assuming that the network is realized in series-parallel connection, it contains no more elements than the network by the Bott-Duffin synthesis procedure.*

Proof. Since $Z(s)$ is non-regular, it implies that $B \neq 0$. Based on Lemma 3.21, $Z(s)$ is realizable as the configuration in Fig. 3.20, through the method in the proof of Lemma 3.21, if $|B| \leq a_1 b_1$.

Therefore, assume that the condition of Lemma 3.21 does not hold. If $\Delta_a \geq 0$ and $B > 0$, then $Z(s)$ can be expressed in the form of (3.86), where H, β_0, $\beta_1 > 0$, $z_2 \geq z_1 > 0$, $z_1 z_2 > \beta_0$, and (3.87) is satisfied. Through the method in Theorem 3.14 (outlined right after its proof), the series-parallel network shown in Fig. 3.21(a) containing no more than nine elements is obtained. Specifically, the number of elements is no more than eight when $Z(s)$ is a minimum function[5]. Since the Bott-Duffin synthesis procedure can make a non-regular positive-real biquadratic impedance realizable with no more than nine elements, when the network is in series-parallel connection (the number is eight for a minimum function), the realization contains no more elements than the network by the Bott-Duffin synthesis procedure.

---

[5]A nonzero real-rational function $H(s)$ is called a *minimum function* if $H(s)$ is positive-real with no zero and no pole on $j\mathbb{R} \cup \infty$, and $\Re(H(j\omega_0)) = 0$ with $\Im(H(j\omega_0)) = X_0 \neq 0$ for a finite $\omega_0 > 0$ (see [Balabanian (1958), pp. 90–91]).

By the principle of duality and frequency inversion, the other cases can also be proved. □

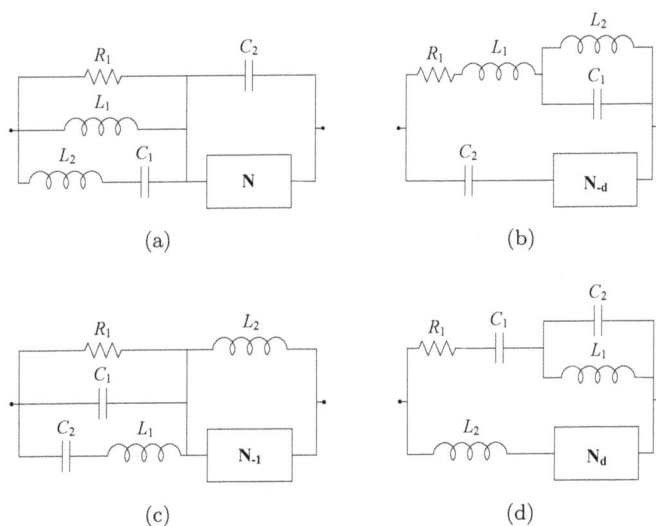

(a)   (b)

(c)   (d)

Fig. 3.21   The configurations obtained by the alternative realization method, where N, N_d, N_{-1}, and N_{-d} are series-parallel networks containing no more than four elements as discussed in Lemma 3.19. The realization configurations of N, N_d, N_{-1}, and N_{-d} are shown in Fig. 3.19 (some elements may be short- or open-circuited).

3.6.3   *Further Generalization to General Biquadratic Impedances*

In the remaining part of this section, the above results will be further generalized to arbitrary biquadratic impedances.

By the principle of duality and frequency inversion, it suffices to consider $Z(s)$ in the form of

$$Z(s) = H \frac{s^2 + \alpha_1 s + \alpha_0}{s^2 + \beta_1 s + \beta_0}, \tag{3.90}$$

where H, α_1, α_0, β_1, $\beta_0 > 0$, and $\alpha_0 > \beta_0$. Since $Z(s)$ is positive-real, assume that the condition of Lemma 3.21 does not hold. The coefficients of $Z(s)$ should further satisfy

$$\frac{(\sqrt{\alpha_0} - \sqrt{\beta_0})^2}{\alpha_1} \leq \beta_1 < \frac{\alpha_0 - \beta_0}{\alpha_1}. \tag{3.91}$$

Then, the realization of this class of impedances is presented in the next theorem.

Theorem 3.16. *Consider a biquadratic impedance $Z(s)$ in the form of (3.90), where H, α_1, α_0, β_1, $\beta_0 > 0$, $\alpha_0 > \beta_0$, and (3.91) holds. Then, the synthesis procedure in the proof of Theorem 3.14 can be generalized to realize $Z(s)$ as a series-parallel network containing no more than nine elements, where the realization is the series-parallel configuration in Fig. 3.21(a). In particular, if $Z(s)$ is a minimum function, then the number of elements is no more than eight.*

Proof. Since the case of $\alpha_1^2 - 4\alpha_0 \geq 0$ has been discussed in Theorem 3.14, it is assumed that $\alpha_1^2 - 4\alpha_0 < 0$ in the proof of this theorem. Multiplying the numerator and denominator of $Z(s)$ by a common factor $(s+q)$, $q > 0$, yields

$$Z(s) = H\frac{s^3 + (q+\alpha_1)s^2 + (\alpha_1 q + \alpha_0)s + \alpha_0 q}{s^3 + (q+\beta_1)s^2 + (\beta_1 q + \beta_0)s + \beta_0 q} =: H\frac{m_1 + n_1}{m_2 + n_2},$$

where $m_1 = (q+\alpha_1)s^2 + \alpha_0 q$, $n_1 = s^3 + (\alpha_1 q + \alpha_0)s$, $m_2 = (q+\beta_1)s^2 + \beta_0 q$, and $n_2 = s^3 + (\beta_1 q + \beta_0)s$. Decompose $Z(s)$ as $Z(s) = Z_1(s) + Z_2(s)$, where $Z_1(s) = Hn_2/(m_2 + n_2) = H(s^3 + (\beta_1 q + \beta_0)s)/(s^3 + (q+\beta_1)s^2 + (\beta_1 q + \beta_0)s + \beta_0 q)$ and $Z_2(s) = H(m_1 + n_1 - n_2)/(m_2 + n_2) = H((q+\alpha_1)s^2 + ((\alpha_1 - \beta_1)q + (\alpha_0 - \beta_0))s + \alpha_0 q)/(s^3 + (q+\beta_1)s^2 + (\beta_1 q + \beta_0)s + \beta_0 q)$. As shown in the proof of Theorem 3.14, $Z_1(s)$ is realizable as a resistor in parallel with a lossless subnetwork containing two inductors and one capacitor, which comprises the left part of the configuration in Fig. 3.21(a). Removing the pole of $Z_2^{-1}(s)$ at $s = \infty$, one obtains $Z_2^{-1}(s) = s/(H(q+\alpha_1)) + Z_3^{-1}(s)$, where $Z_3(s)$ is in the form of (3.88) with $\alpha_2' = H(q+\alpha_1)^2$, $\alpha_1' = H((\alpha_1 - \beta_1)q + (\alpha_0 - \beta_0))(q+\alpha_1)$, $\alpha_0' = H\alpha_0 q(q+\alpha_1)$, $\beta_2' = q^2 + 2\beta_1 q + (\alpha_1 \beta_1 + \beta_0 - \alpha_0)$, $\beta_1' = \beta_1 q^2 + (\alpha_1 \beta_1 + \beta_0 - \alpha_0)q + \alpha_1 \beta_0$, and $\beta_0' = \beta_0 q(q+\alpha_1)$. Let $\alpha_2'\beta_0' = \alpha_0'\beta_2'$. Then, the following equation in q is obtained:

$$(\alpha_0 - \beta_0)q^2 + 2(\alpha_0\beta_1 - \alpha_1\beta_0)q + (\alpha_0\alpha_1\beta_1 + \alpha_0\beta_0 - \alpha_0^2 - \alpha_1^2\beta_0) = 0. \quad (3.92)$$

Similarly to the proof of Theorem 3.14, it can be proved that (3.92) has one and only one positive root q_0 because $\alpha_0 > \beta_0$ and using (3.91). It is clear that α_2', α_0', $\beta_0' > 0$ when $q = q_0$. The equation $\beta_2' = 0$ in q has one and only one positive root because $\alpha_1\beta_1 + \beta_0 - \alpha_0 < 0$. To prove that $\beta_2' > 0$, it suffices to show that $q_1 < q_0$, where q_1 is the unique positive root of $\beta_2' = 0$ in q. Substituting $q_1^2 + 2\beta_1 q_1 + (\alpha_1\beta_1 + \beta_0 - \alpha_0) = 0$ into the left-hand side of

(3.92), that is, $W(q_1) := (\alpha_0 - \beta_0)q_1^2 + 2(\alpha_0\beta_1 - \alpha_1\beta_0)q_1 + (\alpha_0\alpha_1\beta_1 + \alpha_0\beta_0 - \alpha_0^2 - \alpha_1^2\beta_0)$, one obtains that $W(q_1) = 2\beta_0(\beta_1 - \alpha_1)q_1 + \beta_0(\alpha_1\beta_1 + \beta_0 - \alpha_0 - \alpha_1^2)$ and $W(q_1) = (\beta_0/\beta_1)(\alpha_1 - \beta_1)q_1^2 - (\alpha_1\beta_0/\beta_1)(\alpha_0 - \beta_0)$. Therefore, $W(q_1) < 0$, which proves that $q_1 < q_0$.

Regarding $\beta_1' = \beta_1 q^2 + (\alpha_1\beta_1 + \beta_0 - \alpha_0)q + \alpha_1\beta_0$ as a quadratic function of q, and using a similar method to the proof of Theorem 3.14, it can be proved that $\beta_1' > 0$ when $\beta_1 \in (x_1, x_2)$ and $q = q_0$; $\beta_1' = 0$ when $\beta_1 = x_1$ and $q = q_0$. Hence, $Z(s)$ is a minimum function.

Now, to complete the proof of this theorem, it suffices to show that $\alpha_1' > 0$ when $q = q_0$. For the case of $\alpha_1 \geq \beta_1$, $\alpha_1' > 0$ when $q = q_0$ since $\alpha_0 > \beta_0$. For the case of $\alpha_1 < \beta_1$, let $q_m := (\alpha_0 - \beta_0)/(\beta_1 - \alpha_1)$. It is obvious that $\alpha_1' > 0$ when $q = q_0$, if and only if the left-hand side of (3.92) is positive, that is, $W_1(q) = (\alpha_0 - \beta_0)q^2 + 2(\alpha_0\beta_1 - \alpha_1\beta_0)q + (\alpha_0\alpha_1\beta_1 + \alpha_0\beta_0 - \alpha_0^2 - \alpha_1^2\beta_0) > 0$, when $q = q_m$. This is equivalent to $W_1(\beta_1) = (\alpha_1\beta_1 + \alpha_0 - \alpha_1^2 - \beta_0)W_2(\beta_1) > 0$, where $W_2(\beta_1) = \alpha_0\beta_1^2 - \alpha_1(\alpha_0 + \beta_0)\beta_1 + (\alpha_0^2 - 2\alpha_0\beta_0 + \alpha_1^2\beta_0 + \beta_0^2)$, whose discriminant is $(\alpha_0 - \beta_0)^2(\alpha_1^2 - 4\alpha_0)$. Since it is assumed that $\alpha_1^2 - 4\alpha_0 < 0$, the equation $W_2(\beta_1) = 0$ in β_1 has no real root. Therefore, the equation $W_1(\beta_1) = 0$ in β_1 has one and only one real root, which is $(\alpha_1^2 + \beta_0 - \alpha_0)/\alpha_1$. Thus, $W_1(\beta_1) > 0$ holds if and only if $\beta_1 > (\alpha_1^2 + \beta_0 - \alpha_0)/\alpha_1$ if and only if $\alpha_0 - \beta_0 > \alpha_1(\alpha_1 - \beta_1)$, which always holds because of $\alpha_1 < \beta_1$ and $\alpha_0 > \beta_0$. Thus, $Z_3(s)$ is realizable with no more than four elements (three elements for a minimum function). $\qquad\square$

Theorem 3.17. *A non-regular positive-real biquadratic impedance $Z(s)$ in the form of (3.1), where $a_i > 0$, $i = 0,1,2$, and $b_j > 0$, $j = 0,1,2$, is realizable as the configuration in Fig. 3.20, through the method in the proof of Lemma 3.21, or as the configuration in Fig. 3.21 using the procedure outlined in Theorem 3.14 (or its dual, or frequency inverse). Furthermore, if the networks are realized in series-parallel connections, then the realization contains no more elements than the network by the Bott-Duffin synthesis procedure.*

Proof. Combining Lemma 3.21 and Theorem 3.16, this theorem can be proved together with the principle of duality and the principle of frequency inversion. $\qquad\square$

Finally, two examples are presented for illustration.

Example 3.5. Consider a non-regular positive-real biquadratic impedance $Z(s)$ in the form of (3.1), where $a_2 = 1$, $a_1 = 1$, $a_0 = 3$, $b_2 = 1$, $b_1 = 1$, and $b_0 = 1$. Following the realization procedure discussed in this section, $Z(s)$ is realizable as the configuration in Fig. 3.22(a), with the values of the elements satisfying $R_1 = 1$ Ω, $R_2 = 3$ Ω, $R_3 = 3(11 - 6\sqrt{3})/(6\sqrt{3} - 9)$ Ω, $L_1 = \sqrt{3}/(\sqrt{3} - 1)$ H, $L_2 = 3/(4\sqrt{3} - 3)$ H, $L_3 = 3(11 - 6\sqrt{3})/(3\sqrt{3} - 3)$ H, $C_1 = \sqrt{3}(4\sqrt{3} - 3)/9$ F, $C_2 = \sqrt{3}/3$ F, and $C_3 = 1/(11\sqrt{3} - 18)$ F, where the common factor is calculated to be $(s + \sqrt{3} - 1)$. Through the Bott-Duffin synthesis procedure, $Z(s)$ is realizable as the configuration in Fig. 3.22(b), with the values of the elements satisfying $R_1' = 2/3$ Ω, $R_2' = 8/3$ Ω, $R_3' = 1/3$ Ω, $L_1' = 4/3$ H, $L_2' = 4/9$ H, $L_3' = 2$ H, $C_1' = 9/8$ F, $C_2' = 3/4$ F, and $C_3' = 1/4$ F.

It is noted that these two realizations contain the same number of elements.

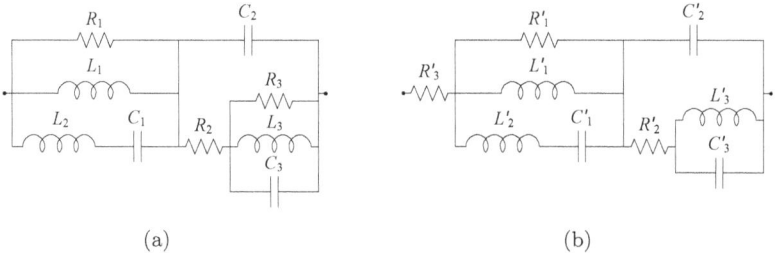

(a)   (b)

Fig. 3.22   The realization configurations discussed in Example 3.5.

The number of elements by the realization method in this section may be smaller than that by the Bott-Duffin synthesis procedure, as shown by the following example.

Example 3.6. Consider a non-regular positive-real biquadratic impedance $Z(s)$ in the form of (3.1), where $a_2 = 1$, $a_1 = 1$, $a_0 = 4$, $b_2 = 1$, $b_1 = 2$, and $b_0 = 1$. Following the realization procedure discussed in this section, $Z(s)$ is realizable as the configuration in Fig. 3.23(a), with the values of the elements satisfying $R_1 = 1$ Ω, $R_2 = 4$ Ω, $L_1 = 5$ H, $L_2 = 15/32$ H, $C_1 = 32/25$ F, and $C_2 = 3/4$ F, where the common factor is $(s + 2)$. Through the Bott-Duffin synthesis procedure, $Z(s)$ is realizable as the configuration in Fig. 3.23(b), with the values of the elements satisfying $R_1' = 25/32$ Ω, $R_2' = 121/32$ Ω, $R_3' = 7/32$ Ω, $L_1' = 275/96$ H, $L_2' = 1375/4096$ H, $L_3' = 4$ H, $C_1' = 4096/3025$ F, $C_2' = 32/33$ F, and $C_3' = 5/44$ F. It is obvious that the number of elements in Fig. 3.23(a) is smaller than that in Fig. 3.23(b).

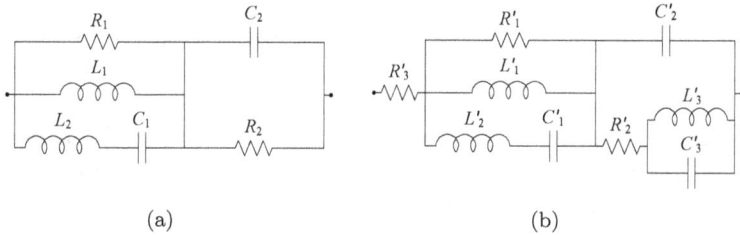

(a)   (b)

Fig. 3.23   The configurations discussed in Example 3.6.

3.7   A Generalized Theorem of Reichert for Biquadratic Minimum Functions

Through the Foster preamble, a minimum function can always be obtained from any positive-real function, and its McMillan degree can be further reduced through a Bott-Duffin cycle without loss of positive-realness. Therefore, the minimum function plays an important role in passive network synthesis. A *biquadratic minimum function* is a minimum function whose McMillan degree is two. Utilizing Pantell's modified Bott-Duffin synthesis procedure [Pantell (1954)] (called *Reza-Pantell-Fialkow-Gerst procedure* in [Hughes (2014)]), any biquadratic minimum impedance function is realizable as a seven-element configuration as shown in Fig. 3.24. The minimal realization problem of biquadratic minimum functions in terms of the total number of elements has been solved in [Foster (1963); Seshu (1959)], showing that a biquadratic minimum function is realizable with fewer than seven elements if and only if it is realizable as the impedance of configurations in [Foster (1963), Fig. 1].

This section presents a generalized theorem of Reichert [Reichert (1969)] for biquadratic minimum functions, which states that a biquadratic minimum impedance function realizable with n reactive elements and an arbitrary number of resistors is always realizable with n reactive elements and two resistors. First, a series of constraints on networks realizing minimum impedance functions are presented. Furthermore, by investigating possible resistor edges incident with vertices of reactive-element graphs under constraints, it is shown that any minimum function realizable as the impedance of networks with three reactive elements and an arbitrary number of resistors can always be realized with three reactive elements and two resistors, verifying the case of $n = 3$. Furthermore, realization configurations are presented, which cover all the cases with four reactive elements and more than two resistors to realize a minimum function. By discussing the realization problem of a biquadratic minimum function as these configurations, the case of $n = 4$ is verified. Finally, the generalized theorem of Reichert for biquadratic minimum functions is obtained.

Some results of this section are present in [Hughes (2014), Section 3.5], which investigate the realization problem of minimum functions with the least number of reactive elements. However, this section presents independent and distinct proofs of these results, and restate these results as the realization problem of biquadratic minimum functions with the least number of resistors. In the investigation in [Hughes (2014), Section 3.5], by showing that the realization can always be a bridge structure in [Hughes (2014), Figure 32] and by investigating the properties of its subnetworks, configurations for realizations of minimal impedances with no more than four reactive elements are obtained.

In this section, based on the topological constraints on realizations of minimum impedances with no more than four reactive elements, possible reactive-element graphs are formulated and possible resistor edges incident with vertices of them to determine configurations are discussed containing more than two resistors that may be able to realize a minimum impedance. The methodology of this section appears to be simpler and more intuitive, expectedly providing some new insight for relevant investigations especially the synthesis of networks with low complexity.

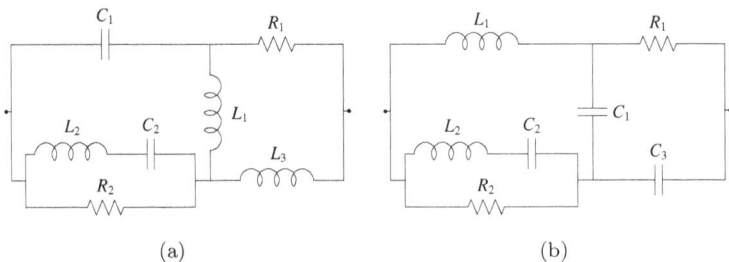

Fig. 3.24   Realizations of a biquadratic minimum impedance function using Pantell's modified Bott-Duffin synthesis procedure [Pantell (1954)] (called *Reza-Pantell-Fialkow-Gerst procedure* in [Hughes (2014)]), where (a) is for $X_0 > 0$, and (b) is for $X_0 < 0$ [Seshu (1959)].

Consider a biquadratic impedance function in the form of (3.1), that is,

$$Z(s) = \frac{a_2 s^2 + a_1 s + a_0}{b_2 s^2 + b_1 s + b_0},$$

where $a_i \geq 0$, $i = 0, 1, 2$, and $b_j \geq 0$, $j = 0, 1, 2$. As stated in Lemma 3.1, $Z(s)$ is positive-real if and only if $a_1 b_1 \geq (\sqrt{a_0 b_2} - \sqrt{a_2 b_0})^2$. As shown in Lemma 3.4, $Z(s)$ is a minimum function if and only if $a_i > 0$, $i = 0, 1, 2$, $b_j > 0$, $j = 0, 1, 2$, and $(\sqrt{a_0 b_2} - \sqrt{a_2 b_0})^2 = a_1 b_1$.

Lemma 3.22. *[Seshu (1959), pg. 346] The network graph of a network N realizing a minimum impedance function cannot contain any path $\mathcal{P}(a, a')$ or cut-set $\mathcal{C}(a, a')$ consisting of only one kind of elements, where vertices a and a' correspond to terminals a and a' of N.*

Lemma 3.23. *[Seshu (1959), pg. 346] The network graph of a network N realizing a minimum impedance function must contain $\mathcal{P}(a, a')$ and $\mathcal{C}(a, a')$ consisting of any two kinds of elements, where vertices a and a' correspond to terminals a and a' of N.*

Lemma 3.24. *[Seshu (1959), pg. 346] A minimum function cannot be realized as the impedance of a network that is the series or parallel connection of two subnetworks, one of which contains only two kinds of elements.*

Lemma 3.25. *If a minimum function is realizable as the impedance of a network N such that the reactive elements cannot be fewer, then the reactive-element graph of N cannot contain the graph in Fig. 3.25 as a subgraph, where vertices a and a' correspond to terminals a and a' of N.*

Proof. This lemma follows directly from Lemma 3.22. $\square$

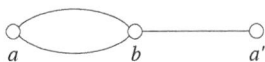

Fig. 3.25 A subgraph, which the reactive-element graph of networks realizing a minimum function cannot contain, where vertices a and a' correspond to terminals a and a'.

Consider a minimum function $H(s)$, realizable as the impedance of a network N with two terminals a and a', where $\Re(H(j\omega_0)) = 0$ and $\Im(H(j\omega_0)) = X_0 \neq 0$ for a finite $\omega_0 > 0$. As shown in [Seshu (1959), pg. 346], all the AC steady-state currents at $\omega = \omega_0$ in the resistors must be zero, as well as the voltages. Furthermore, as pointed out in [Seshu (1959), pg. 347], one can regard N as a two-terminal-pair network by replacing any resistor with a new port, where the point $s = j\omega_0$ must be a zero of the transmission between the input terminals (a, a') and the two new terminals. This means that, at $s = j\omega_0$, the voltage across the new port must be zero if the new port is open-circuited. Moreover, the two-terminal-pair network can only contain reactive elements at $s = j\omega_0$, since all the other resistors can be open-circuited or short-circuited, because of their zero currents and zero voltages. The above property is a constraint on realizations of a minimum function, which is called the *zero-of-transmission requirement* [Seshu (1959), pg. 347].

Lemma 3.26. *A minimum function cannot be realized as the impedance of a network, whose network graph is shown in Fig. 3.26, where (i) any LC-$\mathcal{P}(a, a')$ contains $\mathcal{P}_1(b_1, b_{k+1})$ with $k \geq 1$, (ii) all the edges in $\mathcal{P}_1(b_1, b_{k+1})$ correspond to the same kind of reactive elements, and (iii) any edge $\mathcal{E}(i, j) \in \mathcal{P}_1(b_1, b_{k+1}) \cup \mathcal{P}_1(b_{k+1}, d) \cup \mathcal{P}_1(d, c) \cup \mathcal{P}_1(c, b_1)$ does not belong to any loop[6] whose edges only correspond to reactive elements (LC-loop).*

Proof. By replacing the resistor connecting vertices c and d with a new port, short-circuiting all the resistors whose corresponding edges are in $\mathcal{P}_1(c, b_1) \cup \mathcal{P}_1(b_{k+1}, d)$, and open-circuiting all the other resistors, one obtains a two-terminal-pair (two-port) network containing only reactive elements. Based on the zero-of-transmission requirement, the voltage across the new port (c, d) must be zero, provided that it is open-circuited. This contradicts the fact that all the edges in $\mathcal{P}_1(b_1, b_{k+1})$ correspond to the same kind of reactive elements.   □

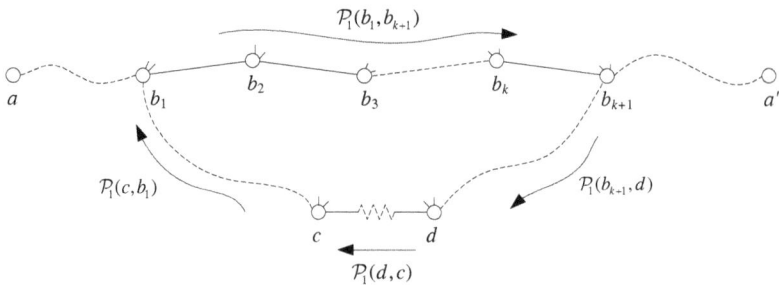

Fig. 3.26   The network graph mentioned in Lemma 3.26, where vertices a and a' correspond to two terminals a and a', a solid line segment denotes a reactive-element edge, a line segment with a resistor notation denotes a resistor edge, a dashed curve segment connecting two vertices (such as vertices b_1 and c) means that there exists at least one path between the two vertices, or these two vertices coincide with each other.

---

[6]The definition of *loop* (or *circuit*) can be referred to Definition 2.8.

Lemma 3.27. *A minimum function cannot be realized as the impedance of a network, whose network graph is shown in Fig. 3.27, where (i) any LC-$\mathcal{P}(a, a')$ contains one of edges incident with vertices b_1 and b_2, and (ii) any edge $\mathcal{E}(i, j) \in \mathcal{P}_1(b_2, d) \cup \mathcal{P}_1(d, c) \cup \mathcal{P}_1(c, b_1)$ does not belong to any loop, whose edges only correspond to reactive elements (LC-loop).*

Proof. Similarly to the proof of Lemma 3.26, this lemma can be proved, which also follows from the zero-of-transmission requirement. □

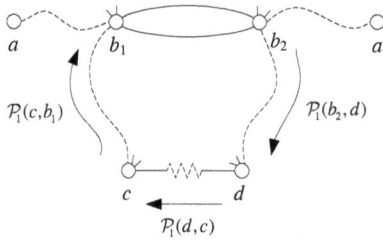

Fig. 3.27   The network graph mentioned in Lemma 3.27, where vertices a and a' correspond to two terminals a and a', a solid curve segment denotes a reactive-element edge, a line segment with a resistor notation denotes a resistor edge, a dashed curve segment connecting two vertices (such as vertices b_1 and c) means that there exists at least one path between the two vertices, or these two vertices coincide with each other.

Theorem 3.18. *Any minimum function, which is realizable as the impedance of a network containing three reactive elements and an arbitrary number of resistors, can always be realized as the impedance of a network containing three reactive elements and two resistors.*

Proof. By Lemma 3.23, there exists LC-$\mathcal{P}(a, a')$, any one of which contains at least two edges. Furthermore, together with Lemma 3.25, all the possible reactive-element graphs are shown in Fig. 3.28, where vertices a and a' correspond to terminals a and a'. Since the generalized star-mesh transformation can eliminate any vertex incident with only resistors [Versfeld (1970)], it suffices to consider vertices of the reactive-element graph. Moreover, it suffices to consider the case where there is no more than one resistor edge R-$\mathcal{E}(i, j)$ incident with any two vertices i and j. As it is stated in [Seshu (1959), Theorem 9] that at least three reactive elements are needed to realize a minimum function, one cannot consider the configuration that is always equivalent to one containing fewer than three reactive elements.

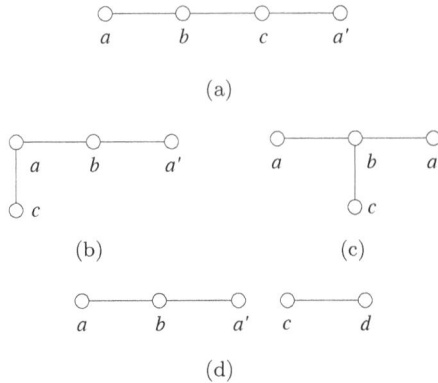

Fig. 3.28   Possible reactive-element graphs of three-reactive-element network realizing a minimum function, where vertices a and a' correspond to terminals a and a', respectively.

For Fig. 3.28(a), since the number of vertices is four, there are no more than six resistor edges. Furthermore, there is no R-$\mathcal{E}(a, a')$ by Lemma 3.22, and there is no R-$\mathcal{E}(a, b)$, R-$\mathcal{E}(b, c)$, or R-$\mathcal{E}(c, a')$ by Lemma 3.26. Therefore, there are no more than two resistors in the possible realizations.

For Fig. 3.28(b), the number of resistor edges cannot exceed six because there are four vertices. There is no R-$\mathcal{E}(a, a')$ by Lemma 3.22, and there is no R-$\mathcal{E}(a, b)$, R-$\mathcal{E}(b, a')$, or R-$\mathcal{E}(b, c)$ by Lemma 3.26. Since R-$\mathcal{E}(a, c)$ and R-$\mathcal{E}(c, a')$ cannot simultaneously exist, there is no more than one resistor in the possible realizations.

For Fig. 3.28(c), based on Lemma 3.26, vertex c cannot be adjacent to vertex a or a' by a resistor edge. This means that the configuration realizing a minimum impedance function can always be equivalent to one containing two reactive elements, which is impossible.

For Fig. 3.28(d), it suffices to consider the case where each of vertices c and d is adjacent to at least two of vertices a, b and a' by resistor edges. Otherwise, the network can be equivalent to one whose reactive-element graph is shown in Fig. 3.28(b) or Fig. 3.28(c). Since R-$\mathcal{E}(a, c)$ and R-$\mathcal{E}(c, a')$ cannot simultaneously exist by Lemma 3.22, there must simultaneously exist R-$\mathcal{E}(a, c)$ and R-$\mathcal{E}(b, c)$ or simultaneously exist R-$\mathcal{E}(c, a')$ and R-$\mathcal{E}(b, c)$. Assume that the network graph contains R-$\mathcal{E}(a, c)$ and R-$\mathcal{E}(b, c)$ (resp., R-$\mathcal{E}(c, a')$ and R-$\mathcal{E}(b, c)$). Then, by Lemma 3.26, vertex d (resp., vertex c) cannot be adjacent to a, b, or a' by a resistor edge. Therefore, the configuration realizing a minimum impedance function can always be equivalent to one containing two reactive elements, which is impossible.   □

Lemma 3.28. *Any minimum function that is realizable as the impedance of a network containing four reactive elements and an arbitrary number of resistors can always be realized as the impedance of a network with four reactive elements and two resistors, or one of the configurations in Fig. 3.29.*

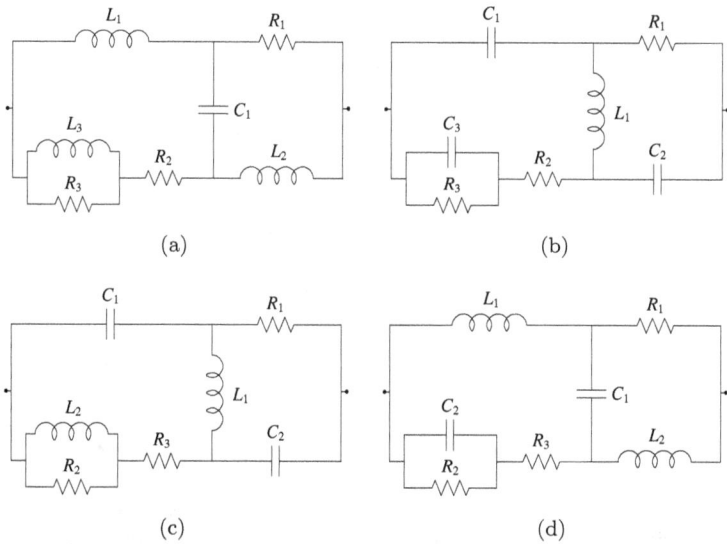

Fig. 3.29 Configurations containing four reactive elements and three resistors, which are mentioned in Lemma 3.28.

Proof. As discussed in the proof of Theorem 3.18, the reactive-element graph for any possible realization of any minimum function must contain at least one LC-$\mathcal{P}(a, a')$, and any LC-$\mathcal{P}(a, a')$ must contain four, three, or two edges. If the reactive-element graph is *connected* [Seshu and Reed (1961), pg. 15] and the shortest LC-$\mathcal{P}(a, a')$ contains four or three edges, then all the possible cases are listed in Fig. 3.30, where vertices a and a' correspond to terminals a and a'. If the reactive-element graph is connected and the shortest LC-$\mathcal{P}(a, a')$ contains two edges, then by Lemma 3.25 the number of vertices must be four or five, which correspond to Figs. 3.31 and 3.32, respectively, where vertices a and a' correspond to terminals a and a'. If the reactive-element graph is not connected, then together with Lemma 3.25 all the possible cases are listed in Fig. 3.33, where vertices a and a' correspond to terminals a and a'.

As discussed in the proof of Theorem 3.18, it is assumed that no vertex incident with only resistor edges exists and there is no more than one resistor edge incident with any two vertices. Moreover, since in Theorem 3.18 one has already considered the configuration realizing a minimum function that can be equivalent to one containing fewer than four reactive elements, this case will not be discussed in the following.

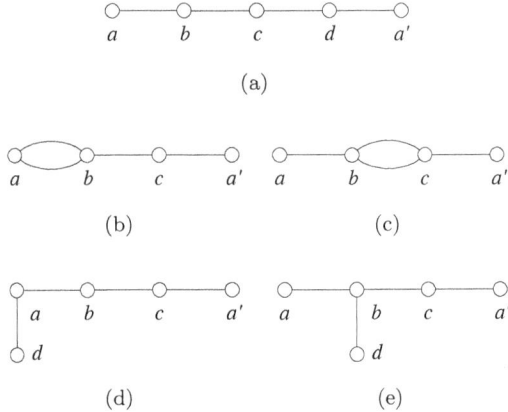

Fig. 3.30 Possible connected reactive-element graphs of networks with four reactive elements realizing a minimum function, where vertices a and a' correspond to terminals a and a' and the shortest LC-$\mathcal{P}(a, a')$ contains four or three edges.

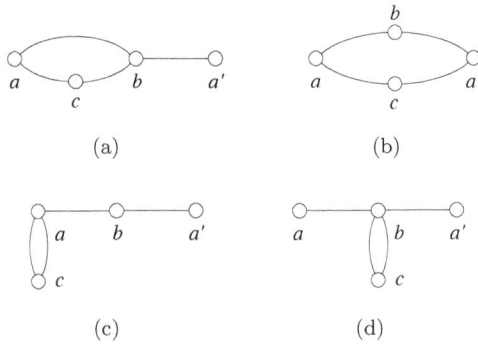

Fig. 3.31 Possible connected four-vertex reactive-element graphs of networks with four reactive elements realizing a minimum function, where vertices a and a' correspond to terminals a and a' and the shortest LC-$\mathcal{P}(a, a')$ contains two edges.

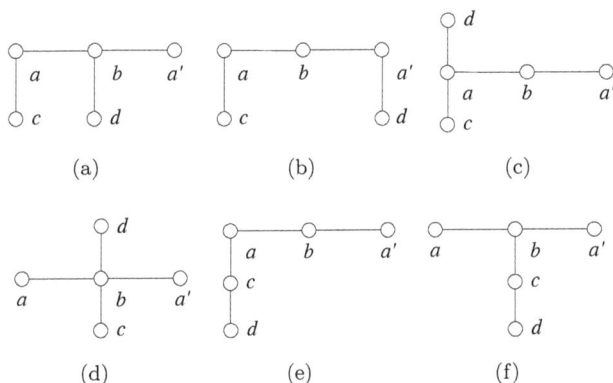

Fig. 3.32   Possible connected five-vertex reactive-element graphs of networks with four reactive elements realizing a minimum function, where vertices a and a' correspond to terminals a and a' and the shortest $LC\text{-}\mathcal{P}(a, a')$ contains two edges.

For Fig. 3.30(a), each of vertices a and a' must be incident with at least one resistor edge based on Lemma 3.22. By Lemma 3.26, there cannot exist $R\text{-}\mathcal{E}(a, b)$, $R\text{-}\mathcal{E}(b, c)$, $R\text{-}\mathcal{E}(c, d)$, or $R\text{-}\mathcal{E}(d, a')$. By Lemma 3.22, there is no $R\text{-}\mathcal{E}(a, a')$. Therefore, there should exist $R\text{-}\mathcal{E}(a, c)$ or $R\text{-}\mathcal{E}(a, d)$. Together with Lemma 3.22, the resistor edges for the possible realizations containing at least three resistors are (I) $R\text{-}\mathcal{E}(a, c)$, $R\text{-}\mathcal{E}(b, d)$, and $R\text{-}\mathcal{E}(b, a')$; (II) $R\text{-}\mathcal{E}(a, c)$, $R\text{-}\mathcal{E}(a, d)$, and $R\text{-}\mathcal{E}(b, a')$. Using Lemma 3.26, all the possible configurations corresponding to the above two cases can be obtained, where one of the configurations corresponding to Case (I) is shown in Fig. 3.34, and one of the configurations corresponding to Case (II) is shown in Fig. 3.35. For the configuration in Fig. 3.34, by short-circuiting R_2, open-circuiting R_3 and replacing R_1 by a new port, it is noted that the zero-of-transmission requirement cannot be satisfied. For the configuration in Fig. 3.35, by open-circuiting R_1, short-circuiting R_2 and replacing R_3 by a new port, it is noted that the zero-of-transmission requirement is not satisfied. One can similarly show that other possible configurations cannot realize minimum functions, either.

For Fig. 3.30(b), since the graph contains four vertices, the total number of resistor edges cannot exceed six. Lemma 3.22 implies that there is no $R\text{-}\mathcal{E}(a, a')$, Lemma 3.26 implies that there is no $R\text{-}\mathcal{E}(b, c)$ or $R\text{-}\mathcal{E}(c, a')$, and Lemma 3.27 implies that there is no $R\text{-}\mathcal{E}(a, b)$. Therefore, all possible realizations contain no more than two resistors. A similar discussion can be carried out for Fig. 3.30(c).

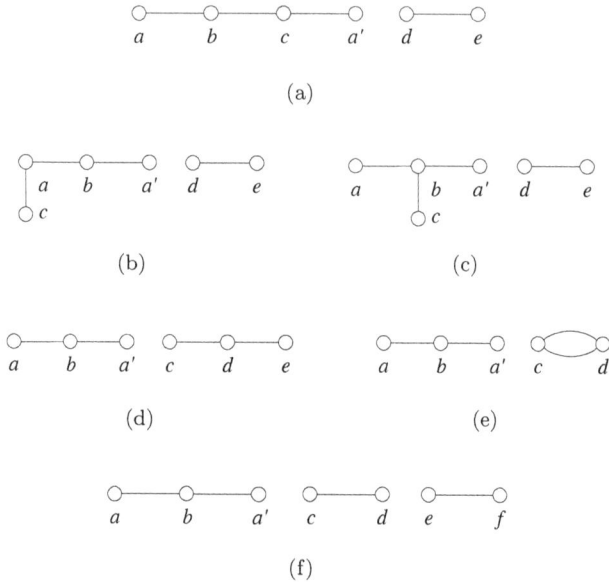

Fig. 3.33 Possible unconnected reactive-element graphs of networks with four reactive elements realizing a minimum function, where vertices a and a' correspond to terminals a and a'.

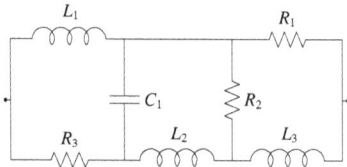

Fig. 3.34 A configuration, whose reactive-element graph is shown in Fig. 3.30(a), with resistor edges R-$\mathcal{E}(a, c)$, R-$\mathcal{E}(b, d)$, and R-$\mathcal{E}(b, a')$.

For Fig. 3.30(d), it is clear that vertex d must be adjacent to at least one of vertices b, c and a' by resistor edges. Lemma 3.22 implies that there is no R-$\mathcal{E}(a, a')$, and Lemma 3.26 implies that there is no R-$\mathcal{E}(a, b)$, R-$\mathcal{E}(b, c)$, R-$\mathcal{E}(c, a')$, or R-$\mathcal{E}(b, d)$. Therefore, there should exist R-$\mathcal{E}(c, d)$ or R-$\mathcal{E}(d, a')$. Moreover, by Lemma 3.22, there exists R-$\mathcal{E}(b, a')$ if R-$\mathcal{E}(d, a')$ does not exist, and there cannot exist R-$\mathcal{E}(a, d)$ if R-$\mathcal{E}(d, a')$ exists. Furthermore, the resistor edges for the possible realizations containing at least three resistors are (I) R-$\mathcal{E}(a, d)$, R-$\mathcal{E}(c, d)$ and R-$\mathcal{E}(b, a')$; (II) R-$\mathcal{E}(a, c)$, R-$\mathcal{E}(c, d)$

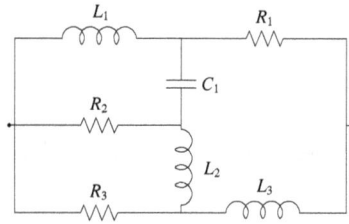

Fig. 3.35 A configuration, whose reactive-element graph is shown in Fig. 3.30(a), with resistor edges R-$\mathcal{E}(a, c)$, R-$\mathcal{E}(a, d)$, and R-$\mathcal{E}(b, a')$.

and R-$\mathcal{E}(b, a')$; (III) R-$\mathcal{E}(a, c)$, R-$\mathcal{E}(a, d)$, R-$\mathcal{E}(c, d)$ and R-$\mathcal{E}(b, a')$; (IV) R-$\mathcal{E}(b, a')$, R-$\mathcal{E}(c, d)$ and R-$\mathcal{E}(d, a')$; (V) R-$\mathcal{E}(a, c)$, R-$\mathcal{E}(b, a')$ and R-$\mathcal{E}(d, a')$. By Lemmas 3.22 and 3.26, one can obtain all the possible configurations corresponding to Cases (I)–(III). All the possible configurations corresponding to Case (I) are shown in Fig. 3.29. Based on the equivalence in [Jiang and Smith (2011), Lemma 11], it is noted that any configuration corresponding to Cases (II) and (III) can always be equivalent to the configuration in Fig. 3.29. By Lemma 3.26, all the configurations corresponding to Case (IV) are shown in Fig. 3.36. It is noted that the zero-of-transmission requirement is not satisfied for Fig. 3.36(a) by open-circuiting R_1, short-circuiting R_2, and replacing R_3 by a new port. A similar discussion can be carried out for Fig. 3.36(b). By Lemma 3.24, any configuration corresponding to Case (V) cannot realize a minimum function.

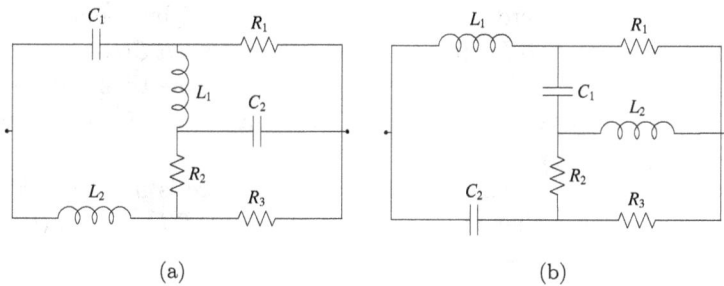

(a) (b)

Fig. 3.36 Configurations, whose reactive-element graphs are the same as Fig. 3.30(d), with resistor edges R-$\mathcal{E}(b, a')$, R-$\mathcal{E}(c, d)$, and R-$\mathcal{E}(d, a')$.

For Fig. 3.30(e), based on Lemma 3.26, vertex d can only be adjacent to vertex a' by a resistor edge among vertices a, c, and a'. Therefore, any possible realization can always be equivalent to the configuration whose reactive-element graph is shown in Fig. 3.30(d).

For Fig. 3.31(a), there exists R-$\mathcal{E}(c,a')$ by Lemmas 3.22 and 3.26. As a consequence, there is no R-$\mathcal{E}(a,c)$, and R-$\mathcal{E}(a,b)$ and R-$\mathcal{E}(b,c)$ cannot simultaneously exist. Therefore, all possible realizations contain no more than two resistors.

For Fig. 3.31(b), there is no R-$\mathcal{E}(a,a')$ by Lemma 3.22, and R-$\mathcal{E}(a,b)$ and R-$\mathcal{E}(b,a')$ (resp., R-$\mathcal{E}(a,c)$ and R-$\mathcal{E}(c,a')$) cannot simultaneously exist. Therefore, the resistor edges for the possible realizations containing at least three resistors are R-$\mathcal{E}(a,b)$, R-$\mathcal{E}(b,c)$ and R-$\mathcal{E}(a,c)$, which correspond to the configuration in Fig. 3.37. It is noted that the zero-of-transmission is not satisfied for Fig. 3.37 by open-circuiting R_2, short-circuiting R_3, and replacing R_1 by a new port.

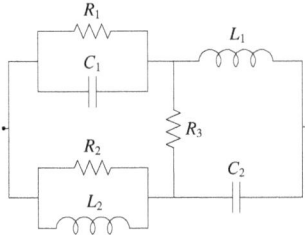

Fig. 3.37   Configurations, whose reactive-element graphs are shown in Fig. 3.31(b), with resistor edges R-$\mathcal{E}(a,b)$, R-$\mathcal{E}(b,c)$, and R-$\mathcal{E}(a,c)$.

For Fig. 3.31(c), there is no R-$\mathcal{E}(a,b)$ or R-$\mathcal{E}(b,a')$ by Lemma 3.26. By Lemma 3.22, there is no R-$\mathcal{E}(a,a')$, which implies that R-$\mathcal{E}(c,a')$ must exist and there is no R-$\mathcal{E}(a,c)$. Therefore, there are no more than two resistors in all possible realizations. A similar discussion can be carried out for Fig. 3.31(d).

For Fig. 3.32(a), Lemma 3.22 implies that there is no R-$\mathcal{E}(a,a')$ and Lemma 3.26 implies that there is no R-$\mathcal{E}(a,b)$, R-$\mathcal{E}(b,a')$, R-$\mathcal{E}(b,c)$, R-$\mathcal{E}(a,d)$, R-$\mathcal{E}(d,a')$, or R-$\mathcal{E}(c,d)$. Therefore, vertex d can only be incident with R-$\mathcal{E}(b,d)$. This means that there are always fewer than four reactive elements, as discussed in Theorem 3.18.

For Fig. 3.32(b), Lemma 3.22 implies that vertex b cannot be incident with any resistor edge. Therefore, any configuration whose reactive-element graph is Fig. 3.32(b) cannot realize a minimum function by Lemma 3.24. Similar discussions can be carried out for Figs. 3.32(c) and 3.32(e).

For Fig. 3.32(d), vertex a cannot be incident with any resistor edge by Lemmas 3.22 and 3.26. This means that there exists a C-$\mathcal{C}(a, a')$ or L-$\mathcal{C}(a, a')$. By Lemma 3.22, any configuration whose reactive-element graph is Fig. 3.32(d) cannot realize a minimum function. A similar discussion can be carried out for Fig. 3.32(f).

For Fig. 3.33(a), there is no R-$\mathcal{E}(a, a')$ by Lemma 3.22 and there is no R-$\mathcal{E}(a, b)$, R-$\mathcal{E}(b, c)$ or R-$\mathcal{E}(c, a')$ by Lemma 3.26. To avoid discussing the configuration that can always be equivalent to one whose reactive-element graph is connected, one considers the case where each of vertices d and e is adjacent to at least two of vertices a, b, c and a' by resistor edges. Together with Lemma 3.22, the resistor edges for the possible realizations containing at least three resistors are (I) R-$\mathcal{E}(b, a')$, R-$\mathcal{E}(a, d)$, R-$\mathcal{E}(c, d)$, R-$\mathcal{E}(a, e)$ and R-$\mathcal{E}(c, e)$; (II) R-$\mathcal{E}(a, c)$, R-$\mathcal{E}(b, a')$, R-$\mathcal{E}(a, d)$, R-$\mathcal{E}(c, d)$, R-$\mathcal{E}(a, e)$ and R-$\mathcal{E}(c, e)$; (III) R-$\mathcal{E}(b, a')$, R-$\mathcal{E}(a, d)$, R-$\mathcal{E}(c, d)$, R-$\mathcal{E}(a, e)$, R-$\mathcal{E}(c, e)$ and R-$\mathcal{E}(d, e)$; (IV) R-$\mathcal{E}(a, c)$, R-$\mathcal{E}(b, a')$, R-$\mathcal{E}(a, d)$, R-$\mathcal{E}(c, d)$, R-$\mathcal{E}(a, e)$, R-$\mathcal{E}(c, e)$ and R-$\mathcal{E}(d, e)$. It is noted that any configuration corresponding to the above cases can be equivalent to the network whose reactive-element graph is shown in Figs. 3.30(d) or 3.30(e).

For Fig. 3.33(b), there does not exist R-$\mathcal{E}(a, b)$, R-$\mathcal{E}(b, c)$, or R-$\mathcal{E}(b, a')$ by Lemma 3.26. One only needs to consider the case where each of vertices d and e is adjacent to at least two of vertices a, b, c and a' by resistor edges. Based on Lemma 3.24, there exists R-$\mathcal{E}(b, d)$ or R-$\mathcal{E}(b, e)$. If there exists R-$\mathcal{E}(b, d)$ (resp., R-$\mathcal{E}(b, e)$), then based on Lemma 3.26 vertex e (resp., vertex d) cannot be adjacent to vertex a, c, or a' by a resistor edge. This means that any possible realization can always be equivalent to one with fewer than four reactive elements.

For Fig. 3.33(c), by Lemmas 3.22 and 3.26, vertex a cannot be adjacent to vertex b, c, and a' by a resistor edge. To avoid a C-$\mathcal{C}(a, a')$ or L-$\mathcal{C}(a, a')$, there exists R-$\mathcal{E}(a, d)$ or R-$\mathcal{E}(a, e)$. One only needs to consider the case where each of vertices d and e is adjacent to at least two of vertices a, b, c and a' by resistor edges. If there exists R-$\mathcal{E}(a, d)$ (resp., R-$\mathcal{E}(a, e)$), then based on Lemma 3.26 vertex e (resp., vertex d) must be adjacent to vertices a and a' by resistor edges. It means that there exists an R-$\mathcal{P}(a, a')$, which is impossible by Lemma 3.22.

For Fig. 3.33(d), it suffices to consider the case where each of vertices c and e is adjacent to at least two of vertices a, b and a' by resistor edges. By Lemma 3.22, there must simultaneously exist R-$\mathcal{E}(a, c)$ and R-$\mathcal{E}(b, c)$ or simultaneously exist R-$\mathcal{E}(a', c)$ and R-$\mathcal{E}(b, c)$. If there simultaneously exist R-$\mathcal{E}(a, c)$ and R-$\mathcal{E}(b, c)$ (resp., R-$\mathcal{E}(a', c)$ and R-$\mathcal{E}(b, c)$), then vertex

e cannot be adjacent to vertex a, b or a' by a resistor edge. It means that any possible realization can always be equivalent to one with fewer than four reactive elements.

For Fig. 3.33(e), one only needs to consider the case where each of vertices c and d is adjacent to at least two of vertices a, b and a' by resistor edges. By Lemma 3.22, there must simultaneously exist $R\text{-}\mathcal{E}(a,c)$ and $R\text{-}\mathcal{E}(b,c)$ or simultaneously exist $R\text{-}\mathcal{E}(a',c)$ and $R\text{-}\mathcal{E}(b,c)$. If there simultaneously exist $R\text{-}\mathcal{E}(a,c)$ and $R\text{-}\mathcal{E}(b,c)$ (resp., $R\text{-}\mathcal{E}(a',c)$ and $R\text{-}\mathcal{E}(b,c)$), then there will exist $R\text{-}\mathcal{P}(a,a')$, $L\text{-}\mathcal{C}(a,a')$, or $C\text{-}\mathcal{C}(a,a')$, which is impossible.

For Fig. 3.33(f), one only needs to consider the case where each of vertices c and d is adjacent to at least two of vertices a, b, a', e and f by resistor edges, and each of vertices e and f is adjacent to at least two of vertices a, b, a', c and d by resistor edges. Since there is no $R\text{-}\mathcal{E}(a,b)$ or $R\text{-}\mathcal{E}(a,a')$ by Lemmas 3.22 and 3.26, vertex a must be adjacent to at least one of vertices c, d, e and f by resistor edges. Without loss of generality, assume that there exists $R\text{-}\mathcal{E}(a,c)$. Then, there exists $R\text{-}\mathcal{E}(b,c)$, $R\text{-}\mathcal{E}(c,f)$ or $R\text{-}\mathcal{E}(c,e)$. For the case where there exists $R\text{-}\mathcal{E}(b,c)$, there is no $R\text{-}\mathcal{E}(a,d)$, $R\text{-}\mathcal{E}(b,d)$ or $R\text{-}\mathcal{E}(d,a')$ by Lemma 3.26. Therefore, $R\text{-}\mathcal{E}(d,e)$ and $R\text{-}\mathcal{E}(d,f)$ must exist. Based on Lemmas 3.22 and 3.26, vertex a' cannot be incident with any resistor edge, which is impossible. For the case where there exists $R\text{-}\mathcal{E}(c,f)$, vertex d must be incident with at least two resistor edges of $R\text{-}\mathcal{E}(a,d)$, $R\text{-}\mathcal{E}(d,e)$, $R\text{-}\mathcal{E}(d,f)$, and $R\text{-}\mathcal{E}(d,a')$. If there exist $R\text{-}\mathcal{E}(a,d)$ and $R\text{-}\mathcal{E}(d,e)$, then vertex a' cannot be incident with any resistor edge by Lemmas 3.22 and 3.26, which implies that a $C\text{-}\mathcal{C}(a,a')$ or $L\text{-}\mathcal{C}(a,a')$ exists. If there exist $R\text{-}\mathcal{E}(a,d)$ and $R\text{-}\mathcal{E}(d,f)$, then $R\text{-}\mathcal{E}(e,a')$ must exist. By Lemmas 3.22 and 3.26, vertex e cannot be adjacent to vertex a, b, c or d by a resistor edge, which will not be considered. If there exist $R\text{-}\mathcal{E}(d,e)$ and $R\text{-}\mathcal{E}(d,f)$, then vertex a' cannot be incident with any resistor edge by Lemmas 3.22 and 3.26. It implies that a $C\text{-}\mathcal{C}(a,a')$ or $L\text{-}\mathcal{C}(a,a')$ must exist. If there exist $R\text{-}\mathcal{E}(d,e)$ and $R\text{-}\mathcal{E}(d,a')$, then vertex b cannot be incident with any resistor edge by Lemma 3.26. It means that the realizations must contain a parallel subnetwork consisting of two elements, which is impossible by Lemma 3.24. If there exist $R\text{-}\mathcal{E}(a,d)$ and $R\text{-}\mathcal{E}(d,a')$ or exist $R\text{-}\mathcal{E}(d,f)$ and $R\text{-}\mathcal{E}(d,a')$, an $R\text{-}\mathcal{P}(a,a')$ exists, which is impossible by Lemma 3.22. For the case where there exists $R\text{-}\mathcal{E}(c,e)$, a similar discussion can be carried out. $\square$

Lemma 3.29. *Any biquadratic minimum impedance function $Z(s)$ cannot be realized as the configuration in Fig. 3.29(a).*

Proof. It is shown in [Seshu (1959)] that any biquadratic minimum function can be written as

$$Z(s) = R\frac{s^2 + (1-k)(\omega_0 X_0/(kR))s + k\omega_0^2}{s^2 + (1-k)(\omega_0 R/X_0)s + (\omega_0^2/k)}, \tag{3.93}$$

where $k \in (0,1)$ and $Z(j\omega_0) = jX_0$. It is obvious that $Z(0) = k^2R$ and $Z(\infty) = R$. Assume that $Z(s)$ is realizable as the configuration in Fig. 3.29(a). This lemma will be proved by contradiction.

When $s = 0$, the inductors L_1, L_2, and L_3 are short-circuited and the capacitor C_1 is open-circuited, which implies that

$$k^2R = \frac{R_1 R_2}{R_1 + R_2}. \tag{3.94}$$

When $s = \infty$, the inductors L_1, L_2, and L_3 are open-circuited and the capacitor C_1 is short-circuited, which implies that

$$R = R_1 + R_2 + R_3. \tag{3.95}$$

When $s = j\omega_0$, with open-circuiting resistors R_2 and R_3, the fact that both voltages and currents of R_1 are zero yields

$$\omega_0^2 = \frac{1}{C_1 L_2}. \tag{3.96}$$

When $s = j\omega_0$, with open-circuiting resistors R_1 and R_3, the fact that both voltages and currents of R_2 are zero yields

$$\omega_0^2 = \frac{1}{C_1 L_1}. \tag{3.97}$$

Based on the impedance at $s = j\omega_0$, it follows that

$$Z(j\omega_0) = L_1 s|_{s=j\omega_0} = j\omega_0 L_1 = jX_0, \tag{3.98}$$

by short-circuiting R_1 and open-circuiting R_2 and R_3. From (3.96)–(3.98), one obtains $L_1 = L_2 = X_0/\omega_0$ and $C_1 = 1/(\omega_0 X_0)$. Therefore, the impedance of the configuration in Fig. 3.29(a) is expressed as $Z_1(s) = m_1(s)/n_1(s)$, where $m_1(s) = L_3 X_0^2(R_1 + R_2 + R_3)s^4 + X_0(\omega_0 L_3 X_0^2 + (R_1 + R_2)R_3 X_0 + 2\omega_0 L_3 R_1(R_2 + R_3))s^3 + \omega_0 X_0(R_3 X_0^2 + \omega_0 L_3(R_1 + R_2 + R_3)X_0 + 2R_1 R_2 R_3)s^2 + \omega_0^2 X_0((R_1 + R_2)R_3 X_0 + \omega_0 L_3 R_1(R_2 + R_3))s + \omega_0^3 X_0 R_1 R_2 R_3$ and $n_1(s) = L_3 X_0^2 s^4 + X_0(R_3 X_0 + \omega_0 L_3(R_1 + R_2 + R_3))s^3 + \omega_0(2\omega_0 L_3 X_0^2 + (R_1 + R_2)R_3 X_0 + \omega_0 L_3 R_1(R_2 + R_3))s^2 + \omega_0^2(2R_3 X_0^2 + \omega_0 L_3(R_1 + R_2 + R_3)X_0 + R_1 R_2 R_3)s + \omega_0^3 X_0(R_1 + R_2)R_3$. The resultant of polynomials $m_1(s)$ and $n_1(s)$ is calculated as $R_0(m_1, n_1, s) = -\omega_0^9 X_0^3 L_3 R_3^2(\omega_0 R_1(R_1 - R_2 - R_3)L_3 - X_0(R_1 - R_2)R_3)^2(\omega_0^2 L_3^2(R_1(R_2 + R_3) + X_0^2)^2 + R_3^2(R_1 R_2 + X_0^2)^2)$. Then, it follows that $R_0(m_1, n_1, s) = 0$, implying that

$$\omega_0 R_1(R_1 - R_2 - R_3)L_3 - X_0(R_1 - R_2)R_3 = 0. \tag{3.99}$$

Based on (3.99), if $R_1 - R_2 - R_3 = 0$, then $R_1 - R_2 = 0$, implying that $R_3 = 0$. This contradicts the assumption that element values are positive and finite. Therefore, $R_1 - R_2 - R_3 \neq 0$ and $R_1 - R_2 \neq 0$. Therefore, (3.99) is equivalent to

$$L_3 = \frac{X_0(R_1 - R_2)R_3}{\omega_0 R_1(R_1 - R_2 - R_3)}. \tag{3.100}$$

Substituting (3.100) into $Z_1(s)$ yields $Z_1(s) = u_1(s)/v_1(s)$, where $u_1(s) = X_0^2(R_1 + R_2 + R_3)(R_1 - R_2)s^3 + \omega_0 X_0(2R_1 R_2(R_1 - R_2 - R_3) - X_0^2(R_1 - R_2))s^2 + \omega_0^2 X_0^2(R_1^2 - R_2(R_2+R_3))s + \omega_0^3 X_0 R_1 R_2(R_1-R_2-R_3)$ and $v_1(s) = X_0^2(R_1-R_2)s^3 + \omega_0 X_0(R_1^2 - R_2(R_2+R_3))s^2 + \omega_0^2(2X_0^2(R_1+R_2) + R_1 R_2(R_1 - R_2 - R_3))s + \omega_0^3 X_0(R_1 + R_2)(R_1 - R_2 - R_3)$. Since $R_1 - R_2 - R_3 \neq 0$ and $R_1 - R_2 \neq 0$, it implies that there exists a common factor $(s+\gamma)$ with $\gamma > 0$ between $u_1(s)$ and $v_1(s)$ such that $Z_1(s)$ is equivalent to $Z(s)$. Based on (3.94) and (3.95), R_2 and R_3 can be expressed in terms of R_1 as

$$R_2 = \frac{k^2 R R_1}{R_1 - k^2 R}, \qquad R_3 = \frac{R R_1 - (R_1^2 + k^2 R^2)}{R_1 - k^2 R}, \tag{3.101}$$

from which $Z_1(s)$ becomes

$$Z_1(s) = R\frac{s^3 + \mu_2 s^2 + \mu_1 s + \mu_0}{s^3 + \nu_2 s^2 + \nu_1 s + \nu_0},$$

where $\mu_2 = \omega_0(2k^2 R(R R_1 + X_0^2 - 2R_1^2) - R_1 X_0^2)/(R X_0(2k^2 R - R_1))$, $\mu_1 = \omega_0^2(kR - R_1)(kR + R_1)/(R(2k^2 R - R_1))$, $\mu_0 = k^2 \omega_0^3 R_1(R - 2R_1)/(X_0(2k^2 R - R_1))$, $\nu_2 = \omega_0(kR - R_1)(kR + R_1)/(X_0(2k^2 R - R_1))$, $\nu_1 = \omega_0^2(k^2 R(R R_1 - 2R_1^2 + 4X_0^2) - 2R_1 X_0^2)/(X_0^2(2k^2 R - R_1))$, and $\nu_0 = \omega_0^3 R_1(R - 2R_1)/(X_0(2k^2 R - R_1))$. Then, one obtains

$$\mu_2 = \frac{\omega_0 X_0 - k\omega_0 X_0 + kR\gamma}{kR}, \tag{3.102}$$

$$\mu_1 = \frac{\omega_0(k^2 \omega_0 R + X_0\gamma - kX_0\gamma)}{kR}, \tag{3.103}$$

$$\mu_0 = k\omega_0^2\gamma, \tag{3.104}$$

$$\nu_2 = \frac{\omega_0 R - k\omega_0 R + X_0\gamma}{X_0}, \tag{3.105}$$

$$\nu_1 = \frac{\omega_0(\omega_0 X_0 + kR\gamma - k^2 R\gamma)}{kX_0}, \tag{3.106}$$

$$\nu_0 = \frac{\omega_0^2\gamma}{k}. \tag{3.107}$$

It follows from (3.104) that

$$\gamma = \frac{k\omega_0 R_1(R - 2R_1)}{X_0(2k^2 R - R_1)}. \tag{3.108}$$

Substituting (3.108) into (3.103), one obtains

$$\frac{\omega_0^2(1 - 2k)(k^2 R^2 - RR_1 + R_1^2)}{R(2k^2 R - R_1)} = 0. \tag{3.109}$$

If $k = 1/2$, then it is obvious that (3.102)–(3.107) hold. Assuming that $k \neq 1/2$, it follows from (3.109) that $k^2 = R_1(R - R_1)/R^2$. Furthermore, from (3.102), one obtains that $R = 2R_1$. This can further imply that $k = 1/2$. Therefore, k can only be equal to $1/2$. From (3.101), it implies that $R_2 R_3 = -(RR_1(2R_1 - R)^2)/(4R_1 - R)^2 \leq 0$, which is obviously impossible. Therefore, this lemma is proved. □

Together with the principle of frequency inversion, it follows from Lemma 3.29 that any biquadratic minimum impedance function $Z(s)$ cannot be realized as the configuration in Fig. 3.29(b), either.

Lemma 3.30. *If a biquadratic minimum impedance function $Z(s)$ is realizable as the configuration in Fig. 3.29(c), then $Z(s)$ is realizable with three reactive elements and two resistors.*

Proof. Similarly to the proof of Lemma 3.29, one can show that k must be equal to 2, indicating that $Z(0) = 4Z(\infty)$. By [Seshu (1959), pg. 348], $Z(s)$ is realizable with three reactive elements and two resistors ([Foster (1963), Fig. 1(f)]). □

Together with the principle of frequency inversion, it follows from Lemma 3.30 that a biquadratic minimum impedance function $Z(s)$ realizable as the configuration in Fig. 3.29(d) can also be realized with three reactive elements and two resistors.

By combining the above results, the following theorem is obtained.

Theorem 3.19. *If a biquadratic minimum function $Z(s)$ is realizable as the impedance of a network containing four reactive elements and an arbitrary number of resistors, then $Z(s)$ is realizable as the impedance of a network with four reactive elements and two resistors.*

Proof. This theorem follows directly from Lemmas 3.28–3.30. □

Finally, the following result is obtained, which is a generalized theorem of Reichert for biquadratic minimum functions.

Theorem 3.20. *If a biquadratic minimum function $Z(s)$ is realizable as the impedance of a network containing n reactive elements and an arbitrary*

*number of resistors, then $Z(s)$ is realizable as the impedance of a network
with n reactive elements and two resistors.*

Proof. As stated in [Seshu (1959), Theorem 9], $n \geq 3$. When $n = 3$, this
theorem is valid by Theorem 3.18. When $n = 4$, this theorem is valid by
Theorem 3.19. Through Pantell's modified Bott-Duffin synthesis procedure
[Pantell (1954)] (called *Reza-Pantell-Fialkow-Gerst procedure* in [Hughes
(2014)]), any biquadratic minimum impedance function is realizable with
five reactive elements and two resistors, which implies that this theorem is
also valid for all $n \geq 5$.   □

3.8   Seven-Element Series-Parallel Realizations of a Specific Class of Biquadratic Impedances

This section investigates the minimal realization problem of a specific biquadratic impedance with double poles and zeros as a class of seven-element
series-parallel networks.

The biquadratic impedances with double poles and zeros can provide
fast transitions, which are expected to have advantages in some mechanical control systems. In [Chang (1969); Tirtoprodjo (1970)], the realization
problem of such an impedance function has been investigated. However,
its minimal realization problem in terms of the number of elements is far
from being solved. Since such a function only contains three parameters, its
realizability conditions can be much neater than the conditions of general
positive-real biquadratic impedances, and the realizability conditions are
always in terms of the pole-zero ratio p/z. In addition, the realization results and the methodology can provide essential guidance on the synthesis
of general biquadratic impedances. Therefore, the investigation on realization problems of such a biquadratic impedance can be a first critical step
towards solving many classical unsolved problems in biquadratic synthesis,
such as the least number of elements (seven or eight) needed to realize any
positive-real biquadratic impedance.

The main results of this section are presented in Theorem 3.26, where
the configurations are shown in Figs. 3.38 and 3.45–3.47.

The biquadratic impedance with double poles and zeros considered here
is in the form of

$$Z(s) = k\frac{(s+z)^2}{(s+p)^2}, \tag{3.110}$$

where k, z, $p > 0$, and $p \neq z$. By a result in [Steiglitz and Zemanian (1962), pg. 271], $Z(s)$ is positive-real if and only if $p/z \in [1/(3 + 2\sqrt{2}), 3 + 2\sqrt{2}]$. For brevity, denote such a class of biquadratic impedances as $\mathcal{Z}_{\delta=2}^{k,p,z}$.

It is clear that (3.110) can be expressed in the form of (3.1), that is,

$$Z(s) = \frac{a_2 s^2 + a_1 s + a_0}{b_2 s^2 + b_1 s + b_0},$$

where

$$a_2 = km, \quad a_1 = 2kzm, \quad a_0 = kz^2 m, \quad b_2 = m, \quad b_1 = 2pm, \quad b_0 = p^2 m,$$
$$(3.111)$$

for any $m > 0$. As a consequence, the realization results of such an impedance as a k-element series-parallel network, for $k = 1, 2, \ldots, 6$, can be derived from the results in Section 3.4 and the results in [Jiang and Smith (2011)].

Theorem 3.21. *Consider a biquadratic impedance $Z(s)$ in the form of (3.110), where k, z, $p > 0$, and $p \neq z$, that is, $Z(s) \in \mathcal{Z}_{\delta=2}^{k,p,z}$. Then, $Z(s)$ cannot be realized with fewer than four elements.*

Theorem 3.22. *Consider a biquadratic impedance $Z(s) \in \mathcal{Z}_{\delta=2}^{k,p,z}$. Then,*

1. *$Z(s)$ is realizable as a four-element series-parallel network if and only if $p/z \in \{1/3, 3\}$;*
2. *$Z(s)$ is realizable as a five-element series-parallel network if and only if $p/z \in [1/3, 3]$ or $p/z \in \{1/(2 + \sqrt{2}), 2 + \sqrt{2}\}$;*
3. *$Z(s)$ is realizable as a six-element series-parallel network if and only if $p/z \in [1/\eta, \eta]$ or $p/z \in \{1/(2 + \sqrt{3}), 2 + \sqrt{3}, 1/(2 + \sqrt{5}), 2 + \sqrt{5}, 2/(5 + \sqrt{17}), (5 + \sqrt{17})/2, 1/\eta_1, \eta_1\}$, where η is the unique distinct root of the equation $19x^3 - 82x^2 + 59x - 16 = 0$ for $x \in (2 + \sqrt{2}, 2 + \sqrt{3})$, and η_1 is the unique distinct root of the equation $x^4 - 6x^3 + 6x^2 - 14x + 5 = 0$ for $x \in ((5 + \sqrt{17})/2, +\infty)$.*

To avoid repetition and to guarantee the minimality of the number of elements, it is assumed that any $Z(s) \in \mathcal{Z}_{\delta=2}^{k,p,z}$ does not satisfy the conditions of Theorem 3.22. This assumption means that any network realizing $Z(s)$ contains at least seven elements.

Lemma 3.31. *[Tirtoprodjo (1970)] For a biquadratic impedance $Z(s) \in \mathcal{Z}_{\delta=2}^{k,p,z}$, not satisfying the conditions of Theorem 3.22, if $p/z \in (1/(2 + \sqrt{5}), 2 + \sqrt{5})$, then $Z(s)$ is realizable as one of the four-reactive seven-element series-parallel configurations in Fig. 3.38.*

It is noted that the two configurations in Fig. 3.38 belong to a class of seven-element series-parallel networks, any one of which is the series or parallel connection of two networks: a three-element series-parallel network N_1 and a four-element series-parallel network N_2 (see Fig. 3.39). Moreover, denote the impedances of N_1 and N_2 as $Z_1(s)$ and $Z_2(s)$, respectively. In this section, the objective is to obtain a necessary and sufficient condition for any biquadratic impedance $Z(s) \in \mathcal{Z}_{\delta=2}^{k,p,z}$, not satisfying the conditions of Theorem 3.22, to be realizable as such a network.

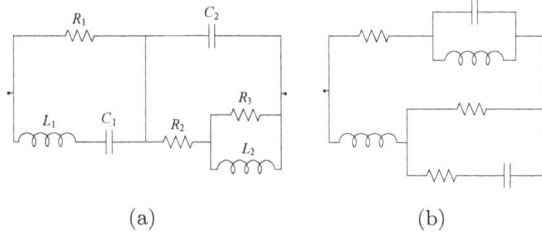

(a)   (b)

Fig. 3.38   Four-reactive seven-element series-parallel configurations in Lemma 3.31, whose one-terminal-pair labeled graphs are $\mathcal{N}_{11a}$ and $\mathcal{N}_{11b}$, respectively, satisfying $\mathcal{N}_{11a} = \mathrm{Dual}(\mathcal{N}_{11b})$.

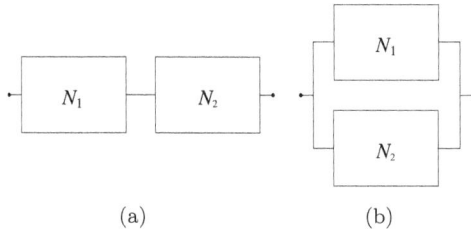

(a)   (b)

Fig. 3.39   A class of seven-element series-parallel networks, where N_1 is a three-element series-parallel network and N_2 is a four-element series-parallel network.

3.8.1   *Preliminary Lemmas*

Lemma 3.32. *Consider a biquadratic impedance $F(s)$ in the form of*

$$F(s) = \frac{\alpha s^2 + \beta s + \gamma}{(s+p)^2}, \qquad (3.112)$$

where α, β, $\gamma \geq 0$, and $p > 0$. Then, $F(s)$ is realizable as a three-element series-parallel network if and only if any one of the following conditions *holds:* 1. $\alpha\gamma = 0$; 2. $\beta = 0$ and $\alpha p^2 - \gamma = 0$; 3. $\gamma = 0$ and $\alpha p - 2\beta = 0$; 4. $\alpha = 0$ and $2\beta p - \gamma = 0$; 5. $\alpha p^2 - \beta p + \gamma = 0$.

Proof. Since $F(s)$ is obviously not a *reactance function* [Baher (1984), Definition 3.1], its realization must contain at least one resistor. Therefore, no more than two reactive elements are needed.

When the number of reactive elements is at most one, the McMillan degree of $F(s)$ cannot exceed one according to [Anderson and Vongpanitlerd (1973), Theorem 4.4.3]. Therefore, there must exist at least one common factor between $(\alpha s^2 + \beta s + \gamma)$ and $(s + p)^2$, which holds if and only if Condition 5 is satisfied.

When the number of reactive elements is two, there is one resistor. Conditions 1–4 can be obtained by enumerating all the possible realizations. □

Lemma 3.33. *Consider a biquadratic impedance $F(s)$ in the form of (3.112), where α, β, $\gamma \geq 0$, $p > 0$, and the condition of Lemma 3.32 does not hold. Then, $F(s)$ is realizable as a four-element series-parallel network if and only if at least one of the following six conditions holds:* 1. $\alpha = 0$ *and* $\gamma < 2\beta p$; 2. $\gamma = 0$ *and* $\alpha p < 2\beta$; 3. α, β, $\gamma > 0$ *and* $\alpha p^2 - \gamma = 0$; 4. α, β, $\gamma > 0$, $\alpha p^2 < \gamma$, *and* $(3\alpha p^2 + \gamma - 2\beta p)(\beta^2 p^2 + \gamma^2 - \alpha\gamma p^2 - 2\beta\gamma p) = 0$; 5. α, β, $\gamma > 0$, $\alpha p^2 > \gamma$, *and* $(\alpha p^2 + 3\gamma - 2\beta p)(\alpha^2 p^2 + \beta^2 - 2\alpha\beta p - \alpha\gamma) = 0$; 6. α, β, $\gamma > 0$ *and* $\alpha^2 p^4 - 2\alpha\beta p^3 + 6\alpha\gamma p^2 - 2\beta\gamma p + \gamma^2 = 0$. *Moreover, if one of the above six conditions holds, then $F(s)$ is realizable as a two-reactive four-element series-parallel network.*

Proof. $F(s)$ can also be expressed as a general biquadratic impedance in (3.1) by letting $a_2 = \alpha m$, $a_1 = \beta m$, $a_0 = \gamma m$, $b_2 = m$, $b_1 = 2pm$, and $b_0 = p^2 m$ for any $m > 0$. Thus, the positive-real condition of $F(s)$ can be directly derived from that of the general biquadratic impedance $((\sqrt{a_2 b_0}) - \sqrt{a_0 b_2})^2 \leq a_1 b_1$ [Chen and Smith (2009a); Foster (1962)]). Since any positive-real biquadratic impedance with zero coefficients is realizable as a two-reactive four-element series-parallel network [Jiang and Smith (2011), Lemma 8], Conditions 1 and 2 of this lemma are obtained. For the case of α, β, $\gamma > 0$, one obtains Conditions 3–6 from [Wang *et al.* (2014), Theorem 5]. □

Lemma 3.34. *Consider a biquadratic impedance $F(s)$ in the form of (3.112), where α, β, γ, $p > 0$, and neither the condition of Lemmas 3.32* *nor the condition of Lemma 3.33 holds. Then, $F(s)$ is realizable as a two-reactive five-element series-parallel network if and only if any one of the following conditions holds: 1. $\alpha p^2 > \gamma$ and $\alpha p^2 + 3\gamma - 2\beta p < 0$; 2. $\alpha p^2 > \gamma$ and $\alpha^2 p^2 + \beta^2 - 2\alpha\beta p - \alpha\gamma < 0$; 3. $\alpha p^2 < \gamma$ and $3\alpha p^2 + \gamma - 2\beta p < 0$; 4. $\alpha p^2 < \gamma$ and $\beta^2 p^2 + \gamma^2 - \alpha\gamma p^2 - 2\beta\gamma p < 0$.*

Proof. This lemma can be proved by (3.111) and [Jiang and Smith (2011), Theorem 1]. □

3.8.2 Realizations as Three-Reactive Seven-Element Series-Parallel Networks

In this subsection, the three-reactive-element case is considered.

Lemma 3.35. *For a biquadratic impedance $Z(s) \in \mathcal{Z}_{\delta=2}^{k,p,z}$, not satisfying the conditions of Theorem 3.22, if $Z(s)$ is realizable as a three-reactive seven-element series-parallel network in Fig. 3.39(a), where N_1 is a three-element series-parallel network and N_2 is a four-element series-parallel network, then $Z(s)$ is also realizable as the configuration in Fig. 3.40, where N_b is a two-reactive five-element series-parallel network.*

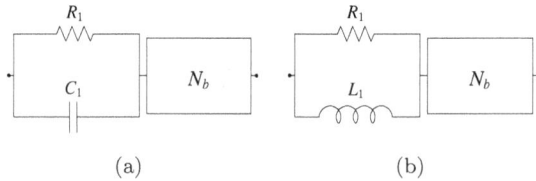

Fig. 3.40   A class of three-reactive seven-element series-parallel networks, where N_b is a two-reactive five-element series-parallel network.

Proof. Together with some results in Section 3.4 and [Anderson and Vongpanitlerd (1973), pg. 370], one can prove that N_1 must be a one-reactive three-element network and N_2 must be a two-reactive four-element network. Otherwise, the assumption that $Z(s)$ cannot be realized with fewer than seven elements is contradictory. Thus, the impedance $Z_1(s)$ of N_1 must be of degree one, which implies that N_1 is equivalent to one of the configurations in Fig. 3.41. Regarding the series connection of R_2 and N_2 as N_b, the lemma is proved. □

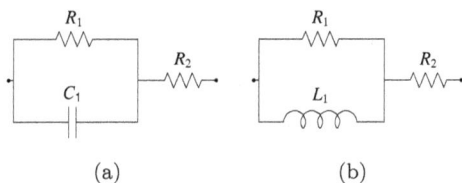

Fig. 3.41   Configurations for N_1 in the proof of Lemma 3.35.

Theorem 3.23. *Any biquadratic impedance $Z(s) \in \mathcal{Z}_{\delta=2}^{k,p,z}$, not satisfying the conditions of Theorem 3.22, cannot be realized as a three-reactive seven-element series-parallel network in Fig. 3.39, where N_1 is a three-element series-parallel network and N_2 is a four-element series-parallel network.*

Proof. It will be shown that $Z(s) \in \mathcal{Z}_{\delta=2}^{k,p,z}$ cannot be realized as the networks in Fig. 3.40, where N_b is a two-reactive five-element series-parallel network. By Lemma 3.35, this implies that $Z(s)$ cannot be realized as a three-reactive seven-element series-parallel network in Fig. 3.39(a), where N_1 is a three-element series-parallel network and N_2 is a four-element series-parallel network. Since any series-parallel network in Fig. 3.39(b) can be a dual network to the case of Fig. 3.39(a), the proof can be completed by the principle of duality.

Assume that $Z(s)$ is realizable as a network in Fig. 3.40(a), where N_b is a two-reactive five-element series-parallel network. Let $Z(s) = Z_a(s) + Z_b(s)$, where $Z_a(s)$ is the impedance of the parallel connection of R_1 and C_1, and $Z_b(s)$ is the impedance of N_b. It is clear that $Z_a(s)$ is in the form of $Z_a(s) = m/(s+p)$, where $m > 0$. It implies that N_b cannot be equivalent to a network containing fewer than five elements, since the conditions of Theorem 3.22 do not hold.

Therefore, the McMillan degree of $Z_b(s)$ cannot be fewer than two, and it can be written as $Z_b(s) = (\alpha s^2 + \beta s + \gamma)/(s+p)^2$, where $\alpha, \beta, \gamma > 0$, so the condition of Lemma 3.34 holds. Since

$$Z_a(s) + Z_b(s) = \frac{\alpha s^2 + (\beta + m)s + (\gamma + mp)}{(s+p)^2},$$

one obtains $\alpha s^2 + (\beta + m)s + (\gamma + mp) = k(s+z)^2$. Then,

$$\beta = 2\alpha z - m, \qquad \gamma = \alpha z^2 - mp. \tag{3.113}$$

Since $\beta, \gamma > 0$, it follows from (3.113) that

$$\alpha > \max\left\{\frac{m}{2z}, \frac{mp}{z^2}\right\}. \tag{3.114}$$

If $Z_b(s)$ satisfies Condition 1 of Lemma 3.34, then $\alpha p^2 > \gamma$ and $\alpha p^2 + 3\gamma - 2\beta p < 0$. Together with (3.113), one obtains $(p - z)(p + z)\alpha + mp > 0$ and

$$(p - z)(p - 3z)\alpha - mp < 0. \tag{3.115}$$

If $p/z \in (1, 3]$, then the condition of Theorem 3.22 holds. If $p/z \in (0, 1) \cup (3, +\infty)$, then $\alpha < mp/((p - z)(p - 3z))$ by (3.115). Together with (3.114), it implies that $mp/z^2 < mp/((p - z)(p - 3z))$, which indicates that $p/z \in (2 - \sqrt{2}, 2 + \sqrt{2})$. Therefore, the conditions of Theorem 3.22 hold.

When $Z_b(s)$ satisfies one of Conditions 2–4 in Lemma 3.34, it can be proved that the conditions of Theorem 3.22 also hold. The details are omitted for brevity.

As a conclusion, $Z(s)$ cannot be realized as the configuration in Fig. 3.40(a), where N_b is a two-reactive five-element series-parallel network. Note that any network in Fig. 3.40(b) can be the frequency inverse network of a network in Fig. 3.40(a). By the principle of frequency inversion, $Z(s)$ cannot be realized as the configuration in Fig. 3.40(b), where N_b is a two-reactive five-element series-parallel network. Thus, as discussed at the beginning of this proof, the lemma is established.   □

3.8.3   *Realizations as Four-Reactive Seven-Element Series-Parallel Networks*

Lemma 3.36. *Consider a biquadratic impedance $Z(s)$ in the form of (3.1), where $a_i > 0$, $i = 0, 1, 2$, and $b_j > 0$, $j = 0, 1, 2$, which cannot be realized as a series-parallel network containing fewer than seven elements. If $Z(s)$ is realizable as a four-reactive seven-element series-parallel network in Fig. 3.39(a), where N_1 is a two-reactive three-element series-parallel network and N_2 is a two-reactive four-element series-parallel network, then N_1 can be one of the configurations in Fig. 3.42.*

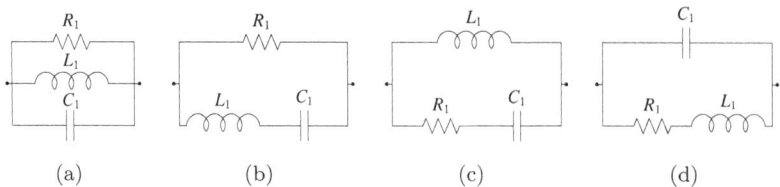

(a)   (b)   (c)   (d)

Fig. 3.42   Two-reactive three-element configurations for the network N_1 mentioned in Lemma 3.36.

Proof. By Lemma 3.2, for any realization of $Z(s)$ there is no cut-set $\mathcal{C}(a, a')$ corresponding to only one kind of reactive elements, where a and a' denote two terminals. By the method of enumeration, the lemma is proved. $\square$

Lemma 3.37. *Any biquadratic impedance $Z(s) \in \mathcal{Z}_{\delta=2}^{k,p,z}$, not satisfying the conditions of Theorem 3.22, cannot be realized as the configuration in Fig. 3.39(a), where N_1 is the configuration in Fig. 3.42(a) and N_2 is a two-reactive four-element series-parallel network.*

Proof. The impedance of the configuration in Fig. 3.42(a) is calculated as

$$Z_1(s) = \frac{R_1 L_1 s}{R_1 L_1 C_1 s^2 + L_1 s + R_1}.$$

Considering the assumption that the condition of Theorem 3.22 does not hold, N_2 cannot be equivalent to a series-parallel network containing fewer than four elements. Thus, the impedance of N_1 is in the form of

$$Z_1(s) = \frac{ms}{(s+p)^2},$$

where $m > 0$, and the impedance of N_2 is in the form of

$$Z_2(s) = \frac{\alpha s^2 + \beta s + \gamma}{(s+p)^2}, \tag{3.116}$$

where α, β, $\gamma > 0$, and the condition of Lemma 3.33 holds. Since $\alpha s^2 + (\beta + m)s + \gamma = k(s+z)^2$, it follows that

$$\beta + m = 2z\alpha, \tag{3.117}$$

$$\gamma = z^2 \alpha. \tag{3.118}$$

Since α, $\gamma > 0$, $Z_2(s)$ satisfies neither Condition 1 nor Condition 2 of Lemma 3.33.

If $Z_2(s)$ satisfies Condition 3 of Lemma 3.33, then $\alpha p^2 - \gamma = 0$. Together with (3.118), it implies that $p = z$, which contradicts the assumption.

If $Z_2(s)$ satisfies Condition 4 of Lemma 3.33, then $\alpha p^2 < \gamma$ and either $3\alpha p^2 + \gamma - 2\beta p = 0$ or $\beta^2 p^2 + \gamma^2 - \alpha\gamma p^2 - 2\beta\gamma p = 0$ holds. For the case of $3\alpha p^2 + \gamma - 2\beta p = 0$, together with (3.117) and (3.118), one obtains $\alpha = -2mp/((3p - z)(p - z))$, $\beta = -m(3p^2 + z^2)/((3p - z)(p - z))$, and $\gamma = -2mz^2 p/((3p - z)(p - z))$, which indicates that $p/z \in (1/3, 1)$ for α, β, $\gamma > 0$. This contradicts the assumption that the condition of Theorem 3.22 does not hold. For the case of $\beta^2 p^2 + \gamma^2 - \alpha\gamma p^2 - 2\beta\gamma p = 0$, together with (3.117) and (3.118), one obtains

$$\alpha = \frac{mp}{z(3p - z)}, \quad \beta = -\frac{m(p - z)}{3p - z}, \quad \gamma = \frac{mzp}{3p - z}, \tag{3.119}$$

or

$$\alpha = \frac{mp}{z(p-z)}, \quad \beta = \frac{m(p+z)}{p-z}, \quad \gamma = \frac{mzp}{p-z}. \tag{3.120}$$

Because α, β, $\gamma > 0$, it follows from (3.119) that $p/z \in (1/3, 1)$, which satisfies the condition of Theorem 3.22. Substituting (3.120) into $\alpha p^2 < \gamma$ yields $(p+z)pm/z < 0$, which is impossible.

Similarly, one can prove the case where Condition 5 or Condition 6 of Lemma 3.33 holds.

The lemma is thus proved. $\qquad\square$

Lemma 3.38. *If a biquadratic impedance* $Z(s) \in \mathcal{Z}_{\delta=2}^{k,p,z}$, *not satisfying the conditions of Theorem 3.22, is realizable as the configuration in Fig. 3.39(a), where N_1 is one of the configurations in Figs. 3.42(b)–3.42(d) and N_2 is a two-reactive four-element series-parallel network, then the condition of Lemma 3.31 holds.*

Proof. Consider only the case where N_1 is a configuration in Fig. 3.42(b) and N_2 is a two-reactive four-element series-parallel network. Thus, other two cases can be similarly proved.

The impedance of the configuration in Fig. 3.42(b) can be calculated as

$$Z_1(s) = \frac{R_1 L_1 C_1 s^2 + R_1}{L_1 C_1 s^2 + R_1 C_1 s + 1}. \tag{3.121}$$

Since it is assumed that the conditions of Theorem 3.22 do not hold, N_2 cannot be equivalent to a series-parallel network containing fewer than four elements. Thus, the impedance of N_1 is in the form of

$$Z_1(s) = \frac{m(s^2 + p^2)}{(s+p)^2},$$

where $m > 0$, and the impedance of N_2 is in the form of (3.116), where $\beta > 0$, α, $\gamma \geq 0$, and the condition of Lemma 3.33 holds. Since $(m+\alpha)s^2 + \beta s + (mp^2 + \gamma) = k(s+z)^2$, it follows that

$$\beta = 2z(m+\alpha), \tag{3.122}$$

$$mp^2 + \gamma = z^2(m+\alpha). \tag{3.123}$$

If $Z_2(s)$ satisfies Condition 1 of Lemma 3.33, then β, $\gamma > 0$, $\alpha = 0$, and $\gamma < 2\beta p$. Together with (3.122) and (3.123), one obtains

$$\beta = 2mz, \quad \gamma = -m(p-z)(p+z), \tag{3.124}$$

which implies that $p < z$ by β, $\gamma > 0$. Substituting (3.124) into $\gamma < 2\beta p$ yields $m(p^2 + 4zp - z^2) > 0$. This implies that $p/z \in (1/(2+\sqrt{5}), 1)$, which satisfies the condition of Lemma 3.31.

If $Z_2(s)$ satisfies Condition 6 of Lemma 3.33, then α, β, $\gamma > 0$ and $\alpha^2 p^4 + 6\alpha\gamma p^2 + \gamma^2 - 2\alpha\beta p^3 - 2\beta\gamma p = 0$. Together with (3.122) and (3.123), one obtains

$$\alpha = \frac{m(3p^4 - 2z^2 p^2 + 4z^3 p - z^4 + 2p^2\sqrt{2p^4 + 2z^4})}{(p - z)^4},$$

$$\beta = \frac{4mzp^2(2p^2 - 2zp + 2z^2 + \sqrt{2p^4 + 2z^4})}{(p - z)^4}, \qquad (3.125)$$

$$\gamma = \frac{mp^2(-p^4 + 4zp^3 - 2z^2 p^2 + 3z^4 + 2z^2\sqrt{2p^4 + 2z^4})}{(p - z)^4},$$

or

$$\alpha = \frac{m(3p^4 - 2z^2 p^2 + 4z^3 p - z^4 - 2p^2\sqrt{2p^4 + 2z^4})}{(p - z)^4},$$

$$\beta = \frac{4mzp^2(2p^2 - 2zp + 2z^2 - \sqrt{2p^4 + 2z^4})}{(p - z)^4}, \qquad (3.126)$$

$$\gamma = \frac{mp^2(-p^4 + 4zp^3 - 2z^2 p^2 + 3z^4 - 2z^2\sqrt{2p^4 + 2z^4})}{(p - z)^4}.$$

Consider the solutions in (3.125). Assume that $p/z \in [2 + \sqrt{5}, +\infty)$. Then, $\gamma < 0$ since $-p^4 + 4zp^3 - 2z^2 p^2 + 3z^4 < 0$ and $(2z^2\sqrt{2p^4 + 2z^4})^2 - (-p^4 + 4zp^3 - 2z^2 p^2 + 3z^4)^2 = -(p + z)(p^2 - 4zp - z^2)(p - z)^5 \leq 0$. This contradicts the assumption. Assume that $p/z \in (0, 1/(2 + \sqrt{5})]$. Then, $\alpha < 0$ since $3p^4 - 2z^2 p^2 + 4z^3 p - z^4 < 0$ and $(2p^2\sqrt{2p^4 + 2z^4})^2 - (3p^4 - 2z^2 p^2 + 4z^3 p - z^4)^2 = -(p + z)(p^2 + 4zp - z^2)(p - z)^5 \leq 0$. This also contradicts the assumption.

Consider the solutions in (3.126). Assume that $p/z \in [2 + \sqrt{5}, +\infty)$. Then, $\gamma < 0$ because of $-p^4 + 4zp^3 - 2z^2 p^2 + 3z^4 < 0$. This contradicts the assumption. Assume that $p/z \in (0, 1/(2 + \sqrt{5})]$. Then, $\alpha < 0$ because of $3p^4 - 2z^2 p^2 + 4z^3 p - z^4 < 0$. This also contradicts the assumption.

Similarly, the case where one of Conditions 2–5 of Lemma 3.33 holds can be proved. $\qquad \square$

Lemma 3.39. *Consider a biquadratic impedance $Z(s)$ in the form of (3.1), where $a_i > 0$, $i = 0, 1, 2$, and $b_j > 0$, $j = 0, 1, 2$, which cannot be realized as a series-parallel network containing fewer than seven elements. If $Z(s)$ is realizable as a four-reactive seven-element series-parallel network in Fig. 3.39(a), where N_1 is a one-reactive three-element series-parallel network and N_2 is a three-reactive four-element series-parallel network, then N_1 can be equivalent to one of the configurations in Fig. 3.43, and N_2 can be equivalent to one of the configurations in Fig. 3.44.*

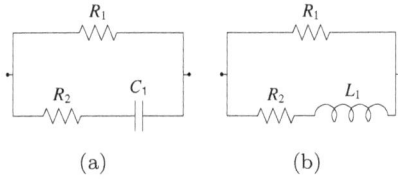

Fig. 3.43   One-reactive three-element series-parallel configurations for the network N_1 in Lemma 3.39.

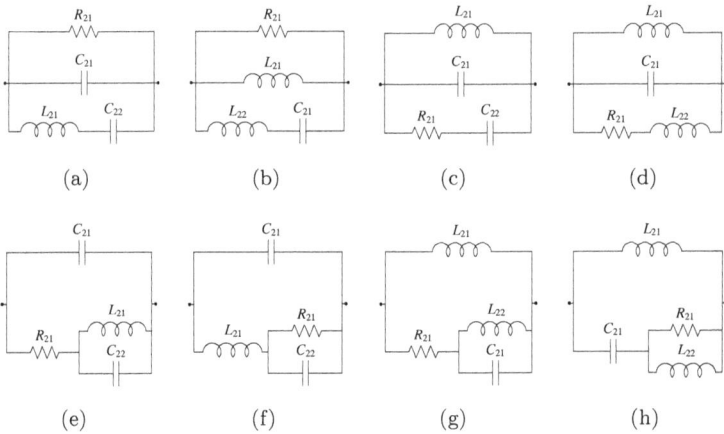

Fig. 3.44   Three-reactive four-element series-parallel configurations for the network N_2 in Lemma 3.39.

Proof. Using a similar method to the proof of Lemma 3.36, this lemma can be proved.   □

Lemma 3.40. *Any biquadratic impedance* $Z(s) \in \mathcal{Z}_{\delta=2}^{k,p,z}$, *not satisfying the conditions of Theorem 3.22, cannot be realized as the configuration in Fig. 3.39(a), where* N_1 *is the configuration in Fig. 3.43(a) and* N_2 *is one of the configurations in Figs. 3.44(a) and 3.44(c)–3.44(e).*

Proof. The impedance of the configuration in Fig. 3.44(a) is calculated to be

$$Z_2(s) = \frac{R_{21}L_{21}C_{22}s^2 + R_{21}}{R_{21}L_{21}C_{21}C_{22}s^3 + L_{21}C_{22}s^2 + R_{21}(C_{21} + C_{22})s + 1}. \quad (3.127)$$

If $Z(s)$ is realizable as the configuration in Fig. 3.39(a), where N_1 is the configuration in Fig. 3.43(a) and N_2 is the configuration in Fig. 3.44(a), then the impedance of N_1 is in the form of

$$Z_1(s) = \frac{ms + q}{s + p_1},$$

(3.128)

where m, q, $p_1 > 0$ and

$$q - mp_1 > 0$$

(3.129)

holds, and the impedance of N_2 is in the form of

$$Z_2(s) = \frac{\alpha s^2 + \gamma}{(s + p_1)(s + p)^2},$$

(3.130)

where α, $\gamma > 0$. Thus, it follows from (3.127) and (3.130) that

$$\frac{1}{C_{21}} = \alpha,$$

(3.131)

$$\frac{1}{L_{21}C_{21}C_{22}} = \gamma,$$

(3.132)

$$\frac{1}{R_{21}C_{21}} = 2p + p_1,$$

(3.133)

$$\frac{C_{21} + C_{22}}{L_{21}C_{21}C_{22}} = p(p + 2p_1),$$

(3.134)

$$\frac{1}{R_{21}L_{21}C_{21}C_{22}} = p_1 p^2.$$

(3.135)

By (3.131), one obtains

$$C_{21} = \frac{1}{\alpha}.$$

(3.136)

It follows from (3.133) and (3.136) that

$$R_{21} = \frac{\alpha}{2p + p_1}.$$

(3.137)

By (3.132), (3.134), and (3.136), one obtains

$$C_{22} = \frac{\alpha p^2 + 2\alpha p_1 p - \gamma}{\alpha \gamma}.$$

(3.138)

By (3.132), (3.136), and (3.138), one obtains

$$L_{21} = \frac{\alpha^2}{\alpha p^2 + 2\alpha p_1 p - \gamma}.$$

(3.139)

Substituting (3.136)–(3.139) into (3.135) yields

$$\alpha p_1 p^2 - 2\gamma p - \gamma p_1 = 0.$$

(3.140)

The assumption that $C_{22} > 0$ and $L_{21} > 0$ implies that

$$\alpha p^2 + 2\alpha p_1 p - \gamma > 0.$$

(3.141)

Based on (3.128) and (3.130), calculation yields

$$Z(s) = Z_1(s) + Z_2(s) = \frac{ms^3 + (2mp + \alpha + q)s^2 + p(mp + 2q)s + (qp^2 + \gamma)}{(s + p_1)(s + p)^2}.$$

$$(3.142)$$

Comparing (3.110) with (3.142) gives

$$m = k, \tag{3.143}$$

$$2mp + \alpha + q = k(p_1 + 2z), \tag{3.144}$$

$$p(mp + 2q) = kz(z + 2p_1), \tag{3.145}$$

$$qp^2 + \gamma = kp_1 z^2. \tag{3.146}$$

Then, (3.143) and (3.145) together yield

$$q = \frac{-k(p^2 - 2p_1 z - z^2)}{2p}. \tag{3.147}$$

By (3.143), (3.144), and (3.147), one obtains

$$\alpha = \frac{-k(p - z)(3p - z - 2p_1)}{2p}. \tag{3.148}$$

By (3.143) and (3.146), one obtains

$$\gamma = \frac{1}{2}k(p - z)(p^2 + zp - 2zp_1). \tag{3.149}$$

Together with (3.143) and (3.147)–(3.149), condition (3.129) is equivalent to $p < z$, and (3.140) and (3.141) are equivalent to

$$(p + z)p_1^2 - 2p(p - z)p_1 - p^2(p + z) = 0, \tag{3.150}$$

$$p^2 + p_1 p - (p_1^2 + zp_1) > 0, \tag{3.151}$$

respectively. Thus, it follows from (3.150) and (3.151) that $-p_1(p-z)^2/(p+z) > 0$, which is impossible.

Therefore, $Z(s)$ cannot be realized as the configuration in Fig. 3.39(a), where N_1 is the configuration in Fig. 3.43(a), and N_2 is the configurations in Fig. 3.44(a).

The other cases can be similarly proved. □

Lemma 3.41. *If a biquadratic impedance* $Z(s) \in \mathcal{Z}_{\delta=2}^{k,p,z}$, *not satisfying the conditions of Theorem 3.22, is realizable as the configuration in Fig. 3.39(a), where N_1 is the configuration in Fig. 3.43(a) and N_2 is one of the configurations in Figs. 3.44(b) and 3.44(f), then the condition of Lemma 3.31 holds.*

Proof. This lemma can be proved using a similar method to the proof of Lemma 3.40. □

Lemma 3.42. *A biquadratic impedance* $Z(s) \in \mathcal{Z}_{\delta=2}^{k,p,z}$, *not satisfying the conditions of Theorem 3.22, is realizable as the configuration in Fig. 3.39(a), where* N_1 *is the configuration in Fig. 3.43(a) and* N_2 *is the configuration in Fig. 3.44(g) (that is, the configuration in Fig. 3.45(a)), if and only if*

$$(p - z)(p - 3z) > 0, \tag{3.152}$$

$$p^4 - 6zp^3 + 6z^2p^2 - 14z^3p + 5z^4 < 0. \tag{3.153}$$

Proof. By (3.111), this realizability condition can be obtained from [Jiang and Zhang (2014), Table I], where the realizability condition of a general biquadratic impedance as such a configuration is presented. Moreover, the element values can be derived as $R_1 = q/p_1$, $R_2 = mq/(q - mp_1)$, $C_1 = (q - mp_1)/q^2$, $R_{21} = \alpha$, $L_{21} = \alpha/(2p + p_1)$, $L_{22} = \alpha\beta/\gamma$, and $C_{21} = 1/\beta$, where $q = kz^2p_1/p^2$, $m = k(2z(p - z)p_1^2 - p(p - z)(p - 3z)p_1 + 2z^2p^2)/(2p^4)$, $\alpha = k(p - z)(2p + p_1)(p^2 + zp - 2zp_1)/(2p^4)$, $\beta = 2k(p - z)(-zp_1^2 + p(p - z)p_1 + zp^2)/p^3$, $\gamma = kp_1(p - z)(p^2 + zp - 2zp_1)/(2p^2)$, and p_1 is a positive root of $(3z - p)p_1^2 - 2p(p - z)p_1 + p^2(p - 3z) = 0$. □

Lemma 3.43. *If a biquadratic impedance* $Z(s) \in \mathcal{Z}_{\delta=2}^{k,p,z}$, *not satisfying the conditions of Theorem 3.22, is realizable as the configuration in Fig. 3.39(a), where* N_1 *is the configuration in Fig. 3.43(a) and* N_2 *is the configuration in Fig. 3.44(h), then the condition of Lemma 3.42 holds.*

Proof. The detail of the proof is omitted for brevity, which is similar to that of Lemma 3.40. □

Theorem 3.24. *A biquadratic impedance* $Z(s) \in \mathcal{Z}_{\delta=2}^{k,p,z}$, *not satisfying the conditions of Theorem 3.22, is realizable as a four-reactive seven-element series-parallel network as shown in Fig. 3.39, where* N_1 *is a three-element series-parallel network and* N_2 *is a four-element series-parallel network, if and only if* $p/z \in [1/\eta_1, \eta_1]$, *where* η_1 *is the unique distinct root of the equation* $x^4 - 6x^3 + 6x^2 - 14x + 5 = 0$ *for* $x \in (2 + \sqrt{5}, +\infty)$. *Moreover, if the condition holds, then* $Z(s)$ *is realizable as one of the configurations in Figs. 3.38 and 3.45.*

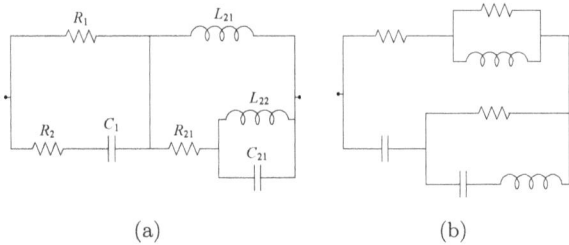

Fig. 3.45   Four-reactive seven-element series-parallel configurations realizing $Z(s) \in \mathcal{Z}_{\delta=2}^{k,p,z}$, whose one-terminal-pair labeled graphs are $\mathcal{N}_{12a}$ and $\mathcal{N}_{12b}$, respectively, satisfying $\mathcal{N}_{12a} = \mathrm{Dual}(\mathcal{N}_{12b})$.

Proof. Since any network in Fig. 3.39(b) can be the dual network of another one in Fig. 3.39(a), based on the principle of duality, one only needs to discuss the case of Fig. 3.39(a).

By Lemma 3.6, $Z(s) \in \mathcal{Z}_{\delta=2}^{k,p,z}$ cannot be realized as a series connection of two networks, one of which contains only reactive elements. Therefore, N_1 in Fig. 3.39(a) can only contain one or two reactive elements.

When N_1 is a two-reactive three-element network and N_2 is a two-reactive four-element network, N_1 is one of the four configurations in Fig. 3.42 by Lemma 3.36, which implies that the condition of Lemma 3.31 is satisfied by Lemmas 3.37 and 3.38. This means that the condition of this theorem holds.

When N_1 is a one-reactive three-element network and N_2 is a three-reactive four-element network, N_1 can be equivalent to one of the configurations in Fig. 3.43 and N_2 can be equivalent to one of the configurations in Fig. 3.44 by Lemma 3.39. This implies that the realizations can be equivalent to the configurations whose one-terminal-pair labeled graphs are $\mathcal{N}_{12a}$ (Fig. 3.45(a)) and $\mathrm{Inv}(\mathcal{N}_{12a})$ by Lemmas 3.40–3.43. Based on the principle of duality and the principle of frequency inversion, the realizability condition of $Z(s) \in \mathcal{Z}_{\delta=2}^{k,p,z}$, as a configuration whose one-terminal-pair labeled graph is $\mathrm{Inv}(\mathcal{N}_{12a})$, is the same as that of $\mathrm{Dual}(\mathcal{N}_{12a})$ (Fig. 3.45(b)). This can be derived from Lemma 3.42 by $p \leftrightarrow p^{-1}$ and $z \leftrightarrow z^{-1}$ (or $p \leftrightarrow z$).

The combination of the condition of Lemma 3.31 and the realizability conditions of $Z(s) \in \mathcal{Z}_{\delta=2}^{k,p,z}$ as the configurations in Fig. 3.45 proves the theorem.   □

3.8.4 Realizations as Five-Reactive Seven-Element Series-Parallel Networks

Lemma 3.44. *Consider a biquadratic impedance $Z(s)$ in the form of (3.1), where $a_i > 0$, $i = 0, 1, 2$, and $b_j > 0$, $j = 0, 1, 2$, which cannot be realized as a series-parallel network containing fewer than seven elements. If $Z(s)$ is realizable as a five-reactive seven-element series-parallel network as the configuration in Fig. 3.39(a), where N_1 is a three-element series-parallel network and N_2 is a four-element series-parallel network, then N_1 is one of the configurations in Figs. 3.42(b)–3.42(d) and N_2 is equivalent to one of the configurations in Fig. 3.44.*

Proof. By Lemmas 3.5 and 3.6, this lemma can be proved based on the method of enumeration. □

Lemma 3.45. *Any biquadratic impedance $Z(s) \in \mathcal{Z}_{\delta=2}^{k,p,z}$, not satisfying the condition of Theorem 3.24, cannot be realized as the configuration in Fig. 3.39(a), where N_1 is the configuration in Fig. 3.42(b) and N_2 is one of the configurations in Figs. 3.44(a)–3.44(d), 3.44(f), and 3.44(h).*

Proof. The proof of this lemma is similar to that of Lemma 3.40, therefore is omitted for brevity. □

Lemma 3.46. *A biquadratic impedance $Z(s) \in \mathcal{Z}_{\delta=2}^{k,p,z}$, not satisfying the condition of Theorem 3.24, is realizable as the configuration in Fig. 3.39(a), where N_1 is the configuration in Fig. 3.42(b) and N_2 is the configuration in Fig. 3.44(e) (that is, the configuration in Fig. 3.46(a) whose one-terminal-pair labeled graph is $\mathcal{N}_{13a}$), if and only if*

$$16p^4 - 40zp^3 + 31z^2p^2 - 10z^3p + z^4 = 0 \qquad (3.154)$$

and $p < z/(2 + \sqrt{5})$.

Proof. *Necessity.* It can be verified that the impedance of the configuration in Fig. 3.42(b) is in the form of (3.121), and the impedance of the configuration in Fig. 3.44(e) is in the form of

$$Z_2(s) = \frac{R_{21}L_{21}C_{22}s^2 + L_{21}s + R_{21}}{R_{21}L_{21}C_{21}C_{22}s^3 + L_{21}(C_{21} + C_{22})s^2 + R_{21}C_{21}s + 1}. \qquad (3.155)$$

Since it is assumed that the condition of Theorem 3.24 does not hold, $Z(s)$ cannot be realized with fewer than five reactive elements. If $Z(s)$ is realizable as the configuration in Fig. 3.39(a), where N_1 is the configuration in

Fig. 3.42(b) and N_2 is the configuration in Fig. 3.44(e), then the impedance of N_1 is of degree two and is in the form of

$$Z_1(s) = \frac{m(s^2 + pp_1)}{(s + p_1)(s + p)},$$

(3.156)

where m, p_1, $p > 0$, and the impedance of N_2 is of degree three and is in the form of

$$Z_2(s) = \frac{\alpha s^2 + \beta s + \gamma}{(s + p_1)(s + p)^2},$$

(3.157)

where α, $\beta > 0$ and

$$\alpha p^2 + 2\alpha p_1 p - \gamma = 0,$$

(3.158)

$$\alpha^2 p_1 p^2 - 2\alpha\gamma p - \gamma(\alpha p_1 - \beta) = 0,$$

(3.159)

$$2\alpha p + \alpha p_1 - \beta > 0.$$

(3.160)

Furthermore, one obtains

$$Z(s) = Z_1(s) + Z_2(s) = \frac{ms^3 + (mp + \alpha)s^2 + (mp_1 p + \beta)s + (mp_1 p^2 + \gamma)}{(s + p_1)(s + p)^2}.$$

(3.161)

Comparing (3.110) with (3.161) yields

$$m = k,$$

(3.162)

$$mp + \alpha = k(p_1 + 2z),$$

(3.163)

$$mp_1 p + \beta = kz(2p_1 + z),$$

(3.164)

$$mp_1 p^2 + \gamma = kz^2 p_1.$$

(3.165)

Then, (3.162) and (3.163) yield

$$\alpha = -k(p - p_1 - 2z).$$

(3.166)

By (3.162) and (3.164), one obtains

$$\beta = -k(p_1 p - 2zp_1 - z^2).$$

(3.167)

It follows from (3.162) and (3.165) that

$$\gamma = -kp_1(p - z)(p + z).$$

(3.168)

Substituting (3.162) and (3.166)–(3.168) into (3.158)–(3.160) yields

$$2pp_1^2 + z(4p - z)p_1 - p^2(p - 2z) = 0, \quad (3.169)$$

$$(2p^2 - z^2)p_1^2 + 2pz(2p - z)p_1 - (p^4 - 5z^2 p^2 + 4z^3 p - z^4) = 0, \quad (3.170)$$

$$p_1^2 + 2pp_1 - (2p^2 - 4zp + z^2) > 0. \quad (3.171)$$

respectively. It is verified that the resultant of (3.169) and (3.170) in p_1 is $z^2(p+z)(p-z)^3(16p^4 - 40zp^3 + 31z^2p^2 - 10z^3p + z^4)$. Since the condition of Lemma 3 does not hold, there exists at least one common root between (3.169) and (3.170) in p_1 if and only if (3.154) holds with $p < z/(2 + \sqrt{5})$. It implies that $p_1 > 0$. By (3.169), it follows that (3.171) is equivalent to

$$p_1 > \frac{p(3p^2 - 6zp + 2z^2)}{(2p - z)^2}. \tag{3.172}$$

From (3.170), it follows that (3.172) is equivalent to $(p^2 + 2zp - z^2)(p - z)^3(2p^3 + 10zp^2 - 7z^2p + z^3) > 0$, which can be verified to hold.

Sufficiency. Based on the discussion in the necessity part, there exists $p_1 > 0$ such that (3.169)–(3.171) hold. Let m, α, β, and γ satisfy (3.162) and (3.166)–(3.168), which obviously implies that α, β, γ, $m > 0$. Therefore, (3.158)–(3.160) and (3.163)–(3.165) hold. Consequently, $Z(s)$ can be written in the form of (3.161). Decompose $Z(s)$ as $Z(s) = Z_1(s) + Z_2(s)$, where $Z_1(s)$ is in the form of (3.156) with m, p_1, $p > 0$ and $Z_2(s)$ is in the form of (3.157) with α, $\beta > 0$. By letting $R_1 = m$, $L_1 = m/(p + p_1)$, and $C_1 = (p+p_1)/(mpp_1)$, $Z_1(s)$ is realizable as the configuration in Fig. 3.42(b). Let $C_{21} = 1/\alpha$, $C_{22} = (2\alpha p + \alpha p_1 - \beta)/(\alpha\beta)$, $R_{21} = \alpha^2/(2\alpha p + \alpha p_1 - \beta)$, and $L_{21} = \alpha^2\beta/(\gamma(2\alpha p + \alpha p_1 - \beta))$. Since (3.160) holds, C_{21}, C_{22}, R_{21}, $L_{21} > 0$. Thus, it can be verified that $1/C_{21} = \alpha$, $1/(R_{21}C_{21}C_{22}) = \beta$, $1/(L_{21}C_{21}C_{22}) = \gamma$, $(C_{21} + C_{22})/(R_{21}C_{21}C_{22}) = 2p + p_1$, $1/(L_{21}C_{22}) = p(p+2p_1)$, and $1/(R_{21}L_{21}C_{21}C_{22}) = p_1p^2$ hold because (3.158) and (3.159) are satisfied. Therefore, $Z_2(s)$ can be realized as the configuration in Fig. 3.44(e).

The lemma is thus proved. $\qquad\square$

Remark 3.4. It can be verified that there is only one distinct root of the equation $16x^4 - 40x^3 + 31x^2 - 10x + 1 = 0$ for $x \in (0, 1/(2 + \sqrt{5}))$.

Lemma 3.47. *Any biquadratic impedance $Z(s) \in \mathcal{Z}_{\delta=2}^{k,p,z}$, not satisfying the condition of Theorem 3.24, cannot be realized as the configuration in Fig. 3.39(a), where N_1 is the configuration in Fig. 3.42(c) and N_2 is one of the configurations in Figs. 3.44(a) and 3.44(f).*

Proof. The detail of the proof is omitted for brevity, which is similar to that of Lemma 3.40. $\qquad\square$

Lemma 3.48. *A biquadratic impedance $Z(s) \in \mathcal{Z}_{\delta=2}^{k,p,z}$, not satisfying the condition of Theorem 3.24, is realizable as the configuration in Fig. 3.39(a), where N_1 is the configuration in Fig. 3.42(c) and N_2 is the configuration in*

Fig. 3.44(e) (that is, the configuration in Fig. 3.47(a) whose one-terminal-pair labeled graph is N_{5a}), if and only if $p < z/(2+\sqrt{5})$ and $p^{10} - 16zp^9 + 118z^2p^8 - 476z^3p^7 + 1066z^4p^6 - 1372z^5p^5 + 1064z^6p^4 - 524z^7p^3 + 161z^8p^2 - 28z^9p + 2z^{10} = 0$.

Proof. The detail of the proof is omitted for brevity, which is similar to that of Lemma 3.46. $\qquad\square$

Remark 3.5. It can be verified that the equation $x^{10} - 16x^9 + 118x^8 - 476x^7 + 1066x^6 - 1372x^5 + 1064x^4 - 524x^3 + 161x^2 - 28x + 2 = 0$ has only one distinct root for $x \in (0, 1/(2+\sqrt{5}))$.

Theorem 3.25. *A biquadratic impedance $Z(s) \in \mathcal{Z}_{\delta=2}^{k,p,z}$, not satisfying the condition of Theorem 3.24, is realizable as a five-reactive seven-element series-parallel network as the configuration in Fig. 3.39, where N_1 is a three-element series-parallel network and N_2 is a four-element series-parallel network, if and only if $Z(s)$ satisfies $p/z \in \{\eta_2, 1/\eta_2, \eta_3, 1/\eta_3\}$, where η_2 is the unique distinct root of the equation $16x^4 - 40x^3 + 31x^2 - 10x + 1 = 0$ for $x \in (0, 1/(2+\sqrt{5}))$, and η_3 is the unique distinct root of the equation $x^{10} - 16x^9 + 118x^8 - 476x^7 + 1066x^6 - 1372x^5 + 1064x^4 - 524x^3 + 161x^2 - 28x + 2 = 0$ for $x \in (0, 1/(2+\sqrt{5}))$. Moreover, if $p/z = \eta_2$ (resp., $p/z = 1/\eta_2$), then $Z(s)$ is realizable as the configuration in Fig. 3.46(a) (resp., Fig. 3.46(b)); if $p/z = \eta_3$ (resp., $p/z = 1/\eta_3$), then $Z(s)$ is realizable as the configuration in Fig. 3.47(a) (resp., Fig. 3.47(b)).*

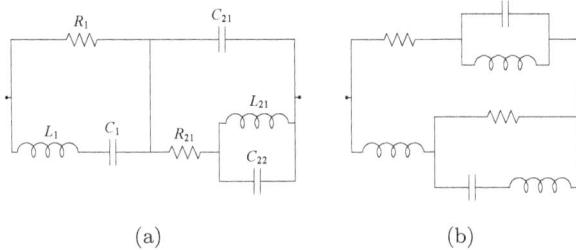

(a)   (b)

Fig. 3.46 Five-reactive seven-element series-parallel configurations realizing $Z(s) \in \mathcal{Z}_{\delta=2}^{k,p,z}$, whose one-terminal-pair labeled graphs are $\mathcal{N}_{13a}$ and $\mathcal{N}_{13b}$, respectively, satisfying $\mathcal{N}_{13a} = \mathrm{Dual}(\mathcal{N}_{13b})$.

Proof. Since any network in Fig. 3.39(b) can be the dual network of another one in Fig. 3.39(a), it suffices to discuss the case of Fig. 3.39(a) by the principle of duality.

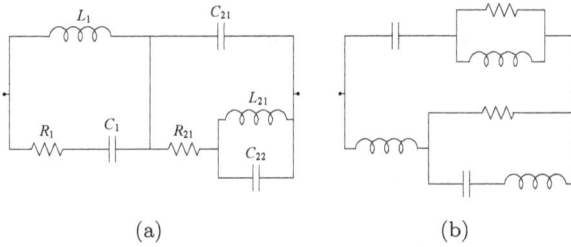

(a)   (b)

Fig. 3.47 Five-reactive seven-element series-parallel configurations realizing $Z(s) \in \mathcal{Z}_{\delta=2}^{k,p,z}$, whose one-terminal-pair labeled graphs are $\mathcal{N}_{14a}$ and $\mathcal{N}_{14b}$, respectively, satisfying $\mathcal{N}_{14a} = \mathrm{Dual}(\mathcal{N}_{14b})$.

By Lemma 3.44, N_1 is one of the configurations in Figs. 3.42(b)–3.42(d) and N_2 can be equivalent to one of the configurations in Fig. 3.44.

First, it is to discuss the case where N_1 is the configuration in Fig. 3.42(b). Under the assumption that the condition of Theorem 3.24 does not hold, by Lemma 3.45, N_2 cannot be any of the configurations in Figs. 3.44(a)–3.44(d), 3.44(f), and 3.44(h). When N_2 is the configuration in Fig. 3.44(e), a necessary and sufficient condition for the realization as such a network is presented in Lemma 3.46 ($p/z = \eta_2$), corresponding to the configuration in Fig. 3.46(a) whose one-terminal-pair labeled graph is $\mathcal{N}_{13a}$. When N_2 is the configuration in Fig. 3.44(g), the one-terminal-pair labeled graph of the realization configuration is $\mathrm{Inv}(\mathcal{N}_{13a})$. Based on the principle of duality and the principle of frequency inversion, the realizability condition of $Z(s)$, as the configuration whose one-terminal-pair labeled graph is $\mathrm{Inv}(\mathcal{N}_{13a})$, is the same as those of $\mathrm{Dual}(\mathcal{N}_{13a}) = \mathcal{N}_{13b}$ (Fig. 3.46(b)). This can be derived from Lemma 3.46 as $p/z = 1/\eta_2$ by $p \leftrightarrow p^{-1}$ and $z \leftrightarrow z^{-1}$ (or $p \leftrightarrow z$). The combination of the realizability conditions for the configurations in Fig. 3.46 implies that $p/z \in \{\eta_2, 1/\eta_2\}$.

Next, consider the case where N_1 is the configuration in Fig. 3.42(c). By Lemma 3.5, $Z(s)$ cannot be realized as a network whose network graph contains a path $\mathcal{P}(a, a')$ corresponding to only one kind of reactive elements. This implies that N_2 cannot be any of the configurations in Figs. 3.44(b)–3.44(d), 3.44(g), and 3.44(h). Under the assumption that the condition of Theorem 3.24 does not hold, by Lemma 3.47, N_2 cannot be any of the configurations in Figs. 3.44(a) and 3.44(f). When N_2 is the configuration in Fig. 3.44(e), a necessary and sufficient condition for the realization as such a network is presented in Lemma 3.48 ($p/z = \eta_3$), corresponding to the configuration in Fig. 3.47(a) whose one-terminal-pair labeled graph is $\mathcal{N}_{14a}$.

Based on the principle of duality, the realizability condition of $Z(s)$, as the configuration whose one-terminal-pair labeled graph is $\text{Dual}(\mathcal{N}_{14a}) = \mathcal{N}_{14b}$, can be derived from Lemma 3.48 as $p/z = 1/\eta_3$ by $p \leftrightarrow z$. The combination of the realizability conditions for the configurations in Fig. 3.47 implies that $p/z \in \{\eta_3, 1/\eta_3\}$.

It is clear that any network in Fig. 3.39(a), with N_1 being the configuration in Fig. 3.42(d), can be the frequency inverse network of another one in Fig. 3.39(a), with N_1 being the configuration in Fig. 3.42(c). Based on the principle of frequency inversion, $p/z \in \{\eta_3, 1/\eta_3\}$ also holds.

Thus, the theorem is proved. $\square$

3.8.5 *Main Results*

Theorem 3.26. *A biquadratic impedance $Z(s) \in \mathcal{Z}_{\delta=2}^{k,p,z}$, not satisfying the conditions of Theorem 3.22, is realizable as a seven-element series-parallel network as shown in Fig. 3.39, where N_1 is a three-element series-parallel network and N_2 is a four-element series-parallel network, if and only if the condition of Theorem 3.24 or Theorem 3.25 holds. Moreover, the realization configurations are shown in Figs. 3.38, 3.45–3.47.*

Proof. It is known from Lin's results [Lin (1965)] that any two-reactive-element series-parallel network realizing a biquadratic impedance can be equivalent to a two-reactive five-element series-parallel network. Together with the result in Theorem 3.23, under the assumption that the conditions of Theorem 3.22 do not hold, the number of reactive elements for seven-element realizations must be at least four. When the number of reactive elements is six, there is only one resistor, which implies that $Z(0) = Z(\infty)$, contradicting the assumption of $p \neq z$. Therefore, the number of reactive elements is four or five. Finally, the combination of Theorems 3.24 and 3.25 proves the theorem. $\square$

Chapter 4

Synthesis of n-Port Resistive Networks

4.1 Introduction

An n-port resistive network is a passive transformerless circuit consisting of only passive resistors. The passivity of such a network implies that its admittance (resp., impedance) matrix is necessarily an $n \times n$ non-negative definite matrix. However, unlike n-port resistor-transformer networks, a non-negative definite matrix may not be realizable as an n-port resistive network, and the realization problem of such networks is still unsolved. The realization of resistive n-port networks is a critical step towards the synthesis of n-port RLC networks. Through the approach of analogy, their realizations as n-port networks containing other one-kind-of-elements can be directly obtained. Furthermore, the minimal realization problem of one-port networks can be connected with n-port resistive network synthesis by element extraction (see the next chapter for details).

This chapter first presents a brief review of the development on n-port resistive network synthesis (see [Chen *et al.* (2014b)]) in Section 4.2. Then, some recent results [Chen *et al.* (2015b); Wang and Chen (2015)] on this topic are presented in Sections 4.3 and 4.4.

In this chapter, the *augmented graph*, *port graph*, and *network graph* of a network is denoted as $\mathcal{G}$, $\mathcal{G}_p$, and $\mathcal{G}_e$, respectively. Note that $\mathcal{G} = \mathcal{G}_p \cup \mathcal{G}_e$. Let $\mathcal{T}$ denote a tree of $\mathcal{G}$, $\mathcal{G}_{pt}$ denote a port graph that is also a tree of $\mathcal{G}$, and $\mathcal{G}_{ec}$ denote a network graph that can be regarded as a complete graph. The *fundamental cut-set matrix* and *fundamental circuit matrix* are denoted as Q_f and B_f, respectively. Let W and L respectively denote the submatrices of Q_f and B_f corresponding to $\mathcal{G}_e$. Also, let I_n denotes an nth-order unit matrix and let $E_{n,m}$ denote a matrix containing the first m columns of I_n. The vectors of port currents and voltages are denoted as $\hat{I}$

and $\hat{V}$, and the vectors of element currents and voltages are denoted as $\hat{J}$ and $\hat{U}$, respectively. Furthermore, let G_d (resp., D_d) denote the diagonal element admittance (resp., impedance) matrix, which is the diagonal matrix with diagonal entries being the admittances of the elements of the n-port network. Denote the admittance matrix and the impedance matrix of an n-port network as $Y_{n \times n}$ and $Z_{n \times n}$. In addition, let $\mathfrak{W}_{ij}$ denote the *tree transformation matrix* from $\mathcal{G}_{pt_i}$ to $\mathcal{G}_{pt_j}$, and $\dot{+}$ denote the direct sum of two matrices. Finally, $\mathbb{S}^n$ denotes the set of real symmetric $n \times n$ matrices, and $\mathbb{R}^{p \times q}$ denotes the set of real $p \times q$ matrices. On the other hand, the classical graph theory is referred to [Seshu and Reed (1961); Harary (1969)].

In this chapter, the following assumption is made without loss of generality, since any n-port resistive network can be reformulated to a network satisfying this assumption.

Assumption 4.1. For an n-port resistive network, there is no internal node and its augmented graph is connected.

4.2 A Review of n-Port Resistive Network Synthesis

4.2.1 *Realizations with $n \leq 3$*

This subsection briefly reviews a reworking of the proof [Chen (2007), Appendix A] given in [Tellegen (1952)], where the realization problems of resistive two-port and three-port networks are solved. Tellegen's original work was in Dutch and is not available to most readers.

The *paramountcy* of a real symmetric matrix is an important concept in n-port resistive networks, which is defined as follows.

Definition 4.1. [Slepian and Weinberg (1958)] A real symmetric matrix is *paramount* if each of its principal minors is not less than the absolute value of any other minor built from the same rows.

As shown in [Slepian and Weinberg (1958)], a paramount matrix is necessarily non-negative definite. Another concept, the *cross-sign change* of a matrix is defined as follows.

Definition 4.2. [Brown and Reed (1962a)] A *cross-sign change* of a matrix is to change the sign of each non-zero entry in the ith row and the ith column.

It should be noted that a cross-sign change of the impedance matrix (or, admittance matrix) means switching the polarity at a port, while exchanging two rows and the corresponding columns means swapping two ports.

It is proved in [Talbot (1955)] that the absolute value of the voltage v (resp., current i) applied at any port of a resistive network cannot be less than any other response voltages (resp., currents). Based on this result, it can be shown that the impedance matrix $Z_{2\times 2} \in \mathbb{S}^2$ of a two-port resistive network must be paramount. Conversely, any paramount impedance matrix $Z_{2\times 2} \in \mathbb{S}^2$ is realizable as a two-port resistive network whose configuration is one of the configurations in Fig. 4.1, where R_{ij} denotes the ijth entry of $Z_{2\times 2}$, $i, j = 1, 2$. The configuration in Fig. 4.1(b) can be obtained from that in Fig. 4.1(a) by switching the polarity at Port 2 corresponding to a cross-sign change for the second row and the second column.

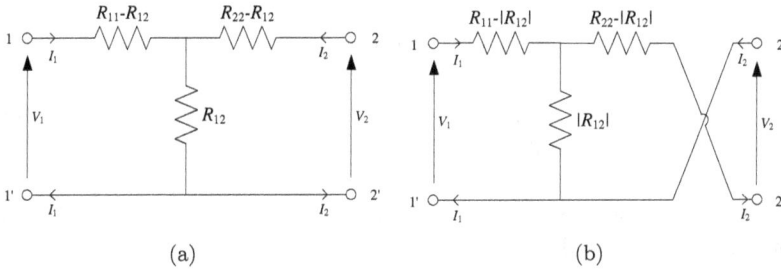

Fig. 4.1 The realization configurations of two-port resistive networks [Chen (2007)], where R_{ij} denotes the ijth entry of $Z_{2\times 2}$, $i, j = 1, 2$. The configuration in (a) corresponds to the case of $R_{12} \geq 0$, and the configuration in (b) corresponds to $R_{12} < 0$.

Through applying the current generator at any port and properly open- or short-circuiting the other two, it can be shown that the impedance matrix $Z_{3\times 3} \in \mathbb{S}^3$ of the resistive three-port network must also be paramount. To show the converse, it suffices to prove that, after a finite number of cross-sign changes and a proper rearrangement of rows and the corresponding columns, any paramount third-order matrix $Z_{3\times 3} \in \mathbb{S}^3$ is realizable as the configuration in Fig. 4.2 with nonnegative element values, where some elements can be open- (infinite resistance) or short- (zero resistance) circuited. The detail of the derivation is referred to [Chen (2007), Appendix A].

Since the augmented graphs of these realizations in Figs. 4.1 and 4.2 are planar, the realization of the admittance matrix can be similarly discussed by the principle of duality. As a summary, the following theorem is presented.

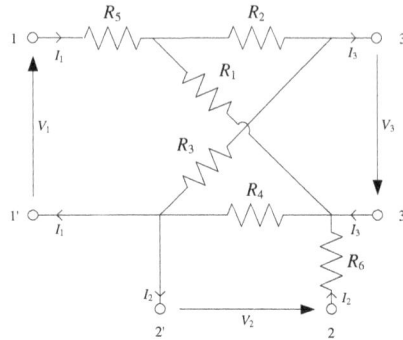

Fig. 4.2   The realization configuration of three-port resistive network [Chen (2007)], where some elements can be open- or short-circuited.

Theorem 4.1. *[Tellegen (1952)] Paramountcy is a necessary and sufficient condition for any second-order (resp., third-order) real symmetric matrix to be realizable as the impedance/admittance matrix of a two-port (resp., three-port) resistive network.*

Theorem 4.1 shows that the realization problem as n-port resistive networks has been solved for $n \leq 3$.

4.2.2 *General Properties of n-Port Resistive Networks*

In addition to the case of $n \leq 3$, the synthesis of n-port resistive networks for $n > 3$ needs to be further explored. Some general properties of n-port resistive networks will be presented. Due to the passivity and reciprocity properties, both the admittance and impedance matrices of n-port resistive networks must be non-negative definite [Anderson and Vongpanitlerd (1973), Section 2.7]. However, as mentioned above, the converse is not always true, and additional conditions are needed because of the transformerless structure.

Utilizing graph theory, Cederbaum [Cederbaum (1958b)] presents a necessary condition for the realization of an impedance (resp., admittance) matrix as an n-port resistive network. Before presenting an overview of the derivation, a concept is introduced.

Definition 4.3. [Cederbaum (1958b); Seshu and Reed (1961)] A matrix with all the subdeterminants equal to ± 1 or 0 is called an *E-matrix*.

Without loss of generality, a mild assumption was made in [Cederbaum (1958b)].

Assumption 4.2. [Cederbaum (1958b)] For an n-port resistive network, the network graph $\mathcal{G}_e$ is connected and contains all the v vertices of the augmented graph $\mathcal{G}$, and the port graph $\mathcal{G}_p$ is part of a tree $\mathcal{T}$ of $\mathcal{G}$.

The above assumption guarantees that $\mathcal{G}_p$ is made part of a tree $\mathcal{T}$ and a co-tree $\mathcal{G} - \mathcal{T}'$, which means by Theorem 2.18 that both the admittance and impedance matrices of the network exist.

Since $\mathcal{G}_p$ is made part of a tree $\mathcal{T}$, the f-cut-set matrix Q_f of $\mathcal{G}$ can be expressed as $Q_f = [I_{v-1}, M] = [E_{v-1,n}, W]$, for $v - 1 \geq n$, whose first n columns correspond to the edges of $\mathcal{G}_p$. Add $v - 1 - n$ new ports in parallel with $\mathcal{T} - \mathcal{G}_p$. Then, the resulting network contains $v - 1$ ports and the f-cut-set matrix becomes $\tilde{Q}_f = [I_{v-1}, W]$. By (2.39), one obtains

$$\begin{bmatrix} \hat{V} \\ \hat{V}_s \end{bmatrix} = (WG_dW^T)^{-1} \begin{bmatrix} \hat{I} \\ \hat{I}_s \end{bmatrix} =: \tilde{Z} \begin{bmatrix} \hat{I} \\ \hat{I}_s \end{bmatrix}.$$

By open-circuiting the $v - 1 - n$ new ports, one can see the impedance matrix $Z_{n \times n}$ is the nth-order principal submatrix of $\tilde{Z}$. Therefore,

$$Q_f \text{ is an E-matrix} \Rightarrow W \text{ is an E-matrix} \quad (Q_f = [E_{v-1,n}, W])$$
$$\Rightarrow WG_dW^T \text{ is paramount} \quad \text{(Binet-Cauchy Theorem)}$$
$$\Rightarrow \tilde{Z} = (WG_dW^T)^{-1} \text{ is paramount} \quad \text{(Jacobi Theorem)}$$
$$\Rightarrow Z_n \text{ is paramount}$$

Since $\mathcal{G}_p$ is made part of the co-tree $\mathcal{G} - \mathcal{T}'$, the f-circuit matrix can be expressed as $B_f = [I_{n+e-v+1}, X] = [E_{n+e-v+1,n}, L]$, for $n + e - v + 1 \geq n$, whose first n columns correspond to the edges of $\mathcal{G}_p$. Add $e - v + 1$ new ports inscribed into $(\mathcal{G} - \mathcal{T} - \mathcal{G}_p)$. Then, the resulting network contains $n + e - v + 1$ ports and the f-cut-set matrix becomes $\tilde{B}_f = [I_{n+e-v+1}, L]$. By (2.38), one obtains

$$\begin{bmatrix} \hat{I} \\ \hat{I}_s \end{bmatrix} = (LD_dL^T)^{-1} \begin{bmatrix} \hat{V} \\ \hat{V}_s \end{bmatrix} =: \tilde{Y} \begin{bmatrix} \hat{V} \\ \hat{V}_s \end{bmatrix}.$$

By short-circuiting the $e - v + 1$ new ports, one can see that the admittance matrix $Y_{n \times n}$ is the nth-order principal submatrix of $\tilde{Y}$. Similarly, one can prove that $Y_{n \times n}$ is paramount.

Therefore, the following result is obtained.

Theorem 4.2. *[Cederbaum (1958b)] The impedance (resp., admittance) matrix of an n-port resistive network is paramount.*

The above mild assumption does not include all the cases. When there exists a circuit in $\mathcal{G}_p$, WG_dW^{T} does not exist because of the nonexistence of $Q_f = [E_{v-1,n}, W]$, whose first n columns correspond to the edges of $\mathcal{G}_p$. An alternative proof of Theorem 4.2 without the assumption was given by Bruno and Weinberg [Bruno and Weinberg (1971)], who show that the admittance and impedance matrices of n-port resistive networks in (2.36) and (2.37) are paramount.

Utilizing the necessary condition of paramountcy, Cederbaum [Cederbaum (1958a)] generalized the result in [Talbot (1955)] to the vector case. If external excitation voltages $\hat{V}_e = [\hat{V}_{e1}, \hat{V}_{e2}, \ldots, \hat{V}_{ek}]^T$ (or currents $\hat{I}_e = [\hat{I}_{e1}, \hat{I}_{e2}, \ldots, \hat{I}_{ek}]^T$) are applied to any k ports of a resistive n-port network, and $\hat{V}_r = [\hat{V}_{r1}, \hat{V}_{r2}, \ldots, \hat{V}_{rk}]^T$ (or, currents $\hat{I}_r = [\hat{I}_{r1}, \hat{I}_{r2}, \ldots, \hat{I}_{rk}]^T$) are voltages (or, currents) of any other k ports (or, elements), then the determinant and all the subdeterminants of the matrix $B \in \mathbb{R}^{k \times k}$ satisfying $\hat{V}_r = B\hat{V}_e$ (or, $\hat{I}_r = B\hat{I}_e$) cannot exceed unity.

In [Cederbaum (1961)], the realization properties of the admittance (or, impedance) matrix with irreducible diagonal entries are investigated, where there is a diagonal entry equal to at least one off-diagonal entry of the same row. The diagonal entry cannot be further reduced because of the paramountcy constraint. Furthermore, the following necessary condition is obtained.

Theorem 4.3. *[Cederbaum (1961)] If an admittance matrix $Y_{n \times n} \in \mathbb{S}^n$ (resp., impedance matrix $Z_{n \times n} \in \mathbb{S}^n$) with all the diagonal entries respectively equal to the absolute value of the entries in a row, that is, $y_{ii} = |y_{ik}|$ (resp., $z_{ii} = |z_{ik}|$), $i = 1, 2, \ldots, n$, is realizable as an n-port resistive network, then $Y_{n \times n}$ (resp., $Z_{n \times n}$) is realizable as an n-port resistive network with $n + 1$ terminals (resp., with n independent circuits.)*

Cederbaum [Cederbaum (1961)] presented a paramount matrix as

$$P = \begin{bmatrix} 3 & 1 & 2 & 3 \\ 1 & 5 & 4 & 5 \\ 2 & 4 & 6 & 6 \\ 3 & 5 & 6 & 53/7 \end{bmatrix},$$

satisfying the condition of Theorem 4.3, and proved that P cannot be realized as the impedance matrix of a four-port resistive network with five terminals and cannot be realized as the admittance matrix of an n-port resistive network with five independent circuits. By Theorem 4.3, the paramount matrix P is realizable as neither the impedance matrix nor the admittance matrix of an n-port resistive network. This means that paramountcy is not always a sufficient condition for realization when $n > 3$.

In addition, by constructing another paramount matrix as

$$
Q = \begin{bmatrix} 3 & 2 & 1 & 3 \\ 2 & 3 & 2 & 3 \\ 1 & 2 & 3 & 3 \\ 3 & 3 & 3 & 5 \end{bmatrix},
$$

which is realizable as the impedance matrix but not the admittance matrix of a four-port resistive network, Cederbaum [Cederbaum (1961)] showed that the realizability conditions of the admittance and impedance matrices of the resistive n-port network are not the same when $n > 3$. In [Foster (1961); Nambiar (1963)], there is also discussion of the difference between the admittance and impedance matrices realizability conditions. In [Loughlin and Slepian (1979)], alternatively it shows that paramountcy is not sufficient, where new criteria guaranteeing the non-realizability of an irreducible paramount matrix are presented.

4.2.3 Realizations of Admittance Matrices with $n + 1$ Terminals

The admittance matrix of any n-port resistive network must be a real symmetric matrix $Y_{n \times n} \in \mathbb{S}^n$ in the form of

$$
Y_{n \times n} = \begin{bmatrix}
y_{11} & y_{12} & y_{13} & \cdots & y_{1,n-1} & y_{1n} \\
y_{12} & y_{22} & y_{23} & \cdots & y_{2,n-1} & y_{2n} \\
y_{13} & y_{23} & y_{33} & \cdots & y_{3,n-1} & y_{3n} \\
\vdots & \vdots & \vdots & \ddots & \vdots & \vdots \\
y_{1,n-1} & y_{2,n-1} & y_{3,n-1} & \cdots & y_{n-1,n-1} & y_{n-1,n} \\
y_{1n} & y_{2n} & y_{3n} & \cdots & y_{n-1,n} & y_{nn}
\end{bmatrix}. \tag{4.1}
$$

By Theorem 2.18, the port graph of any realization must be made part of a tree, provided that the admittance matrix exists. Since any internal node can be eliminated by the generalized star-mesh transformation [Versfeld (1970)], one may assume that all the nodes of n-port resistive networks are terminals. Furthermore, the number of terminals/nodes must range from $n + 1$ to $2n$. For an n-port resistive network with $n + 1$ terminals, the augmented graph $\mathcal{G}$ contains $n + 1$ vertices and the port graph $\mathcal{G}_p$ is precisely a tree of $\mathcal{G}$.

The realization problem of n-port resistive networks with $n + 1$ terminals has been investigated, first by Cederbaum in [Cederbaum (1959)].

By Remark 2.4, the admittance matrix of such a class of networks can be written as the unimodular congruence form, that is, $Y_{n \times n} = WGW^{\mathrm{T}}$, where $L \in \mathbb{R}^{n \times e}$ is an E-matrix and $G \in \mathbb{R}^{e \times e}$ is a diagonal positive definite matrix. A direct algebraic procedure is established in [Cederbaum (1959)] for decomposing any $Y_{n \times n} \in \mathbb{S}^n$ into the unimodular congruence form, which can either generates a unique decomposition or show that the decomposition is impossible.

After obtaining L and G, the technique by Guillemin [Guillemin (1959)] or the Gould algorithm [Gould (1958)] can be utilized for the final realization as an n-port resistive network. Cederbaum's method can also be applied to the realization of the impedance matrix as an n-port resistive network with n independent circuits. The disadvantage of this approach is that the realizability of a given matrix can only be checked by proceeding with the decomposition and the final realization procedure, which is complex for many cases. Therefore, more straightforward methods are needed.

After the work of Cederbaum, Brown *et al.* [Brown and Tokad (1961); Brown and Reed (1962a,b); Guillemin (1960)] have successfully derived necessary and sufficient conditions for admittance matrices to be realizable as n-port resistive networks with $n + 1$ terminals, and derived formulas of the element values in terms of only the matrix entries. The methodology is to establish the relationship between the entries of the matrix and the values of the involved elements. To introduce the relevant results, some terminologies are needed.

Definition 4.4. [Brown and Tokad (1961)] A *path tree* is a tree, in which there are at most two edges incident to each vertex (see Fig. 4.3(a)); a *Lagrangian tree* is a tree, in which each branch is incident to a common vertex (see Fig. 4.3(b)).

Definition 4.5. [Harary (1969)] A *complete graph* is a connected graph, in which there exists one and only one edge between each pair of vertices (see Fig. 4.4(b)).

Definition 4.6. [Cederbaum (1963)] A *dominant matrix* is a real symmetric $n \times n$ matrix, such that each of its main diagonal entries is not less than the sum of the absolute values of all other entries in the same row, that is, $y_{ii} \geq \sum_{j=1,\ldots n, j \neq i} |y_{ij}|$, $i = 1, 2, \ldots, n$.

Definition 4.7. [Cederbaum (1963)] A *uniformly tapered matrix* is a real symmetric $n \times n$ matrix, whose entries satisfy $y_{ij} - y_{i,j+1} \geq y_{i-1,j} - y_{i-1,j+1}$, where $i, j = 1, 2, \ldots, n$, and $i \leq j$, with $y_{0,k} = y_{k,n+1} = 0$, $k = 1, 2, \ldots, n$.

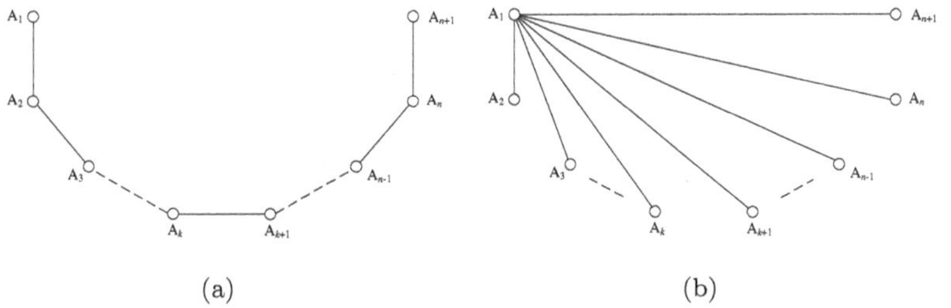

Fig. 4.3   Graphs of (a) path tree and (b) Lagrangian tree containing $n+1$ vertices (see [Brown and Tokad (1961)]).

When investigating n-port resistive networks with $n+1$ terminals, it is assumed in [Brown and Reed (1962a); Brown and Tokad (1961); Brown and Reed (1962b)] that the network graph is a complete graph with each edge corresponding to an element having a non-negative value. Besides, the port graph is also a tree of $\mathcal{G}$. In order to distinguish the $(n+1)$-terminal case with the general cases, denote the network graph and port graph as $\mathcal{G}_{ec}$ and $\mathcal{G}_{pt}$, respectively, which clearly satisfies $\mathcal{G} = \mathcal{G}_{pt} \cup \mathcal{G}_{ec}$. The admittance matrix $Y_{n \times n}$ can be expressed in the form of (2.39), where W is the f-cut-set matrix of $\mathcal{G}_{ec}$. When $\mathcal{G}_{ec}$, $\mathcal{G}_{pt}$, and their orientations are fixed, each element admittance g_{ij} can be linearly and uniquely expressed in terms of the entries of $Y_{n \times n}$.[1] In order to solve the realization problem, it suffices to find conditions such that the values of the above elements are non-negative. In [Brown and Tokad (1961)], the values of the elements are expressed in terms of the entries of $Y_{n \times n}$ when $\mathcal{G}_{pt}$ is a path tree and a Lagrangian tree, respectively.

Lemma 4.1. *[Brown and Tokad (1961); Guillemin (1960)] A matrix $Y_{n \times n} \in \mathbb{S}^n$ is realizable as the admittance matrix of an n-port resistive network with $n+1$ terminals whose port graph is a Lagrangian tree, if and only if after a finite number of cross-sign changes $Y_{n \times n}$ becomes a dominant matrix with all the off-diagonal entries being non-positive. Moreover, if $Y_{n \times n}$ is a dominant matrix with all the off-diagonal entries being non-positive, then all the port edges are oriented towards (or from) the common vertex (see Fig. 4.4(a)), and the admittance values of the elements connecting any two terminals are uniquely determined as $g_{i,j} = -y_{ij}$, $i, j = 1, 2, \ldots, n$ and $i < j$, and $g_{l,n+1} = \sum_{k=1}^{n} y_{lk}$, $l = 1, 2, \ldots, n$ (see Fig. 4.4(b)).*

---

[1] By Remark 2.5, only the orientations of the edges of $\mathcal{G}_{pt}$ can affect the expressions but those of $\mathcal{G}_{ec}$ cannot.

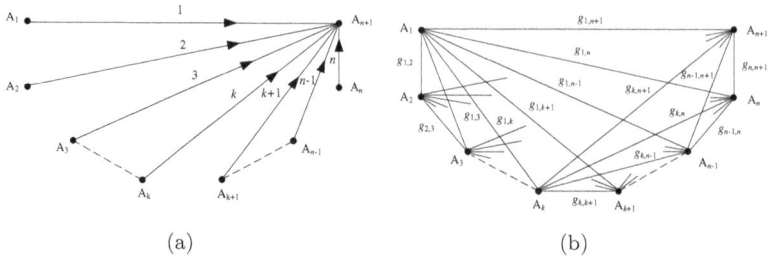

(a)   (b)

Fig. 4.4   (a) The directed port graph $\mathcal{G}_p$, which is a Lagrangian tree with all the port edges oriented towards the common vertex A_{n+1}; (b) the complete graph connecting each two vertices of $\mathcal{G}_p$, where $g_{i,j} \geq 0$, $i, j = 1, 2, ..., n+1$ and $i < j$.

Theorem 4.4. *[Guillemin (1960); Brown and Tokad (1961)] A matrix $Y_{n \times n} \in \mathbb{S}^n$ is realizable as the admittance matrix of an n-port resistive network with $n + 1$ terminals whose port graph is a path tree, if and only if after a finite number of cross-sign changes and a proper rearrangement of rows and columns $Y_{n \times n}$ is a uniformly tapered matrix. Moreover, if $Y_{n \times n}$ is a uniformly tapered matrix, then port edges are ordered and oriented to the same direction (see Fig. 4.5(a)), and the admittance values of the elements connecting any two terminals are uniquely determined as $g_{i,j} = (y_{i,j-1} - y_{ij}) - (y_{i-1,j-1} - y_{i-1,j})$, $i, j = 1, 2, \ldots, n+1$ and $i < j$, where $y_{0,k} = y_{k-1,n+1} = 0$, $k = 1, 2, \ldots, n+1$ (see Fig. 4.5(b)).*

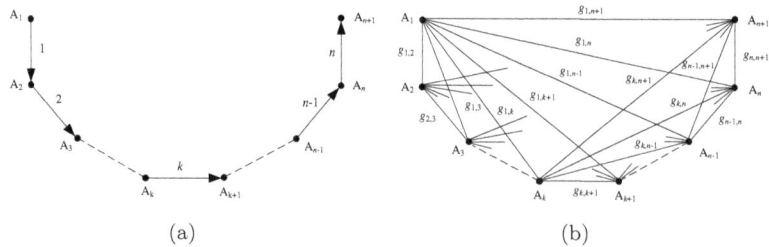

(a)   (b)

Fig. 4.5   (a) The directed port graph $\mathcal{G}_p$, which is a path tree with all the port edges ordered and oriented to the same direction; (b) the complete graph connecting each two vertices of $\mathcal{G}_p$, where $g_{i,j} \geq 0$, $i, j = 1, 2, \ldots, n+1$ and $i < j$.

For the realizations as networks with other kinds of port graphs, it is necessary to convert them to the network with a path tree or a Lagrangian tree. For this purpose, the concept of the *tree transformation matrix* is introduced as follows.

Definition 4.8. [Brown and Reed (1962b)] Consider a graph $\mathcal{P} = \mathcal{G}_{pt_i} \cup \mathcal{G}_{pt_j}$ with $n+1$ vertices, where $\mathcal{G}_{pt_i}$ and $\mathcal{G}_{pt_j}$ are both $(n+1)$-vertex trees. Let $[I_n, \mathfrak{W}_{ij}]$ be an f-cut-set matrix of $\mathcal{P}$, where the columns of I_n correspond to $\mathcal{G}_{pt_j}$ and columns of $\mathfrak{W}_{ij}$ correspond to $\mathcal{G}_{pt_i}$. Then, $\mathfrak{W}_{ij}$ is the *tree transformation matrix* from $\mathcal{G}_{pt_i}$ to $\mathcal{G}_{pt_j}$.

The tree transformation matrix has the following important property.

Theorem 4.5. *[Brown and Reed (1962b)] For the augmented graphs* $\mathcal{G}_i = \mathcal{G}_{pt_i} \cup \mathcal{G}_e$ *and* $\mathcal{G}_j = \mathcal{G}_{pt_j} \cup \mathcal{G}_e$ *of two n-port resistive networks with* $n+1$ *terminals, whose f-cut-set matrices are* $[I_n, W_i]$ *and* $[I_n, W_j]$, *respectively, the following relation is satisfied:* $W_j = \mathfrak{W}_{ij} W_i$.

Therefore, $Y_{n \times n}$ is realizable as an n-port $(n+1)$-terminal resistive network whose port graph is $\mathcal{G}_{pt_i}$ if and only if $\mathfrak{W}_{ij} Y_n \mathfrak{W}_{ij}^T$ is realizable as the one whose port graph is $\mathcal{G}_{pt_j}$.

Let $\mathcal{G}_{pt_j}$ be a Lagrangian tree with all edges oriented from or toward the common vertex. By Theorem 4.1, a necessary and sufficient condition is obtained for the realization as an n-port resistive network with $n+1$ terminals.

Theorem 4.6. *[Brown and Reed (1962b)] A matrix* $Y_{n \times n} \in \mathbb{S}^n$ *is realizable as the admittance matrix of an n-port resistive network with* $n+1$ *terminals whose directed port graph is* $\mathcal{G}_{pt_i}$, *if and only if there exists* $\mathfrak{W}_{iL}$ *such that* $\mathfrak{W}_{iL} Y_n \mathfrak{W}_{iL}^T$ *is a dominant matrix with all the off-diagonal entries being non-positive.*

By enumerating all the possible graphs, Theorem 4.6 can be utilized to test the realizability of any given admittance matrix. However, the test is very complex especially for a large value of n. To resolve this technical issue, Guillemin [Guillemin (1960)] first investigated the relationship between the port graphs and the entries of the admittance matrix $Y_{n \times n}$. Later, Biorci and Civalleri [Biorci and Civalleri (1961a)] proposed a simpler and quicker topological solution to such a problem. Through establishing a set of results relating the properties of a port graph to the *sign matrix* S_n [Biorci (1961)] or to the absolute values of the off-diagonal entries, a unified procedure is developed, which can either determine the possible port graphs including the edge orientations and the edge sequence or show that the given admittance matrix is not realizable. After the determination of the port graphs, Theorem 4.6 can be further utilized to test the realizability. Furthermore, Biorci and Civalleri [Biorci and Civalleri (1961b)] established a more straightforward criterion without going through the actual realization procedure of the port graph topologically. It is shown that if a real symmetric matrix is realizable as the admittance matrix of an n-port resistive network, then its sign matrix S_n must satisfy a special form (called the *F matrix* in [Biorci and Civalleri (1961b)]). Correspondingly, a rapid procedure that can reduce a sign matrix S_n to this form is developed. Finally, the port graph can be directly determined and Theorem 4.6 can be used to test the realizability of the given matrix.

However, the approach in [Biorci and Civalleri (1961b)] is not very efficient when $Y_{n \times n}$ contains many zero entries, since there are 2^k possible sign matrices if k zero entries exist in $Y_{n \times n}$. For this reason, [Halkias *et al.* (1962)] investigated a realization procedure that is particularly convenient when a large number of zero entries exist. In addition to determining possible port graphs of the realization of $Y_{n \times n}$, Boesch and Youla [Boesch and Youla (1965)] focused on directly determining the inverse of the tree transformation matrix $\mathfrak{W}_{iL}$ in Theorem 4.6, that is, $\mathfrak{W}_{iL}^{-1}$. They showed that $\mathfrak{W}_{iL}^{-1}$ satisfies

$$\mathfrak{W}_{iL}^{-1} + \mathfrak{W}_{iL}^{-T} = \frac{\text{sgn } Y_{n \times n} + U_n}{2} + I_n, \tag{4.2}$$

where the entries of sgn $Y_{n \times n}$ are ± 1 and $U_n \in \mathbb{R}^{n \times n}$ is a matrix with only one entry being 1 but all others being 0. By further showing that $Y_{n \times n}$ can always be converted to the canonical form that ensures $\mathfrak{W}_{iL}^{-1}$ to be triangular, and by establishing a rapid determination of the form, $\mathfrak{W}_{iL}^{-1}$ can be easily calculated using (4.2). However, the restriction of this method is that $Y_{n \times n}$ must contain nonzero entries.

As a conclusion, although easy-to-check criteria exist and can be further explored, the realization problem of the admittance matrices as resistive n-port networks containing $n + 1$ terminals has already been solved.

4.2.4 *Realizations of Admittance Matrices with More than* $n + 1$ *Terminals*

Unlike the $(n + 1)$-terminal case, the realization problem of an admittance matrix as an n-port resistive network with more than $n + 1$ terminals is still not completely solved as of today.

One of the main methods is to expand the n-port resistive network containing $n + p$ terminals for $k > 1$ to an $(n + p - 1)$-port network without changing the number of terminals, by properly adding $p - 1$ new ports such that the port graph $\tilde{\mathcal{G}}_p$ of the resulting network is a tree of $\tilde{\mathcal{G}}_p \cup \mathcal{G}_{ec}$. This converts the realization problem into the n'-port $(n'+1)$-terminal case with $n' = n + p - 1$ that has already been solved. This methodology was originally investigated by Guillemin [Guillemin (1961)], without showing any realizability condition. Consequently, Swaminathan and Frisch [Swaminathan and Frisch (1965)] derived a necessary and sufficient condition for the admittance matrix $Y_{n\times n}$ to be realizable as an n-port resistive network containing $n + 2$ terminals, in terms of the entries of $Y_{n\times n}$ and a set of parameters. The following shows the relationship between the admittance matrix $Y_{k\times k} \in \mathbb{S}^k$ of the k-port resistive network N_k containing $k + p$ terminals and the admittance matrix $Y_{(k+1)\times(k+1)} \in \mathbb{S}^{k+1}$ of the $(k + 1)$-port network N_{k+1} obtained by adding a new port to N_k.

Lemma 4.2. *[Swaminathan and Frisch (1965)] If the $(k + 1)$th row and the $(k + 1)$th column of $Y_{(k+1)\times(k+1)}$ correspond to the new port of N_{k+1} and the other correspondences are unchanged, then the relationship between $Y_{k\times k}$ and $Y_{(k+1)\times(k+1)}$ satisfies*

$$Y_{(k+1)\times(k+1)} = (Y_{k\times k} \dotplus 0) + \sigma_k \sigma_k^T, \tag{4.3}$$

where $\sigma_k = [p_1, p_2, \ldots, p_k, p_{k+1}]^T$.

Furthermore, the following conclusion has been established.

Lemma 4.3. *[Swaminathan and Frisch (1965)] Consider two real symmetric matrices, $Y_{k\times k}$ and $Y_{(k+1)\times(k+1)}$, which satisfy the relationship (4.3). If $Y_{(k+1)\times(k+1)}$ is the admittance matrix of a $(k + 1)$-port resistive network, then $Y_{k\times k}$ is the admittance matrix of a k-port resistive network. Moreover, if $Y_{k\times k}$ is the admittance matrix of a k-port resistive network with at least $k + 2$ terminals, then there exists a vector σ_k and a $(k + 1)$-port resistive network whose admittance matrix $Y_{(k+1)\times(k+1)}$ satisfies (4.3).*

Making use of Theorem 4.4 and Lemmas 4.2 and 4.3, the following result can be obtained.

Theorem 4.7. *[Swaminathan and Frisch (1965)] A given admittance matrix $Y_{n\times n} \in \mathbb{S}^n$ is realizable as an n-port resistive network with $n+2$ terminals, if and only if after a proper rearrangement of rows and columns, there exist $\mathfrak{W} = \mathfrak{W}_{1P_1} \dotplus \mathfrak{W}_{2P_2}$ and a set of parameters $p_1, p_2, \ldots, p_n, p$, such that the entries of $\mathfrak{W}Y_n\mathfrak{W}^T$ satisfy*

$$(y_{a,b} - y_{a,b+1}) - (y_{a-1,b} - y_{a-1,b+1}) \geq (p_a - p_{a-1})(p_{b+1} - p_b) \geq 0$$
$$\text{for } a, b = 1, 2, \ldots, m - 2, \text{ and } b \geq a,$$

$$(y_{a,m-1} - y_{a-1,m-1}) \geq (p_a - p_{a-1})(p - p_{m-1}) \geq 0, \ \ for \ a = 1, 2, \ldots, m-1,$$

$$(y_{i,j} - y_{i,j+1}) - (y_{i-1,j} - y_{i-1,j+1}) \geq (p_{i-1} - p_i)(p_j - p_{j+1}) \geq 0$$
$$for \ i, j = m+1, m+2, \ldots, n, \ \ and \ j \geq i,$$

$$(y_{m,j} - y_{m,j+1}) \geq (p - p_m)(p_j - p_{j+1}) \geq 0, \ \ for \ j = m, m+1, \ldots, n,$$

$$(p_a - p_{a-1})(p_j - p_{j+1}) \geq -(y_{a,j} - y_{a,j+1}) + (y_{a-1,j} - y_{a-1,j+1})$$
$$for \ a = 1, 2, \ldots, m-1, \ \ j = m, m+1, \ldots, n,$$

$$(p_a - p_{a-1})(p - p_m) \geq y_{a,m} - y_{a-1,m}, \ \ for \ a = 1, 2, \ldots, m-1,$$

$$(p - p_{m-1})(p - p_m) \geq -y_{m-1,m},$$

$$(p - p_{m-1})(p_j - p_{j+1}) \geq y_{m-1,j} - y_{m-1,j+1}, \ \ for \ j = m, m+1, \ldots, n,$$

where subscripts a and b are utilized to index the rows or columns in the positions indexed by $m-1$ or less, and i and j are used for the rows or columns above $m-1$.

In [Swaminathan and Frisch (1965)], a necessary condition for realization is further derived by eliminating parameters $p_1, p_2, \ldots, p_n, p$, based on Theorem 4.7. Moreover, in [Swaminathan and Frisch (1965)], the necessary condition is extended to the general $(n+p)$-terminal case with $p > 2$ ($\mathcal{G}_p$ consists of multiple subtrees) by short-circuiting all the ports except those corresponding to any two subtrees. However, no necessary and sufficient condition is available. Thereafter, many works such as [Lupo and Halkias (1965); Lupo (1968); Reddy *et al.* (1970); Naidu *et al.* (1976)] investigated the realization of the admittance matrix as an n-port resistive network with more than $n+1$ terminals. After the 1970s, there has not been much effort on investigating the synthesis of resistive n-port networks due to the declining interest, and the realization problem for the $(n+p)$-terminal case with $p \geq 2$ remains not completely solved, even for the case of $p = 2$ [Cederbaum (1984)].

In addition, unlike the $(n+1)$-terminal case (implied from (2.39)), the admittance matrix $Y_{n \times n}$ of an n-port network N containing $n+p$ terminals, which is the parallel connection of two networks N_1 and N_2, may not satisfy $Y_{n \times n} = Y_{n \times n}^{(1)} + Y_{n \times n}^{(2)}$, where $Y_{n \times n}^{(1)}$ and $Y_{n \times n}^{(2)}$ are admittance matrices of N_1 and N_2. Therefore, in [Lempel and Cederbaum (1967)] and [Murti and Thulasiraman (1967)], conditions were investigated for $Y_{n \times n} =$

$Y_{n\times n}^{(1)} + Y_{n\times n}^{(2)}$. As shown in [Cederbaum (1965); Thulasiraman and Murti (1968)], $Y_{n\times n} = W_e G_d W_e^{\mathrm{T}}$, where $W_e = W_1 - W_1 G W_2^{\mathrm{T}} (W_2 G W_2^{\mathrm{T}})^{-1} W_2$. Furthermore, W_e also satisfies $U = W_e^T V$ and $I = W_e J$. Since W_e plays a similar role to the fundamental cut-set matrix for the $(n+1)$-terminal case, W_e is called the *modified cut-set matrix*. If a one-to-one correspondence exists between the ports and the terminals of two n-port networks N_1 and N_2 to make the corresponding ports be incident at corresponding terminals, then N_1 and N_2 are called *compatible* [Lempel and Cederbaum (1967)].

A necessary and sufficient condition of $Y_{n\times n} = Y_{n\times n}^{(1)} + Y_{n\times n}^{(2)}$ is presented in [Lempel and Cederbaum (1967)], which is restated as follows.

Theorem 4.8. *[Lempel and Cederbaum (1967)] The admittance matrix of the network, which is the parallel connection of two compatible n-port resistive networks N_1 and N_2, satisfies $Y_{n\times n} = Y_{n\times n}^{(1)} + Y_{n\times n}^{(2)}$, if and only if their modified cut-set matrices with identical row and column ordering are the same.*

Remark 4.1. Theorem 4.8 is also valid for the networks containing negative elements.

4.3 Synthesis of n-Port Resistive Networks Containing $2n$ Terminals

This section considers the realization problem of an n-port resistive network containing $2n$ terminals.

By utilizing the existing results of $(n + p)$-terminal resistive networks with $p > 1$, a necessary and sufficient condition can be obtained for its realization as an n-port resistive network containing $2n$ terminals. The condition is based on the existence of a parameter matrix, and the element values are also parameterized. Since the condition is based on the existence of a set of parameters, further effort should be made such that the realizability condition can be simpler to apply.

It was shown in Section 4.2 that the realization problem of admittance matrices as n-port resistive networks containing $n + 1$ terminals has been solved [Brown and Reed (1962a,b); Brown and Tokad (1961); Guillemin (1960)], and the realization problem of the $(n+p)$-terminal case with $p > 1$ was studied in [Guillemin (1961); Reddy *et al.* (1970); Reddy and Thulasiraman (1972); Swaminathan and Frisch (1965)], which however remains to be completely solved.

Further investigations are needed to obtain necessary and sufficient conditions for an admittance matrix $Y_{n\times n}$ in the form of (4.19) to be realizable as an n-port resistive network with $2n$ terminals. In addition to other classes of n-port resistive networks containing more than $n+1$ terminals, such a class of networks is essential. If its network graph is described by a complete graph, the symmetric topological property of an n-port network containing $2n$ terminals shows that the realizability condition should be independent of the structures and orientations of the ports, which should be simpler than other cases. Although it may be impossible for an n-port resistive network to be equivalent to one containing $2n$ terminals, it suffices to focus on investigating the networks that do not contain the equivalence.

4.3.1   *A Necessary and Sufficient Condition for Realization*

Theorem 4.9 gives a necessary and sufficient condition for $Y_{n\times n} \in \mathbb{S}^n$ to be realizable as an n-port resistive network with $2n$ terminals. Such a network can be expressed as Fig. 4.6, where the orientation of each port edge is assumed to be in the opposite direction of the port current,[2] the elements are described by a complete graph, and the conductance value of each resistor connecting terminal A_i and A_j satisfies $g_{ij} \geq 0$, $i,j = 1,2,...,n$.

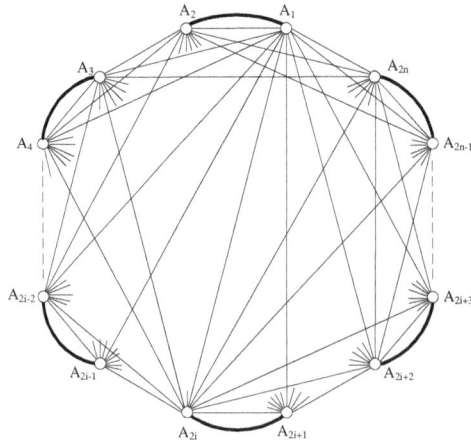

Fig. 4.6   The union of the port graph (edges in bold line segments) and the complete graph of an n-port resistive network containing $2n$ terminals.

---

[2]It can also be in the same direction.

Theorem 4.9. *An admittance matrix $Y_{n \times n} \in \mathbb{S}^n$ in the form of (4.19) is realizable as an n-port resistive network with 2n terminals if and only if there exists a matrix $P \in \mathbb{R}^{(2n-1) \times (n-1)}$ in the form of*

$$P = \begin{bmatrix} \gamma_1 & \gamma_2 & \cdots & \gamma_{n-1} \end{bmatrix} \tag{4.4}$$

with

$$\gamma_k := \begin{bmatrix} p_1^{(k)} & p_{n+1}^{(k)} & \cdots & p_k^{(k)} & p_{n+k}^{(k)} & p_{k+1}^{(k)} & 0 & \cdots p_{n-1}^{(k)} & 0 & p_n^{(k)} \end{bmatrix}^T \tag{4.5}$$

where $k = 1, 2, \ldots, n-1$, such that

$$\max\{ -(\alpha_i - \beta_{i-1})^T(\alpha_j - \beta_j), \ -(\alpha_i - \beta_i)^T(\alpha_j - \beta_{j-1})\} \le y_{ij}$$
$$\le \min\{ -(\alpha_i - \beta_{i-1})^T(\alpha_j - \beta_{j-1}), \ -(\alpha_i - \beta_i)^T(\alpha_j - \beta_j)\} \tag{4.6}$$

where $i, j = 1, 2, \ldots, n$ and $i < j$, and

$$y_{ii} \ge -(\alpha_i - \beta_{i-1})^T(\alpha_i - \beta_i) \tag{4.7}$$

with $i = j = 1, 2, \ldots, n$, where α_l^T is the $(2l-1)$th row of P, $l = 1, 2, \ldots, n$, β_m^T is the 2mth row of P, $m = 1, 2, \ldots, n-1$, and $\beta_0 = \beta_n := 0_{(n-1) \times 1}$.

Proof. It can be seen from Fig. 4.6 that no matter how the port edges of the n-port resistive network N with $2n$ terminals are oriented, it can be augmented to a $(2n-1)$-port resistive network $\overline{N}$ with $2n$ terminals and a linear ordered port graph, by adding $n-1$ new ports. Using Lemma 4.2 for $n-1$ steps, the admittance matrix $\overline{Y}$ of $\overline{N}$ can be written as

$$\overline{Y} = Y_d + Y_p,$$

where

$$Y_d = \begin{bmatrix} y_{11} & 0 & y_{12} & 0 & y_{13} & \cdots & y_{1,n-1} & 0 & y_{1n} \\ 0 & 0 & 0 & 0 & 0 & \cdots & 0 & 0 & 0 \\ y_{12} & 0 & y_{22} & 0 & y_{23} & \cdots & y_{2,n-1} & 0 & y_{2n} \\ 0 & 0 & 0 & 0 & 0 & \cdots & 0 & 0 & 0 \\ y_{13} & 0 & y_{23} & 0 & y_{33} & \cdots & y_{3,n-1} & 0 & y_{3n} \\ \vdots & \vdots & \vdots & \vdots & \vdots & \ddots & \vdots & \vdots & \vdots \\ y_{1,n-1} & 0 & y_{2,n-1} & 0 & y_{3,n-1} & \cdots & y_{n-1,n-1} & 0 & y_{n-1,n} \\ 0 & 0 & 0 & 0 & 0 & \cdots & 0 & 0 & 0 \\ y_{1,n} & 0 & y_{2,n} & 0 & y_{3,n} & \cdots & y_{n-1,n} & 0 & y_{n,n} \end{bmatrix}$$

and

$$Y_p = \sum_{k=1}^{n-1} \gamma_k \gamma_k^T$$

with γ_k defined in (4.5). Therefore, one can express $\overline{Y}$ as

$$\overline{Y} =$$

$$
\begin{bmatrix}
y_{11} + \sum\limits_{k=1}^{n-1} p_1^{(k)} p_1^{(k)} & \sum\limits_{k=1}^{n-1} p_1^{(k)} p_{n+1}^{(k)} & \cdots & \sum\limits_{k=n-1}^{n-1} p_1^{(k)} p_{2n-1}^{(k)} & y_{1n} + \sum\limits_{k=1}^{n-1} p_1^{(k)} p_n^{(k)} \\
* & \sum\limits_{k=1}^{n-1} p_{n+1}^{(k)} p_{n+1}^{(k)} & \cdots & \sum\limits_{k=n-1}^{n-1} p_{n+1}^{(k)} p_{2n-1}^{(k)} & \sum\limits_{k=1}^{n-1} p_{n+1}^{(k)} p_n^{(k)} \\
\vdots & \vdots & \ddots & \vdots & \vdots \\
* & * & \cdots & \sum\limits_{k=n-1}^{n-1} p_{2n-1}^{(k)} p_{2n-1}^{(k)} & \sum\limits_{k=n-1}^{n-1} p_{2n-1}^{(k)} p_n^{(k)} \\
* & * & \cdots & * & y_{nn} + \sum\limits_{k=1}^{n-1} p_n^{(k)} p_n^{(k)}
\end{bmatrix}.
$$

$$(4.8)$$

Using α_i and β_i, $\overline{Y}$ in the form of (4.8) can be further expressed as

$$\overline{Y} =$$

$$
\begin{bmatrix}
y_{11} + \alpha_1^T \alpha_1 & \alpha_1^T \beta_1 & \cdots & y_{1,n-1} + \alpha_1^T \alpha_{n-1} & \alpha_1^T \beta_{n-1} & y_{1n} + \alpha_1^T \alpha_n \\
* & \beta_1^T \beta_1 & \cdots & \beta_1^T \alpha_{n-1} & \beta_1^T \beta_{n-1} & \beta_1^T \alpha_n \\
\vdots & \vdots & \ddots & \vdots & \vdots & \vdots \\
* & * & \cdots & y_{n-1,n-1} + \alpha_{n-1}^T \alpha_{n-1} & \alpha_{n-1}^T \beta_{n-1} & y_{n-1,n} + \alpha_{n-1}^T \alpha_n \\
* & * & \cdots & * & \beta_{n-1}^T \beta_{n-1} & \beta_{n-1}^T \alpha_n \\
* & * & \cdots & * & * & y_{nn} + \alpha_n^T \alpha_n
\end{bmatrix}.
$$

$$(4.9)$$

Using Lemma 4.3, step by step, it can be shown that $Y_{n \times n}$ is realizable as an n-port resistive network with $2n$ terminals if and only if $\overline{Y}$ is realizable as a $(2n-1)$-port resistive network with $2n$ terminals whose port graph is a linear ordered tree. Furthermore, a necessary and sufficient condition for $Y_{n \times n}$ to be realizable as an n-port resistive network with $2n$ terminals is that $\overline{Y}$ is a uniformly tapered matrix, by Theorem 4.4, the condition of which is derived as follows.

When $i, j = 1, 2, \ldots, n$, and $i < j$, define a 3×3 submatrix $A^{(ij)} = [a_{ml}^{(ij)}]_{3 \times 3}$ of $\overline{Y}$ as

$$
A^{(ij)} = \begin{bmatrix} a_{11}^{(ij)} & a_{12}^{(ij)} & a_{13}^{(ij)} \\ a_{21}^{(ij)} & a_{22}^{(ij)} & a_{23}^{(ij)} \\ a_{31}^{(ij)} & a_{32}^{(ij)} & a_{33}^{(ij)} \end{bmatrix} := \begin{bmatrix} \beta_{i-1}^T \beta_{j-1} & \beta_{i-1}^T \alpha_j & \beta_{i-1}^T \beta_j \\ \alpha_i^T \beta_{j-1} & y_{ij} + \alpha_i^T \alpha_j & \alpha_i^T \beta_j \\ \beta_i^T \beta_{j-1} & \beta_i^T \alpha_j & \beta_i^T \beta_j \end{bmatrix}.
$$

When $i = j = 1, 2, \ldots, n$, define a 2×2 submatrix $B^{(i)} = [b_{ml}^{(i)}]_{2 \times 2}$ of $\overline{Y}$ as

$$
B^{(i)} = \begin{bmatrix} b_{11}^{(i)} & b_{12}^{(i)} \\ b_{21}^{(i)} & b_{22}^{(i)} \end{bmatrix} := \begin{bmatrix} \beta_{i-1}^T \alpha_i & \beta_{i-1}^T \beta_i \\ y_{ii} + \alpha_i^T \alpha_i & \alpha_i^T \beta_i \end{bmatrix}.
$$

It is noted that $\overline{Y}$ is uniformly tapered if and only if the entries of $A^{(ij)}$ and $B^{(i)}$ satisfy

$$a_{11}^{(ij)} - a_{12}^{(ij)} \leq a_{21}^{(ij)} - a_{22}^{(ij)} \leq a_{31}^{(ij)} - a_{32}^{(ij)}, \quad 1 \leq i < j \leq n, \quad (4.10)$$

$$a_{12}^{(ij)} - a_{13}^{(ij)} \leq a_{22}^{(ij)} - a_{23}^{(ij)} \leq a_{32}^{(ij)} - a_{33}^{(ij)}, \quad 1 \leq i < j \leq n, \quad (4.11)$$

$$b_{11}^{(i)} - b_{12}^{(i)} \leq b_{21}^{(i)} - b_{22}^{(i)}, \quad 1 \leq i = j \leq n. \quad (4.12)$$

By the definition of A^{ij}, (4.10) is equivalent to

$$
\begin{aligned}
&\beta_{i-1}^T \beta_{j-1} - \beta_{i-1}^T \alpha_j \leq \alpha_i^T \beta_{j-1} - y_{ij} - \alpha_i^T \alpha_j \leq \beta_i^T \beta_{j-1} - \beta_i^T \alpha_j \\
&\Leftrightarrow -(\alpha_i - \beta_i)^T (\alpha_j - \beta_{j-1}) \leq y_{ij} \leq -(\alpha_i - \beta_{i-1})^T (\alpha_j - \beta_{j-1}).
\end{aligned}
\quad (4.13)
$$

Moreover, (4.11) is equivalent to

$$
\begin{aligned}
&\beta_{i-1}^T \alpha_j - \beta_{i-1}^T \beta_j \leq y_{ij} + \alpha_i^T \alpha_j - \alpha_i^T \beta_j \leq \beta_i^T \alpha_j - \beta_i^T \beta_j \\
&\Leftrightarrow -(\alpha_i - \beta_{i-1})^T (\alpha_j - \beta_j) \leq y_{ij} \leq -(\alpha_i - \beta_i)^T (\alpha_j - \beta_j).
\end{aligned}
\quad (4.14)
$$

By the definition of $B^{(i)}$, (4.12) is equivalent to

$$
\begin{aligned}
&y_{ii} + \alpha_i^T \alpha_i - \alpha_i^T \beta_i \geq \beta_{i-1}^T \alpha_i - \beta_{i-1}^T \beta_i, \\
&\Leftrightarrow y_{ii} \geq -(\alpha_i - \beta_{i-1})^T (\alpha_i - \beta_i).
\end{aligned}
\quad (4.15)
$$

Combining (4.13)–(4.15), the condition of the theorem can be obtained. $\square$

Remark 4.2. Although the condition of Theorem 4.9 is concerned with the existence of a parameter matrix, it can contribute to the establishment of alternative necessary and sufficient conditions that may be more transparent and simpler to apply.

4.3.2 Element Value Expressions

The network element values are expressed based on Fig. 4.6 as follows.

Theorem 4.10. *If an admittance matrix $Y_{n \times n} \in \mathbb{S}^n$ in the form of (4.19) is realizable as an n-port resistive network with $2n$ terminals as shown in Fig. 4.6, where the orientation of each port edge is from vertex A_{2k-1} to A_{2k}, $k = 1, 2, \ldots, n$, that is, $Y_{n \times n}$ satisfies the condition of Theorem 4.9, then the conductances of the resistors of the edges are given by*

$$g_{2r-1,2s} = y_{r,s} + (\alpha_r - \beta_{r-1})^T (\alpha_s - \beta_s), \quad 1 \leq r \leq s \leq n,$$

$$g_{2r-1,2s-1} = -y_{r,s} - (\alpha_r - \beta_{r-1})^T (\alpha_s - \beta_{s-1}), \quad 1 \leq r < s \leq n,$$

$$g_{2r,2s-1} = y_{r,s} + (\alpha_r - \beta_r)^T (\alpha_s - \beta_{s-1}), \quad 1 \leq r < s \leq n, \quad (4.16)$$

$$g_{2r,2s} = -y_{r,s} - (\alpha_r - \beta_r)^T (\alpha_s - \beta_s), \quad 1 \leq r < s \leq n,$$

where α_k and β_k are obtained from the parameter matrix P as defined in Theorem 4.9, and $g_{h,l}$ denotes the conductance of the element connecting vertices A_h and A_l, $h, l = 1, 2, \ldots, 2n$, and $h < l$.

Proof. Since the orientation of each port edge is from vertex A_{2k-1} to A_{2k}, $k = 1, 2, \ldots, n$, the newly added ports of the augmented $(2n - 1)$-port network $\overline{N}$ must be in parallel with the edges connecting vertices A_{2k} and A_{2k+1}, $k = 1, 2, \ldots, n - 1$, respectively, and its port graph is a linear ordered tree. By the discussion in Theorem 4.9, a matrix P as defined in the theorem must exist such that (4.6) and (4.7) hold, and the admittance matrix $\overline{Y}$ of $\overline{N}$ is expressed in the form of (4.9).

By Theorem 4.4, element conductances are given as follows, in which

$$
\begin{aligned}
g_{2r-1,2s} &= (\overline{y}_{2r-1,2s-1} - \overline{y}_{2r-1,2s}) - (\overline{y}_{2r-2,2s-1} - \overline{y}_{2r-2,2s}) \\
&= (y_{r,s} + \alpha_r^T \alpha_s - \alpha_r^T \beta_s) - (\beta_{r-1}^T \alpha_s - \beta_{r-1}^T \beta_s) \\
&= y_{r,s} + (\alpha_r - \beta_{r-1})^T (\alpha_s - \beta_s),
\end{aligned}
$$

where $r, s = 1, 2, \ldots, n$, and $r \le s$;

$$
\begin{aligned}
g_{2r-1,2s-1} &= (\overline{y}_{2r-1,2s-2} - \overline{y}_{2r-1,2s-1}) - (\overline{y}_{2r-2,2s-2} - \overline{y}_{2r-2,2s-1}) \\
&= (\alpha_r^T \beta_{s-1} - y_{rs} - \alpha_r^T \alpha_s) - (\beta_{r-1}^T \beta_{s-1} - \beta_{r-1}^T \alpha_s) \\
&= -y_{rs} - (\alpha_r - \beta_{r-1})^T (\alpha_s - \beta_{s-1}),
\end{aligned}
$$

where $r, s = 1, 2, \ldots, n$, and $r < s$;

$$
\begin{aligned}
g_{2r,2s-1} &= (\overline{y}_{2r,2s-2} - \overline{y}_{2r,2s-1}) - (\overline{y}_{2r-1,2s-2} - \overline{y}_{2r-1,2s-1}) \\
&= (\beta_r^T \beta_{s-1} - \beta_r^T \alpha_s) - (\alpha_r^T \beta_{s-1} - y_{rs} - \alpha_r^T \alpha_s) \\
&= y_{rs} + (\alpha_r - \beta_r)^T (\alpha_s - \beta_{s-1}),
\end{aligned}
$$

where $r, s = 1, 2, \ldots, n$, and $r < s$;

$$
\begin{aligned}
g_{2r,2s} &= (\overline{y}_{2r,2s-1} - \overline{y}_{2r,2s}) - (\overline{y}_{2r-1,2s-1} - \overline{y}_{2r-1,2s}) \\
&= (\beta_r^T \alpha_s - \beta_r^T \beta_s) - (y_{rs} + \alpha_r^T \alpha_s - \alpha_r^T \beta_s) \\
&= -y_{rs} - (\alpha_r - \beta_r)^T (\alpha_s - \beta_s),
\end{aligned}
$$

where $r, s = 1, 2, \ldots, n$, and $r < s$. $\qquad\square$

For the parameter matrix P, a necessary condition is given in the following corollary. Let

$$
\begin{aligned}
\Pi_{i,j} &:= \max\{-(\alpha_i - \beta_{i-1})^T (\alpha_j - \beta_j), \ -(\alpha_i - \beta_i)^T (\alpha_j - \beta_{j-1})\}, \\
\Omega_{i,j} &:= \min\{-(\alpha_i - \beta_{i-1})^T (\alpha_j - \beta_{j-1}), \ -(\alpha_i - \beta_i)^T (\alpha_j - \beta_j)\},
\end{aligned}
$$

where $i, j = 1, 2, \ldots, n$, and $i < j$, and

$$
L_i := -(\alpha_i - \beta_{i-1})^T (\alpha_i - \beta_i),
$$

where $i = 1, 2, \ldots, n$.

Corollary 4.1. *If a matrix $Y \in \mathbb{S}^n$ satisfies the condition of Theorem 4.9, then the matrix $P \in \mathbb{R}^{(2n-1)\times(n-1)}$ satisfies*

$$\Omega_{i,j} \geq \Pi_{i,j}, \tag{4.17}$$

$$L_i \geq \max\{ |\Omega_{i,j}|, |\Pi_{i,j}| \}, \tag{4.18}$$

where $i, j = 1, 2, \ldots, n$, and $i < j$.

Proof. Condition (4.17) follows directly from (4.6). To prove (4.18), assume the contrary that there exist i and j such that

$$L_i < \max\{ |\Omega_{i,j}|, |\Pi_{i,j}| \}, \quad i < j.$$

Then, an admittance matrix Y' exists, whose entries are the same as those of $Y_{n\times n}$, except y_{ii} and y_{ij}. For Y', $y'_{ii} \geq L_i$ and $\Pi_{i,j} \leq y'_{ij} \leq \Omega_{i,j}$, but $y'_{ii} < |y'_{ij}|$. It is obvious that Y' also satisfies the condition of Theorem 4.9. Thus, it can be realized as an n-port resistive network with $2n$ terminals, which contradicts the property of paramountcy [Cederbaum (1958b)]. $\quad\square$

4.3.3 Numerical Example

Consider an admittance matrix $Y \in \mathbb{S}^4$ in the form

$$Y_{4\times4} = \begin{bmatrix} 5 & 1 & 1 & 4 \\ 1 & 3 & 2 & 1 \\ 1 & 2 & 4 & 1 \\ 4 & 1 & 1 & 6 \end{bmatrix}.$$

By choosing the parameter matrix P as

$$P = \begin{bmatrix} 1 & 3 & 2 & 0 & 2 & 0 & 1 \\ 1 & 2 & 2 & 5 & 2 & 0 & 1 \\ 1 & 2 & 2 & 2 & 7 & 1 \end{bmatrix}^T,$$

the condition of Theorem 4.9 is satisfied. Therefore, $Y_{4\times4}$ is realizable as a four-port resistive network containing eight terminals. Furthermore, based on (4.16), if the network is described by Fig. 4.6, where A_1, A_3, A_5, and A_7 are at higher potentials, and A_2, A_4, A_6, and A_8 are at lower ones, then the values of the elements are obtained as $g_{12} = 1$, $g_{18} = 7$, $g_{23} = 3$, $g_{27} = 7$, $g_{34} = 1$, $g_{45} = 15$, $g_{56} = 2$, $g_{67} = 35$, $g_{78} = 2$, with other element values being zero.

4.4   Minimal Realization of Three-Port Resistive Networks

This section introduces the minimal realization of any third-order real symmetric admittance matrix as a three-port resistive network.

As shown in Section 4.2, paramountcy is a necessary and sufficient condition for any third-order real symmetric admittance matrix to be realizable as a three-port resistive network. A three-port canonical configuration that can realize any paramount admittance matrix is shown in Fig. 4.7 ([Slepian and Weinberg (1958), Fig. 11(B)]), which contains no more than six elements and is the dual network of Fig. 4.2. It is essential to investigate the realization problem of three-port resistive networks with the least number of elements.

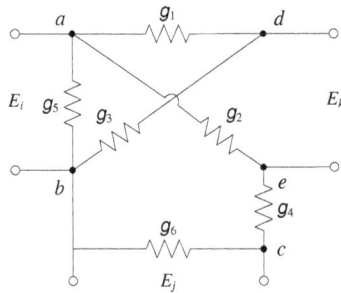

Fig. 4.7   A three-port canonical configuration that can realize a third-order paramount admittance matrix, where the conductances of elements are positive, zero, or infinity [Slepian and Weinberg (1958)].

First, a necessary and sufficient condition is obtained for a third-order real symmetric admittance matrix to be realizable as a three-port four-terminal resistive network containing no more than k elements, $k = 1, 2, \ldots, 5$. Then, it can be proved that the admittance matrix of a three-port resistive network containing at most k elements, $k = 1, 2, 3$, is realizable as the three-port four-terminal network, where the case of $k = 3$ has been presented in [Chen *et al.* (2015c)].

Utilizing topological constraints of the realizations, one can obtain a three-port four-element configuration whose admittance matrix may not be realizable as a three-port four-terminal resistive network containing no more than four elements. After deriving its realizability condition, a necessary and sufficient condition is obtained for a third-order paramount matrix to be realizable with no more than four elements. By similar but more complex arguments, a necessary and sufficient condition can be derived for a paramount admittance matrix to be realizable as a three-port resistive network containing five elements, by determining all the possible five-element configurations whose admittance matrices may not be realizable as three-port four-terminal resistive networks with five elements.

Since the augmented graph of any three-port resistive network containing no more than five elements is planar, the results in this section can be directly transformed to those of the impedance synthesis by the principle of duality.

Consider a matrix $Y_{3\times 3} \in \mathbb{S}^3$ in the form of

$$Y_{3\times 3} = \begin{bmatrix} y_{11} & y_{12} & y_{13} \\ y_{12} & y_{22} & y_{23} \\ y_{13} & y_{23} & y_{33} \end{bmatrix}. \tag{4.19}$$

By Theorem 4.1, if $Y_{3\times 3}$ is realizable as the admittance matrix of a three-port resistive network, then it is necessarily paramount. By the definition of paramountcy (see Definition 4.1), $Y_{3\times 3} \in \mathbb{S}^3$ is paramount if and only if the following inequalities simultaneously hold: $y_{11} \geq \max\{|y_{12}|, |y_{13}|\}$, $y_{22} \geq \max\{|y_{12}|, |y_{23}|\}$, $y_{33} \geq \max\{|y_{13}|, |y_{23}|\}$, $y_{11}y_{22} - y_{12}^2 \geq \max\{|y_{11}y_{23} - y_{12}y_{13}|, |y_{12}y_{23} - y_{13}y_{22}|\}$, $y_{22}y_{33} - y_{23}^2 \geq \max\{|y_{12}y_{33} - y_{13}y_{23}|, |y_{12}y_{23} - y_{13}y_{22}|\}$, and $y_{11}y_{33} - y_{13}^2 \geq \max\{|y_{11}y_{23} - y_{13}y_{12}|, |y_{12}y_{33} - y_{23}y_{13}|\}$.

In this section, the following problem is considered: What are necessary and sufficient conditions for a given paramount admittance matrix $Y_{3\times 3}$ to be realizable as a three-port resistive network with at most k elements, where $k = 1, 2, \ldots, 5$, and what about the covering configurations? It is assumed that all the nodes of the three-port resistive are terminals (Assumption 4.1) and element values are positive and finite.

Consider an n-port RLC network N consisting of three components: N_a, N_b and N_c, each of which has only one common node with another (see Fig. 4.8(a)). Another n-port RLC network N_1 shown in Fig. 4.8(b) can be obtained by exchanging the positions of N_b and N_c without any other alterations.

Lemma 4.4. *The admittance matrices of two n-port RLC networks, N and N_1, whose structures are as shown in Fig. 4.8, are the same.*

Proof. By Theorem 2.18, the admittance matrix $Y_{n\times n}$ of any n-port RLC network is expressed as (2.37), where D_d is a diagonal matrix, in which each diagonal entry is the element impedance, and $B_f = [B_{f1}, B_{f2}]$ is the f-circuit matrix of the augmented graph $\mathcal{G}$, where the columns of

$B_{f1} = [I_{n+e-v+1}, B_{f12}]$ correspond to the edges of the network graph $\mathcal{G}_e$. It is obvious that N and N_1 have the same B_f and D_d. Therefore, the admittance matrices of N and N_1 are the same based on (2.37). $\qquad \square$

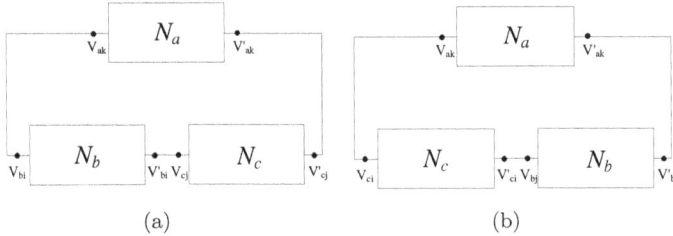

(a)   (b)

Fig. 4.8   The structures of two n-port networks mentioned in Lemma 4.4, where (a) is for N and (b) is for N_1.

The concept of *isomorphism* for graphs can be found in [Seshu and Reed (1961), pg. 13]. Since the augmented graph contains two types of edges: port edges and network edges, it is necessary to define the *isomorphism for augmented graphs*, as follows.

Definition 4.9. Two augmented graphs $\mathcal{G}$ and $\mathcal{G}_1$ are isomorphic if there simultaneously exist a one-to-one correspondence between vertices of $\mathcal{G}$ and $\mathcal{G}_1$, a one-to-one correspondence between port edges of $\mathcal{G}$ and $\mathcal{G}_1$, and a one-to-one correspondence between network edges of $\mathcal{G}$ and $\mathcal{G}_1$, which preserves the incidence relationships.

It is obvious that two isomorphic augmented graphs can correspond to the same n-port resistive network.

4.4.1   *Minimal Realization with Four Terminals*

Lemma 4.5. *An admittance matrix* $Y_{3\times3} \in \mathbb{S}^3$ *is realizable as a three-port four-terminal resistive network containing no more than k elements, $k = 1, 2, \ldots, 5$, if and only if*

1. *when* $y_{12}y_{13}y_{23} \leq 0$, *the following inequalities simultaneously hold with at least* $(6 - k)$ *of the six inequality signs being equality:* $|y_{12}| \geq 0$, $|y_{13}| \geq 0$, $|y_{23}| \geq 0$, $y_{11} - |y_{12}| - |y_{13}| \geq 0$, $y_{22} - |y_{12}| - |y_{23}| \geq 0$, *and* $y_{33} - |y_{13}| - |y_{23}| \geq 0$;

2. *when $y_{12}y_{13}y_{23} \geq 0$, at least one of the following three conditions holds with at least $(6-k)$ of the six inequality signs being equality: i) $-|y_{13}| \leq 0$, $|y_{13}| \leq |y_{12}| \leq y_{11}$, $|y_{13}| \leq |y_{23}| \leq y_{33}$, and $|y_{12}| + |y_{23}| - |y_{13}| \leq y_{22}$; ii) $-|y_{12}| \leq 0$, $|y_{12}| \leq |y_{13}| \leq y_{11}$, $|y_{12}| \leq |y_{23}| \leq y_{22}$, and $|y_{13}| + |y_{23}| - |y_{12}| \leq y_{33}$; iii) $-|y_{23}| \leq 0$, $|y_{23}| \leq |y_{12}| \leq y_{22}$, $|y_{23}| \leq |y_{13}| \leq y_{33}$, and $|y_{12}| + |y_{13}| - |y_{23}| \leq y_{11}$.*

Proof. When the three-port resistive network contains four terminals, the port graph is either a Lagragian tree or a path tree. Therefore, this lemma can be directly proved by using Lemmas 4.1 and 4.4, where Condition 1 corresponds to the Lagrangian-tree case and Condition 2 corresponds to the path-tree case. □

4.4.2 *Realization with at Most Four Elements*

First, a necessary and sufficient condition is presented for the realization with at most k elements, where $k = 1, 2, 3$.

Theorem 4.11. *A paramount admittance matrix $Y_{3\times3} \in \mathbb{S}^3$ is realizable as a three-port resistive network with at most k elements, where $k = 1, 2, 3$, if and only if $Y_{3\times3}$ satisfies the condition of Lemma 4.5.*

Proof. This theorem can be proved using a similar method to those of the four- and five-element cases in the following discussions. The details of this proof is omitted for brevity. Moreover, reference [Chen *et al.* (2015c)] provides an alternative proof for the case of $k = 3$. □

In order to obtain a necessary and sufficient condition for any admittance matrix $Y_{3\times3} \in \mathbb{S}^3$ to be realizable as a three-port resistive network with no more than four elements, it suffices to find the possible configurations that may not always be equivalent to a three-port four-terminal configuration with at most four elements, and to derive their realizability conditions. This is because Lemma 4.5 provides a necessary and sufficient condition for the realization as three-port resistive networks with four terminals and at most four elements when $k = 4$. A series of sufficient conditions have been previously established for the realization as three-port resistive networks containing four terminals and no more than four elements.

Lemma 4.6. *For a paramount admittance matrix $Y_{3\times3} \in \mathbb{S}^3$, if at least two of y_{12}, y_{13}, and y_{23} are zero, then $Y_{3\times3}$ is realizable as a three-port four-terminal resistive network containing no more than four elements.*

Proof. Since at least two of y_{12}, y_{13}, and y_{23} are zero, $Y \in \mathbb{S}^3$ can satisfy $y_{13} = y_{23} = 0$ after a proper rearrangement of rows and the corresponding columns. The paramountcy of $Y_{3\times 3}$ implies that the following inequalities hold: $y_{11} \geq |y_{12}| \geq 0$, $y_{22} \geq |y_{12}| \geq 0$, and $y_{33} \geq 0$. It follows that $y_{12}y_{13}y_{23} = 0$, $|y_{12}| \geq 0$, $|y_{13}| = 0$, $|y_{23}| = 0$, $y_{11} - |y_{12}| - |y_{13}| = y_{11} - |y_{12}| \geq 0$, $y_{22} - |y_{12}| - |y_{23}| = y_{22} - |y_{12}| \geq 0$, and $y_{33} - |y_{13}| - |y_{23}| = y_{33} \geq 0$. Thus, the condition of Lemma 4.5 holds for $k = 4$. Therefore, $Y_{3\times 3}$ is realizable as a three-port four-terminal resistive network containing no more than four elements. □

It is clear that the admittance matrix of a three-port resistive network with a *separable* [Seshu and Reed (1961), pg. 35] augmented graph must contain at least two zero entries among y_{12}, y_{13}, and y_{23}. Therefore, Lemma 4.6 shows that if a three-port resistive network whose admittance matrix exists cannot be equivalent to one containing four terminals and at most four elements, its augmented graph $\mathcal{G}$ must be nonseparable.

Lemma 4.7. *If a paramount admittance matrix $Y_{3\times 3} \in \mathbb{S}^3$ contains two equal rows or two rows for which one row is the negative of the other, then $Y_{3\times 3}$ is realizable as a three-port four-terminal resistive network containing no more than four elements.*

Proof. First, consider the two-equal-row case. Without loss of generality, assume the two rows to be the first and second rows. Due to its symmetry, $Y_{3\times 3}$ is in the form of

$$Y_{3\times 3} = \begin{bmatrix} y_{11} & y_{11} & y_{13} \\ y_{11} & y_{11} & y_{13} \\ y_{13} & y_{13} & y_{33} \end{bmatrix}.$$

Together with the paramountcy of $Y_{3\times 3}$, it follows that $y_{12}y_{13}y_{23} = y_{11}y_{13}^2 \geq 0$, $-|y_{13}| \leq 0$, $y_{11} - |y_{12}| = 0$, $|y_{12}| - |y_{13}| = y_{11} - |y_{13}| \geq 0$, $y_{33} - |y_{23}| = y_{33} - |y_{13}| \geq 0$, $|y_{23}| - |y_{13}| = 0$, and $y_{22} - (|y_{12}| + |y_{23}| - |y_{13}|) = y_{11} - (y_{11} + |y_{13}| - |y_{13}|) = 0$. Therefore, Condition 2 of Lemma 4.5 must hold for $k = 4$. When there is another pair of two equal rows, a similar argument can be applied. Moreover, one can similarly prove the other case when $Y_{3\times 3}$ contains two rows where one row is the negative of the other. □

When a three-port resistive network contains two ports directly in series, it is clear that its admittance matrix must contain two equal rows or two rows for which one row is the negative of the other. Therefore, Lemma 4.7 shows that if a three-port resistive network, whose admittance matrix exists, cannot be equivalent to one containing four terminals and at most four elements, its augmented graph $\mathcal{G} = \mathcal{G}_p \cup \mathcal{G}_e$ cannot contain two port edges directly in series.

Lemma 4.8. *An admittance matrix $Y_{3\times3} \in \mathbb{S}^3$ is realizable as a three-port resistive network containing no more than four elements, if and only if $Y_{3\times3}$ satisfies the condition of Lemma 4.5 for $k = 4$ or $Y_{3\times3}$ is realizable as the configuration in Fig. 4.9.*

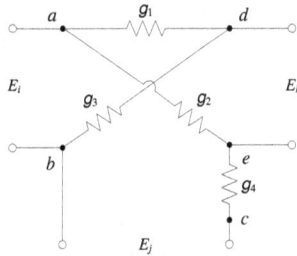

Fig. 4.9   The configuration discussed in Lemma 4.8, where g_1, g_2, g_3, $g_4 > 0$.

Proof. *Sufficiency.* By Lemma 4.5, the sufficiency obviously holds.

Necessity. As discussed above, the network can only contain four, five, or six terminals because of the existence of the admittance matrix $Y_{3\times3}$.

When the number of terminals is four, the condition of Lemma 4.5 holds for $k = 4$.

When the number of terminals is five, the port graph $\mathcal{G}_p$ can only be the one shown in Fig. 4.10(a), containing two *maximal connected subgraphs* (see [Seshu and Reed (1961), pg. 16]). If there are no more than three elements, then the condition of Lemma 4.5 holds for $k = 3$ by Theorem 4.11, which means that the condition of Lemma 4.5 holds for $k = 4$. Therefore, assume that there are four elements.

Next, find out the possible configurations whose admittance matrices do not always satisfy the condition of Lemma 4.5 for $k = 4$. Then, denote the cut-set that separates $\mathcal{G}$ into two parts containing respectively two maximal connected subgraphs of $\mathcal{G}_p$ as $\mathcal{C}_{ep}$. From its definition, one can see that any edge belonging to $\mathcal{C}_{ep}$ is a network edge. By Lemma 4.6, any of vertices d and e is incident by at least one network edge belonging to $\mathcal{C}_{ep}$. Otherwise, $\mathcal{G}$ is a separable graph, which implies that the condition of Lemma 4.5 holds for $k = 4$. By Lemma 4.4, if there is only one network edge belonging to $\mathcal{C}_{ep}$ incident at vertex d or e, then one can see that the three-port resistive network can always be equivalent to another three-port four-terminal resistive network containing no more than four elements. This further implies that the condition of Lemma 4.5 holds for $k = 4$. Therefore, one can indicate that each of vertices d and e is incident by two network edges belonging to $\mathcal{C}_{ep}$. By Lemmas 4.6 and 4.7, it also implies that each of vertices a–c is incident by at least one network edge belonging to $\mathcal{C}_{ep}$.

Assuming that two augmented graphs, which are isomorphic with each other, are the same graph, it is not difficult to see that only the augmented graphs $\mathcal{G}$ in Fig. 4.11 are possible, where the bold line segment represents a port edge and the light one represents a network edge. Fig. 4.11(b) can be further eliminated by Lemma 4.4, since the corresponding network can be equivalent to the three-port four-terminal case. The configuration with graph in Fig. 4.11(a) is shown in Fig. 4.9.

When the number of terminals is six, the port graph $\mathcal{G}_p$ must be the one shown in Fig. 4.10(b), consisting of three maximal connected subgraphs. Assume that there exists a network having this port graph. Using a similar argument to the above, it can be shown that at least six elements are needed according to Lemmas 4.4, 4.6 and 4.7, which however is impossible.   □

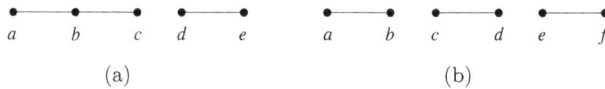

(a)   (b)

Fig. 4.10   (a) The port graph of three-port resistive networks containing five terminals; (b) the port graph of three-port resistive networks containing six terminals.

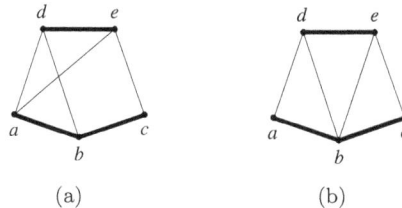

(a)   (b)

Fig. 4.11   Augmented graphs mentioned in the proof of Theorem 4.8, where edges in bold line segments are port edges and those in light line segments are network edges.

Theorem 4.12. *Consider a paramount admittance matrix* $Y_{3\times3} \in \mathbb{S}^3$ *that cannot be realized as a three-port resistive network containing fewer than four elements. Then,* $Y_{3\times3}$ *is realizable as the configuration in Fig. 4.9 if and only if at least one of the following conditions holds: i)* $M_{11} - |M_{12}| = 0$ *and* $M_{33} - |M_{23}| = 0$; *ii)* $M_{22} - |M_{23}| = 0$ *and* $M_{11} - |M_{13}| = 0$; *iii)* $M_{33} - |M_{13}| = 0$ *and* $M_{22} - |M_{12}| = 0$.

Proof. *Necessity.* For Fig. 4.9, assume that Terminals a and b form Port 1 with the port current orientated from a to b, Terminals b and c form Port 2 with the port current orientated from b to c, and Terminals d and e form Port 3 with the port current orientated from e to d. The entries of $Y_{3\times3}$ can be expressed as follows:

$$y_{11} = (g_1 + g_2)(g_3 + g_4)/G, \quad y_{12} = g_4(g_1 + g_2)/G,$$
$$y_{13} = (g_1g_4 - g_2g_3)/G, \quad y_{22} = g_4(g_1 + g_2 + g_3)/G, \qquad (4.20)$$
$$y_{23} = g_4(g_1 + g_3)/G, \quad y_{33} = (g_1 + g_3)(g_2 + g_4)/G,$$

where $G = g_1 + g_2 + g_3 + g_4$. The assumption that $g_1, g_2, g_3, g_4 > 0$ implies that $y_{11}, y_{12}, y_{22}, y_{23}, y_{33} > 0$. Furthermore, it can be verified that $M_{11} - |M_{12}| = 0$ and $M_{33} - |M_{23}| = 0$.

It is noted that properly swapping some of the ports and switching the polarities of some ports can yield any other case. This corresponds to a proper rearrangement of rows and the corresponding columns and a finite number of cross-sign changes. It can be verified that the condition of this theorem also holds.

Sufficiency. Let the element values in Fig. 4.9 satisfy

$$g_1 = \frac{(y_{33} - y_{23} + y_{13})M_{12}}{M_{12} - y_{23}(y_{33} - y_{23})}, \qquad (4.21)$$

$$g_2 = \frac{M_{12}}{y_{23}}, \qquad (4.22)$$

$$g_3 = \frac{(y_{23} - y_{13})M_{12}}{M_{12} - y_{23}(y_{33} - y_{23})}, \qquad (4.23)$$

$$g_4 = \frac{M_{12}}{y_{33} - y_{23}}. \qquad (4.24)$$

Since it is assumed that $Y_{3\times3}$ cannot be realized with fewer than four elements, the condition of Lemma 4.5 does not hold for $k = 3$. Together with the condition of paramountcy, it implies that $y_{11}, y_{22}, y_{33} > 0$.

Consider the case where $y_{12}y_{13}y_{23} \le 0$ and Condition i) holds. Assuming that $y_{12} = 0$ (resp., $y_{23} = 0$), it follows from $M_{33} - |M_{23}| = 0$ (resp.,

$M_{11} - |M_{12}| = 0)$ that $y_{22} - |y_{23}| = 0$ (resp., $y_{22} - |y_{12}| = 0$). Then, together with $M_{11} - |M_{12}| = 0$ (resp., $M_{33} - |M_{23}| = 0$), it implies that $y_{33} - |y_{23}| - |y_{13}| = 0$ (resp., $y_{11} - |y_{12}| - |y_{13}| = 0$). This contradicts the assumption that the condition of Lemma 4.5 does not hold for $k = 3$. Therefore, $y_{12} \neq 0$ and $y_{23} \neq 0$. After some cross-sign changes, one obtains y_{12}, $y_{23} > 0$ and $y_{13} \leq 0$. Then, it follows that $M_{12} > 0$ and $M_{23} > 0$, which implies that $M_{11} - M_{12} = M_{11} - |M_{12}| = 0$ and $M_{33} - M_{23} = M_{33} - |M_{23}| = 0$. As a result, one obtains

$$y_{11}(M_{12} - y_{23}(y_{33} - y_{23})) - y_{33}y_{12}(y_{12} - y_{13}) = 0, \qquad (4.25)$$

$$y_{33}(M_{23} - y_{12}(y_{11} - y_{12})) - y_{11}y_{23}(y_{23} - y_{13}) = 0. \qquad (4.26)$$

Then, it follows that $M_{12} - y_{23}(y_{33} - y_{23}) > 0$ and $M_{23} - y_{12}(y_{11} - y_{12}) > 0$. Moreover, $M_{11} - M_{12} = 0$ and $M_{33} - M_{23} = 0$ also imply $y_{11} - y_{12} + y_{13} \neq 0$ and $y_{33} - y_{23} + y_{13} \neq 0$. Otherwise, $Y_{3\times 3}$ would satisfy the condition of Lemma 4.5 for $k = 3$. Based on the condition of paramountcy, one can prove that $y_{11} - y_{12} + y_{13}$ and $y_{33} - y_{23} + y_{13}$ cannot be both negative. Therefore, if $y_{33} - y_{23} + y_{13} < 0$, then exchanging the first and third rows and the corresponding columns yields $y_{33} - y_{23} + y_{13} > 0$, without altering all the previous conditions. Therefore, $y_{33} - y_{23} > 0$. As a consequence, the element values expressed in (4.21)–(4.24) must be positive and finite. Since $M_{11} - M_{12} = 0$ and by (4.25), it can be verified that (4.20) must hold. Therefore, $Y_{3\times 3}$ is realizable as the required network.

Consider the case where $y_{12}y_{13}y_{23} > 0$ and Condition i) holds. After some cross-sign changes, one obtains $y_{12} > 0$, $y_{13} > 0$, and $y_{23} > 0$. Since M_{22} must be non-negative, it follows that at least one of M_{12} and M_{23} is non-negative. If $M_{12} \geq 0$ (resp., $M_{23} \geq 0$), then $M_{11} - M_{12} = M_{11} - |M_{12}| = 0$ (resp., $M_{33} - M_{23} = M_{33} - |M_{23}| = 0$), which implies that $y_{33}(y_{22} - y_{12}) = y_{23}(y_{23} - y_{13})$ (resp., $y_{11}(y_{22} - y_{23}) = y_{12}(y_{12} - y_{13})$). The condition of paramountcy implies that $y_{23} - y_{13} \geq 0$ (resp., $y_{12} - y_{13} \geq 0$), indicating that $M_{23} \geq 0$ (resp., $M_{12} \geq 0$). Together with the assumption that the condition of Lemma 4.5 does not hold for $k = 3$, one has $y_{12} > y_{13} > 0$, $y_{23} > y_{13} > 0$, $M_{12} > 0$, and $M_{23} > 0$. Therefore, $M_{11} - M_{12} = M_{11} - |M_{12}| = 0$ and $M_{33} - M_{23} = M_{33} - |M_{23}| = 0$, which further implies (4.25). Thus, $M_{12} - y_{23}(y_{33} - y_{23}) > 0$. Assume that $y_{33} - y_{23} = 0$. It follows from $M_{11} - M_{12} = 0$ and $M_{33} - M_{23} = 0$ that $y_{11} = y_{12}$, $y_{33} = y_{23}$, and $(y_{22} - y_{23}) - (y_{12} - y_{13}) = 0$, which contradicts the assumption that the condition of Lemma 4.5 does not hold for $k = 3$. Therefore, $y_{33} - y_{23} > 0$, which means that $y_{33} - y_{23} + y_{13} > 0$. As a result, g_1, g_2, g_3, and g_4 as expressed in (4.21)–(4.24) must have positive and finite values. Since

$M_{11} - M_{12} = 0$ and by (4.25), it can be verified that (4.20) must hold. Therefore, $Y_{3\times3}$ is realizable as the required network.

In addition, if $Y_{3\times3}$ satisfies Condition ii) or iii), then by properly arranging the rows and columns $Y_{3\times3}$ can always yield one of the above two cases. Therefore, this theorem is proved. □

Theorem 4.13. *A paramount admittance matrix $Y_{3\times3} \in \mathbb{S}^3$ is realizable as a three-port resistive network containing no more than four elements, if and only if $Y_{3\times3}$ satisfies the condition of Lemma 4.5 for $k = 4$ or the condition of Theorem 4.12.*

Proof. Combining Lemma 4.8 and Theorems 4.11 and 4.12, this theorem can be directly proved. □

4.4.3 Realization with Five Elements

Now, consider the paramount admittance matrix $Y_{3\times3} \in \mathbb{S}^3$ that cannot be realized as a three-port resistive network containing fewer than five elements, that is, does not satisfy the condition of Theorem 4.13.

Lemma 4.9. *If $Y_{3\times3} \in \mathbb{S}^3$ is the admittance matrix of at least one of the configurations in Fig. 4.12, then $Y_{3\times3}$ is always realizable as the admittance matrix of a three-port four-terminal resistive network containing no more than four elements.*

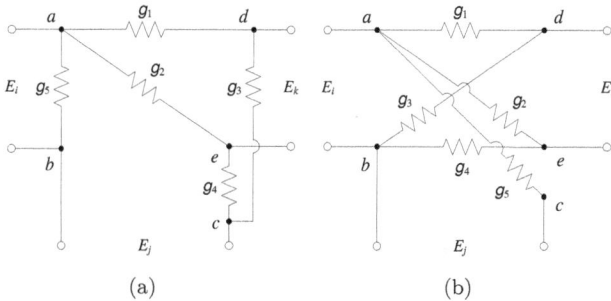

(a)   (b)

Fig. 4.12 Three-port resistive configurations containing five elements mentioned in Lemma 4.9, where $g_1, g_2, g_3, g_4, g_5 > 0$.

Proof. To prove this lemma, it suffices to prove that the admittance matrix of each configuration in Fig. 4.12 satisfies the condition of Lemma 4.5 for $k = 4$. The detail of the proof is omitted for brevity. □

Lemma 4.10. *An admittance matrix $Y_{3\times3} \in \mathbb{S}^3$ not satisfying the condition of Theorem 4.13 is realizable as a three-port resistive network containing five elements, if and only if $Y_{3\times3}$ satisfies the condition of Lemma 4.5 for $k = 5$ or $Y_{3\times3}$ is realizable as one of the configurations in Figs. 4.13 and 4.14.*

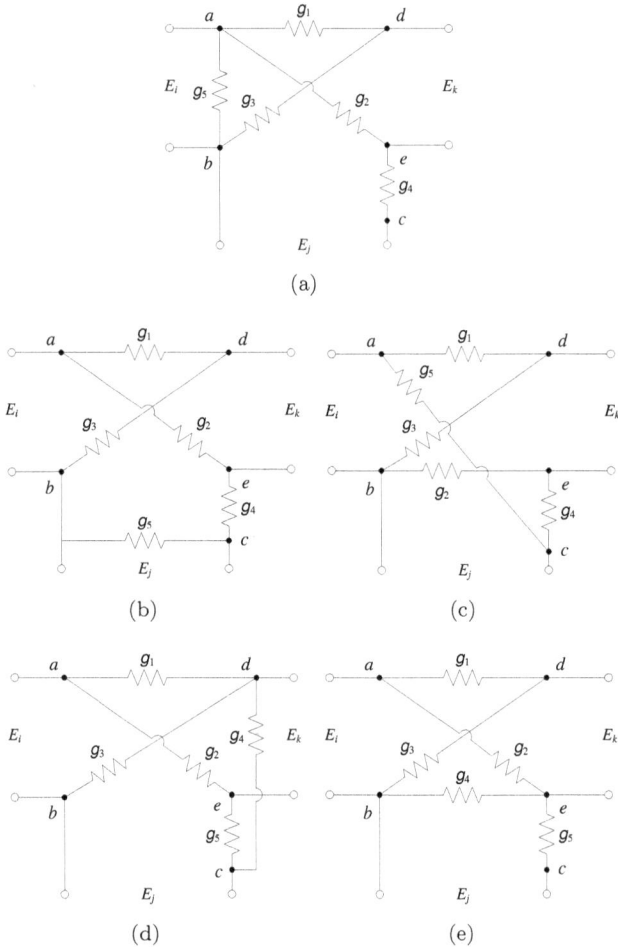

Fig. 4.13   Three-port resistive configurations containing five elements mentioned in Lemma 4.10, where $g_1, g_2, g_3, g_4, g_5 > 0$.

Proof. *Sufficiency.* Together with Lemma 4.5, this sufficiency part can be directly proved.

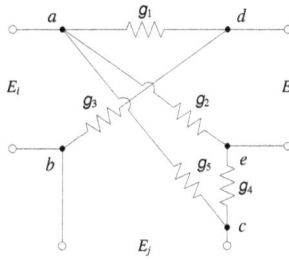

Fig. 4.14   A three-port resistive configuration containing five elements mentioned in Lemma 4.10, where g_1, g_2, g_3, g_4, $g_5 > 0$.

Necessity. The existence of the admittance matrix implies that the number of terminals can only be four, five, or six.

When there are four terminals, the condition of Lemma 4.5 holds for $k = 5$.

When there are five terminals, the port graph $\mathcal{G}_p$ is shown in Fig. 4.10(a), which consists of two maximal connected subgraphs. The assumption that the condition of Theorem 4.13 does not hold means that the network cannot contain fewer than five elements. Then, one will determine all the possible three-port five-element configurations that may not always be equivalent to a three-port four-terminal one containing no more than five elements. As discussed above, Lemma 4.6 implies that the augmented graph $\mathcal{G}$ is nonseparable and Lemma 4.7 means that the augmented graph $\mathcal{G}$ cannot contain two ports directly in series. Denote the cut-set separating $\mathcal{G}$ into two parts containing respectively two components of $\mathcal{G}_p$ as $\mathcal{C}_{ep}$. By Lemma 4.4, each of vertices d and e must be incident by at least two network edges belonging to $\mathcal{C}_{ep}$, and each of vertices a–c must be incident by at least one network edge belonging to $\mathcal{C}_{ep}$. Therefore, there are four or five edges belonging to $\mathcal{C}_{ep}$, which are all network edges. Regarding two augmented graphs that are isomorphic with each other as the same graph, the possible graphs can only be those shown in Figs. 4.15(a)–4.15(f), when $\mathcal{C}_{ep}$ contains four edges. Similarly, it can be verified that only augmented graphs in Figs. 4.15(g) and 4.15(h) are possible when $\mathcal{C}_{ep}$ contains five edges. Furthermore, Figs. 4.15(c) and 4.15(f) are eliminated by Lemma 4.9. As a result, all the possible configurations are shown in Figs. 4.13 and 4.14.

When there are six terminals, the port graph $\mathcal{G}_p$ is shown in Fig. 4.10(b). Assuming that the network cannot be equivalent to one with fewer terminals and no more than five elements, each of the six vertices of $\mathcal{G}_p$ is incident by at least two network edges according to Lemmas 4.4 and 4.6. This implies that at least six network edges are needed, which is impossible. Thus, the lemmas is proved.   □

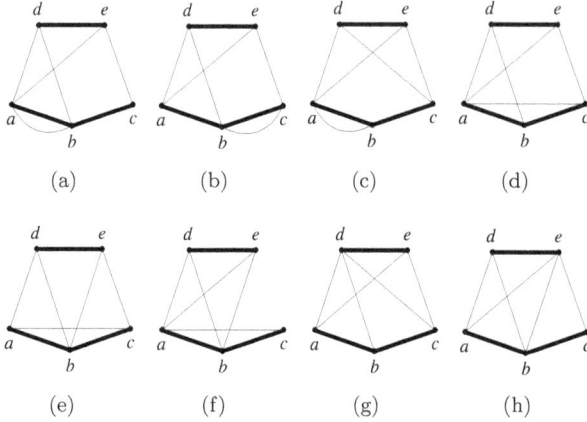

Fig. 4.15   Augmented graphs discussed in the proof of Lemma 4.10, where edges in bold line segments are port edges and those in light line segments are network edges.

Theorem 4.14. *A paramount admittance matrix* $Y_{3\times3} \in \mathbb{S}^3$ *not satisfying the condition of Theorem 4.13 is realizable as the configuration in Fig. 4.13(a), if and only if*

1. *when* $y_{12}y_{13}y_{23} \leq 0$, *at least one of the following conditions holds:*
   i) $y_{33} - |y_{13}| - |y_{23}| > 0$ *and* $(M_{11} - |M_{12}|)(M_{22} - |M_{12}|) = 0$; *ii)* $y_{22} - |y_{12}| - |y_{23}| > 0$ *and* $(M_{11} - |M_{13}|)(M_{33} - |M_{13}|) = 0$; *iii)* $y_{11} - |y_{12}| - |y_{13}| > 0$ *and* $(M_{22} - |M_{23}|)(M_{33} - |M_{23}|) = 0$;
2. *when* $y_{12}y_{13}y_{23} > 0$, *at least one of the following conditions holds:*
   i) $|y_{13}| < \min\{|y_{12}|, |y_{23}|\}$ *and* $(M_{11} - |M_{12}|)(M_{33} - |M_{23}|) = 0$;
   ii) $|y_{23}| < \min\{|y_{12}|, |y_{13}|\}$ *and* $(M_{22} - |M_{12}|)(M_{33} - |M_{13}|) = 0$;
   iii) $|y_{12}| < \min\{|y_{13}|, |y_{23}|\}$ *and* $(M_{11} - |M_{13}|)(M_{22} - |M_{23}|) = 0$.

Proof. *Necessity.* For Fig. 4.13(a), assume that Terminals a and b form Port 1 with the current orientated from a to b, Terminals b and c form Port 2 with the current orientated from b to c, and Terminals d and e

form Port 3 with the current orientated from e to d. Then,

$$y_{11} = ((g_1 + g_2)(g_3 + g_4 + g_5) + g_5(g_3 + g_4))/G,$$
$$y_{12} = g_4(g_1 + g_2)/G, \quad y_{13} = (g_1 g_4 - g_2 g_3)/G,$$
$$y_{22} = g_4(g_1 + g_2 + g_3)/G, \quad y_{23} = g_4(g_1 + g_3)/G,$$
$$y_{33} = (g_1 + g_3)(g_2 + g_4)/G,$$

(4.27)

where $G = g_1 + g_2 + g_3 + g_4$. The assumption that $g_1, g_2, g_3, g_4, g_5 > 0$ implies that $y_{11}, y_{12}, y_{22}, y_{23}, y_{33} > 0$. If $y_{13} \leq 0$, then $y_{12} y_{13} y_{23} \leq 0$, $y_{33} - |y_{13}| - |y_{23}| = y_{33} + y_{13} - y_{23} = g_1(g_2 + g_4)/(g_1 + g_2 + g_3 + g_4) > 0$, $M_{12} = y_{12} y_{33} - y_{13} y_{23} > 0$, and $M_{11} - |M_{12}| = M_{11} - M_{12} = 0$, which implies that Condition 1 holds. If $y_{13} > 0$, then $y_{12} y_{13} y_{23} > 0$, $|y_{12}| - |y_{13}| = y_{12} - y_{13} = g_2(g_3 + g_4)/(g_1 + g_2 + g_3 + g_4) > 0$, $|y_{23}| - |y_{13}| = y_{23} - y_{13} = g_3(g_2 + g_4)/(g_1 + g_2 + g_3 + g_4) > 0$, $M_{12} = g_2 g_4(g_1 + g_3)/(g_1 + g_2 + g_3 + g_4) > 0$, and $M_{11} - |M_{12}| = M_{11} - M_{12} = 0$, which implies that Condition 2 holds.

Properly swapping some of the ports and switching the polarities of some ports can yield any other case, which corresponds to a proper rearrangement of rows and the corresponding columns, with a finite number of cross-sign changes. Note that Condition 1 or 2 also holds for other cases.

Sufficiency. Let the element values in Fig. 4.13(a) satisfy

$$g_1 = \frac{(y_{33} - y_{23} + y_{13})M_{12}}{M_{12} - y_{23}(y_{33} - y_{23})},$$

(4.28)

$$g_2 = \frac{M_{12}}{y_{23}},$$

(4.29)

$$g_3 = \frac{(y_{23} - y_{13})M_{12}}{M_{12} - y_{23}(y_{33} - y_{23})},$$

(4.30)

$$g_4 = \frac{M_{12}}{y_{33} - y_{23}},$$

(4.31)

$$g_5 = \frac{y_{11}(M_{12} - y_{23}(y_{33} - y_{23})) - y_{33} y_{12}(y_{12} - y_{13})}{M_{12} - y_{23}(y_{33} - y_{23})}.$$

(4.32)

Since the condition of Theorem 4.13 does not hold, neither the condition of Lemma 4.5 for $k = 4$ nor the condition of Theorem 4.12 holds. Together with the condition of paramountcy, it implies that $y_{11}, y_{22}, y_{33} > 0$. By Lemma 4.6, it indicates that at most one of y_{12}, y_{13}, and y_{23} is zero.

Consider the case where $y_{12} y_{13} y_{23} \leq 0$ and Condition 1-i) holds. The condition of $M_{11} - |M_{12}| = 0$ can always be guaranteed by properly interchanging the first and second rows and columns of $Y_{3\times3}$. Then, after some cross-sign changes, one obtains $y_{12}, y_{23} \geq 0$, and $y_{13} \leq 0$. Since at most one of y_{12}, y_{13}, and y_{23} is zero, it follows that $y_{12} - y_{13} > 0$,

$y_{23} - y_{13} > 0$, $M_{12} > 0$, $M_{23} > 0$, and $M_{11} - M_{12} = M_{11} - |M_{12}| = 0$. Furthermore, $y_{33} - y_{23} + y_{13} = y_{33} - |y_{13}| - |y_{23}| > 0$, which implies that $y_{33} - y_{23} > 0$. Since the condition of Lemma 4.5 does not hold for $k = 4$, it follows from $M_{11} - M_{12} = 0$ that $y_{23} > 0$. Since the condition of $M_{33} - |M_{23}| = 0$ contradicts the assumption that the condition of Theorem 4.12 does not hold, together with the condition of paramountcy, it implies that $M_{33} - M_{23} = M_{33} - |M_{23}| > 0$. Therefore, $M_{11} - M_{12} = 0$ implies that

$$y_{11}(M_{12} - y_{23}(y_{33} - y_{23})) - y_{33}y_{12}(y_{12} - y_{13}) = y_{33}(M_{33} - M_{23}) > 0,$$
(4.33)

which further implies that $M_{12} - y_{23}(y_{33} - y_{23}) > 0$. As a consequence, the element values as expressed in (4.28)–(4.32) must be positive and finite. By (4.28)–(4.32) and since $M_{11} - |M_{12}| = 0$, it implies that (4.27). Therefore, $Y_{3\times 3}$ is realizable as the required network.

Consider the case where $y_{12}y_{13}y_{23} > 0$ and Condition 2-i) holds. After some cross-sign changes, one can guarantee that $y_{12} > y_{13} > 0$ and $y_{23} > y_{13} > 0$. Moreover, properly interchanging the first and third rows and columns of $Y_{3\times 3}$ yields $M_{11} - |M_{12}| = 0$, which does not alter the assumption. From the condition of paramountcy, it follows that $M_{12} > y_{13}(y_{33} - y_{23}) \geq 0$ and $M_{23} > y_{13}(y_{11} - y_{12}) \geq 0$, implying that $M_{11} - M_{12} = M_{11} - |M_{12}| = 0$. Since the condition of Lemma 4.5 does not hold for $k = 4$, it follows that $y_{33} - y_{23} > 0$, implying that $y_{33} - y_{23} + y_{13} > 0$. Considering that $M_{33} - |M_{23}| = 0$ would contradict the assumption that the condition of Theorem 4.12 does not hold, together with the condition of paramountcy, it follows that $M_{33} - M_{23} = M_{33} - |M_{23}| > 0$. Since $M_{11} - |M_{12}| = 0$ implies (4.33), it follows that $M_{12} - y_{23}(y_{33} - y_{23}) > 0$. As a consequence, the element values as expressed in (4.28)–(4.32) must be positive and finite. Moreover, (4.28)–(4.32) and $M_{11} - |M_{12}| = 0$ together imply (4.27). Therefore, $Y_{3\times 3}$ is realizable as the required network.

Finally, if $Y_{3\times 3}$ satisfies other cases of Condition 1 or 2, then properly arranging the rows and the corresponding columns of $Y_{3\times 3}$ can always generate one of the above two cases.   □

Theorem 4.15. *A paramount matrix $Y_{3\times 3} \in \mathbb{S}^3$ not satisfying the condition of Theorem 4.13 is realizable as the admittance matrix of the configuration in Fig. 4.13(b) if and only if*

1. *when $y_{12}y_{13}y_{23} \leq 0$, at least one of the following conditions holds:*
   i) $y_{33}(M_{33} - |M_{23}|) - y_{11}(M_{11} - |M_{12}|) = 0$; ii) $y_{22}(M_{22} - |M_{12}|) - y_{33}(M_{33} - |M_{13}|) = 0$; iii) $y_{11}(M_{11} - |M_{13}|) - y_{22}(M_{22} - |M_{23}|) = 0$;

2. *when $y_{12}y_{13}y_{23} > 0$, at least one of the following conditions holds:*
   i) $|y_{13}| < \min\{|y_{12}|, |y_{23}|\}$ *and* $y_{33}(M_{33} - |M_{23}|) - y_{11}(M_{11} - |M_{12}|) = 0$;
   ii) $|y_{23}| < \min\{|y_{12}|, |y_{13}|\}$ *and* $y_{22}(M_{22} - |M_{12}|) - y_{33}(M_{33} - |M_{13}|) = 0$; *iii)* $|y_{12}| < \min\{|y_{13}|, |y_{23}|\}$ *and* $y_{11}(M_{11} - |M_{13}|) - y_{22}(M_{22} - |M_{23}|) = 0$.

Proof. This proof is similar to that of Theorem 4.14. $\square$

Theorem 4.16. *A paramount admittance matrix $Y_{3\times3} \in \mathbb{S}^3$ not satisfying the condition of Theorem 4.13 is realizable as the configuration in Fig. 4.13(c) if and only if $y_{12}y_{13}y_{23} > 0$, and at least one of the following conditions holds: i)* $y_{11}y_{22}y_{33} - y_{12}y_{13}y_{23} - y_{33}(|y_{13}|(y_{22} - |y_{12}|) + |y_{23}|(y_{11} - |y_{12}|)) = (y_{11} + y_{22} - |y_{12}|)(y_{33}|y_{12}| - |y_{13}||y_{23}|) > 0$; *ii)* $y_{11}y_{22}y_{33} - y_{12}y_{13}y_{23} - y_{11}(|y_{13}|(y_{22} - |y_{13}|) + |y_{12}|(y_{33} - |y_{23}|)) = (y_{33} + y_{22} - |y_{23}|)(y_{11}|y_{23}| - |y_{13}||y_{12}|) > 0$; *iii)* $y_{11}y_{22}y_{33} - y_{12}y_{13}y_{23} - y_{22}(|y_{12}|(y_{33} - |y_{12}|) + |y_{23}|(y_{11} - |y_{13}|)) = (y_{11} + y_{33} - |y_{13}|)(y_{22}|y_{13}| - |y_{12}||y_{23}|) > 0$.*

Proof. The proof is similar to that of Theorem 4.14. $\square$

Theorem 4.17. *A paramount admittance matrix $Y_{3\times3} \in \mathbb{S}^3$ not satisfying the condition of Theorem 4.13 is realizable as the configuration in Fig. 4.13(d) if and only if*

1. *when $y_{12}y_{13}y_{23} \leq 0$, at least one of the following conditions holds:*
   i) $|y_{12}|(|y_{23}| + |y_{13}|)^2 - (|y_{23}| + |y_{13}|)M_{33} + y_{33}(y_{22} - |y_{12}|)(y_{11} - |y_{12}|) = 0$;
   ii) $|y_{23}|(|y_{12}| + |y_{13}|)^2 - (|y_{12}| + |y_{13}|)M_{11} + y_{11}(y_{22} - |y_{23}|)(y_{33} - |y_{23}|) = 0$; *iii)* $|y_{13}|(|y_{23}| + |y_{12}|)^2 - (|y_{23}| + |y_{12}|)M_{22} + y_{22}(y_{33} - |y_{13}|)(y_{11} - |y_{13}|) = 0$;
2. *when $y_{12}y_{13}y_{23} > 0$, at least one of the following conditions holds:*
   i) $(|y_{12}||y_{23}| - y_{22}|y_{13}|)(y_{11}|y_{23}| - |y_{12}||y_{13}|) > 0$ *and* $|(|y_{23}| - |y_{13}|)M_{33} = |y_{12}|(|y_{23}| - |y_{13}|)^2 + y_{33}(y_{22} - |y_{12}|)(y_{11} - |y_{12}|)$;
   ii) $(|y_{12}||y_{23}| - y_{22}|y_{13}|)(y_{33}|y_{12}| - |y_{23}||y_{13}|) > 0$ *and* $|(|y_{12}| - |y_{13}|)M_{11} = |y_{23}|(|y_{12}| - |y_{13}|)^2 + y_{11}(y_{22} - |y_{23}|)(y_{33} - |y_{23}|)$;
   iii) $(|y_{23}||y_{13}| - y_{33}|y_{12}|)(y_{11}|y_{23}| - |y_{12}||y_{13}|) > 0$ *and* $|(|y_{23}| - |y_{12}|)M_{22} = |y_{13}|(|y_{23}| - |y_{12}|)^2 + y_{22}(y_{33} - |y_{13}|)(y_{11} - |y_{13}|)$.

Proof. The proof is similar to that of Theorem 4.14. $\square$

Theorem 4.18. *A paramount admittance matrix $Y_{3\times3} \in \mathbb{S}^3$ not satisfying the condition of Theorem 4.13 is realizable as the configuration in Fig. 4.13(e) if and only if*

1. when $y_{12}y_{13}y_{23} \leq 0$, at least one of the following conditions holds:
   i) $|y_{12}|(y_{33} - |y_{23}|) - |y_{13}|(y_{22} - |y_{23}|) > 0$ or $|y_{23}|(y_{11} - |y_{12}|) - |y_{13}|(y_{22}-|y_{12}|) > 0$ holds and $y_{11}|y_{23}|(y_{22}-|y_{23}|) - y_{33}|y_{12}|(y_{22}-|y_{12}|) - |y_{12}||y_{23}|(|y_{12}| - |y_{23}|) = 0$; ii) $|y_{12}|(y_{33}-|y_{13}|) - |y_{23}|(y_{11}-|y_{13}|) > 0$ or $|y_{13}|(y_{22} - |y_{12}|) - |y_{23}|(y_{11} - |y_{12}|) > 0$ holds and $y_{22}|y_{13}|(y_{11} - |y_{13}|) - y_{33}|y_{12}|(y_{11} - |y_{12}|) - |y_{12}||y_{13}|(|y_{12}| - |y_{13}|) = 0$; iii) $|y_{13}|(y_{22} - |y_{23}|) - |y_{12}|(y_{33}-|y_{23}|) > 0$ or $|y_{23}|(y_{11}-|y_{13}|) - |y_{12}|(y_{33}-|y_{13}|) > 0$ holds and $y_{11}|y_{23}|(y_{33} - |y_{23}|) - y_{22}|y_{13}|(y_{33} - |y_{13}|) - |y_{13}||y_{23}|(|y_{13}| - |y_{23}|) = 0$;

2. when $y_{12}y_{13}y_{23} > 0$, at least one of the following conditions holds:
   i) $M_{11} + (|y_{12}||y_{23}| - y_{22}|y_{13}|) > y_{33}|y_{12}| - |y_{23}||y_{13}| > 0$ or $M_{33} + (|y_{12}||y_{23}| - y_{22}|y_{13}|) > y_{11}|y_{23}| - |y_{12}||y_{13}| > 0$ holds and $y_{11}|y_{23}|(y_{22} - |y_{23}|) - y_{33}|y_{12}|(y_{22} - |y_{12}|) - |y_{12}||y_{23}|(|y_{12}| - |y_{23}|) = 0$; ii) $M_{22} + (|y_{12}||y_{13}| - y_{11}|y_{23}|) > y_{33}|y_{12}| - |y_{23}||y_{13}| > 0$ or $M_{33} + (|y_{13}||y_{12}| - y_{11}|y_{23}|) > y_{22}|y_{13}| - |y_{12}||y_{23}| > 0$ holds and $y_{22}|y_{13}|(y_{11} - |y_{13}|) - y_{33}|y_{12}|(y_{11} - |y_{12}|) - |y_{12}||y_{13}|(|y_{12}| - |y_{13}|) = 0$; iii) $M_{11} + (|y_{13}||y_{23}| - y_{33}|y_{12}|) > y_{22}|y_{13}| - |y_{23}||y_{12}| > 0$ or $M_{22} + (|y_{23}||y_{13}| - y_{33}|y_{12}|) > y_{11}|y_{23}| - |y_{13}||y_{12}| > 0$ holds and $y_{11}|y_{23}|(y_{33} - |y_{23}|) - y_{22}|y_{13}|(y_{33} - |y_{13}|) - |y_{13}||y_{23}|(|y_{13}| - |y_{23}|) = 0$.

Proof. The proof is similar to that of Theorem 4.14. □

Lemma 4.11. *If a paramount admittance matrix $Y_{3\times 3} \in \mathbb{S}^3$ not satisfying the condition of Theorem 4.13 is realizable as the configuration in Fig. 4.14, then $Y_{3\times 3}$ is also realizable as the configuration in Fig. 4.13(a).*

Proof. It suffices to prove that the admittance matrix of the configuration in Fig. 4.14 always satisfies the condition of Theorem 4.14. The detail is omitted for brevity. □

In summary, the final result is obtained as follows.

Theorem 4.19. *A paramount admittance matrix $Y_{3\times 3} \in \mathbb{S}^3$ not satisfying the condition of Theorem 4.13 is realizable as a three-port resistive network with five elements, if and only if $Y_{3\times 3}$ satisfies the condition of Lemma 4.5 for $k = 5$ or $Y_{3\times 3}$ satisfies at least one of the conditions in Theorems 4.14–4.18.*

Proof. This theorem follows directly from Lemmas 4.10 and 4.11 and Theorems 4.14–4.18. □

Remark 4.3. By Kuratowski's Theorem in [Harary (1969), pg. 109], any graph containing fewer than nine edges must be planar. Therefore, the augmented graphs of three-port resistive networks containing no more than five elements is planar. By [Seshu and Reed (1961), Theorem 3-15], the dual graphs always exist. Based on the principle of duality, the results in Theorems 4.11–4.19 can be directly transformed to those of the impedance case.

Remark 4.4. It follows from the results in this section that at least six elements are needed to realize the entire class of third-order paramount matrices.

Remark 4.5. Recall the theorem of Reichert [Reichert (1969)] assumes that reactive element values can be arbitrary. The reduction of the number of resistors to no more than three is partially due to the variation of reactive element values. The results of this section can be applied to investigate the synthesis of one-port networks containing two reactive elements with restricted values, which has important practical implications.

4.4.4 *Some Examples*

Example 4.1. Consider a paramount admittance matrix $Y_{3\times3} \in \mathbb{S}^3$ satisfying $y_{11} = 2$, $y_{12} = 1$, $y_{13} = -1/3$, $y_{22} = 2$, $y_{23} = 4/3$, $y_{33} = 20/9$, where the condition of Lemma 4.5 does not hold for $k = 4$. For these values, it can be verified that the condition of Theorem 4.12 holds with $y_{12}y_{13}y_{23} < 0$ and Condition i) being also satisfied. Therefore, $Y_{3\times3}$ is realizable as the configuration in Fig. 4.9, where Terminals a and b form Port 1 with the current oriented from a to b, Terminals b and c form Port 2 with the current oriented from b to c, and Terminals d and e form Port 3 with the current oriented from e to d. Moreover, the element values satisfy $g_1 = 1$, $g_2 = 2$, $g_3 = 3$, and $g_4 = 3$.

Example 4.2. Consider a paramount admittance matrix $Y_{3\times3} \in \mathbb{S}^3$ satisfying $y_{11} = 3$, $y_{12} = 1$, $y_{13} = -2$, $y_{22} = 5$, $y_{23} = 3$, and $y_{33} = 6$, where the condition of Theorem 4.13 does not hold. For these values, it can be verified that the condition of Lemma 4.5 holds for $k = 5$. Therefore, $Y_{3\times3}$ is realizable as a three-port four-terminal resistive network containing five elements, whose port graph is a Lagrangian tree.

Example 4.3. Consider a paramount admittance matrix $Y_{3\times3} \in \mathbb{S}^3$ satisfying $y_{11} = 7$, $y_{12} = 5/3$, $y_{13} = 1$, $y_{22} = 20/9$, $y_{23} = 1/3$, and $y_{33} = 2$, where neither the condition of Theorem 4.13 nor the condition of Lemma 4.5 satisfies for $k = 5$. For these values, it can be verified that the condition of Theorem 4.15 holds with $y_{12}y_{13}y_{23} > 0$ and Condition 2-ii) being also satisfied. Therefore, $Y_{3\times3}$ is realizable as the configuration in Fig. 4.13(b), where Terminals b and c form Port 1 with the current oriented from c to b, Terminals a and b form Port 2 with the current oriented from a to b, and Terminals d and e form Port 3 with the current oriented from e to d. Moreover, the element values satisfy $g_1 = 2$, $g_2 = 3$, $g_3 = 1$, $g_4 = 3$, and $g_5 = 5$.

Chapter 5

Mechanical Synthesis of Low-Complexity One-Port Networks

5.1 Introduction

In addition to the classical realization problems for electrical networks, synthesis of mechanical networks has recently been investigated in a series of works [Chen and Smith (2008, 2009b); Chen *et al.* (2013a,b, 2015c)], since mechanical networks have some additional complexity requirements.

Considering the complexity, cost, and weight of elements, one prefers fewer inerters or dampers compared with springs in mechanical systems. Therefore, it is meaningful to solve the realization problem of any positive-real admittance as a mechanical network consisting of m damper, l inerter, and an arbitrary number of springs, $m, l = 1, 2, \ldots$.

In [Chen and Smith (2009b)], Chen and Smith made the first attempt at this unsolved problem, and solved the case where the one-port mechanical network contains only one damper, one inerter, and a finite number of springs. By extracting its damper and inerter, any such a mechanical network N can be formulated as shown in Fig. 5.1, where X denotes the resulting three-port network consisting of only springs. Making a mild assumption that the impedance of X exists, one has

$$\begin{bmatrix} \hat{v}_1 \\ \hat{v}_2 \\ \hat{v}_3 \end{bmatrix} = s \begin{bmatrix} L_1 & L_4 & L_5 \\ L_4 & L_2 & L_6 \\ L_5 & L_6 & L_3 \end{bmatrix} \begin{bmatrix} \hat{F}_1 \\ \hat{F}_2 \\ \hat{F}_3 \end{bmatrix} =: sL \begin{bmatrix} \hat{F}_1 \\ \hat{F}_2 \\ \hat{F}_3 \end{bmatrix}, \qquad (5.1)$$

where sL is the impedance of X and $\hat{\cdot}$ denotes the Laplace transform. By replacing resistors with springs, sL is the impedance of a three-port network consisting of springs if and only if L is the impedance of a three-port resistive network, if and only if L is paramount (see Chapter 4). Therefore, L is necessarily non-negative definite. Since Ports 2 and 3 of X are terminated with the damper and the inerter, respectively, $\hat{F}_2 = -c\hat{v}_2$ and $\hat{F}_3 = -bs\hat{v}_3$

where $b > 0$ and $c > 0$. Therefore, one can calculate the admittance of the one-port network N in Fig. 5.1, and obtains

$$\frac{\hat{F}_1}{\hat{v}_1} = \frac{bc(L_2L_3 - L_6^2)s^3 + bL_3s^2 + cL_2s + 1}{bc\det(L)s^4 + b(L_1L_3 - L_5^2)s^3 + c(L_1L_2 - L_4^2)s^2 + L_1s}. \tag{5.2}$$

It is noted that the admittance expression in (5.2) is concerned with the element values of damper and inerter. In order to further eliminate these two parameters (b and c) in (5.2), the following scalings are carried out: $L_1 \to R_1$, $cL_2 \to R_2$, $bL_3 \to R_3$, $\sqrt{c}L_4 \to R_4$, $\sqrt{b}L_5 \to R_5$, $\sqrt{bc}L_6 \to R_6$ without altering (5.1), which can also be written in the matrix form as

$$R := \begin{bmatrix} R_1 & R_4 & R_5 \\ R_4 & R_2 & R_6 \\ R_5 & R_6 & R_3 \end{bmatrix} = T \begin{bmatrix} L_1 & L_4 & L_5 \\ L_4 & L_2 & L_6 \\ L_5 & L_6 & L_3 \end{bmatrix} T, \tag{5.3}$$

where

$$T = \begin{bmatrix} 1 & 0 & 0 \\ 0 & \sqrt{c} & 0 \\ 0 & 0 & \sqrt{b} \end{bmatrix} \tag{5.4}$$

and R must be non-negative definite. As a consequence, the admittance in (5.2) can be written as

$$Y(s) = \frac{(R_2R_3 - R_6^2)s^3 + R_3s^2 + R_2s + 1}{s(\det Rs^3 + (R_1R_3 - R_5^2)s^2 + (R_1R_2 - R_4^2)s + R_1)}. \tag{5.5}$$

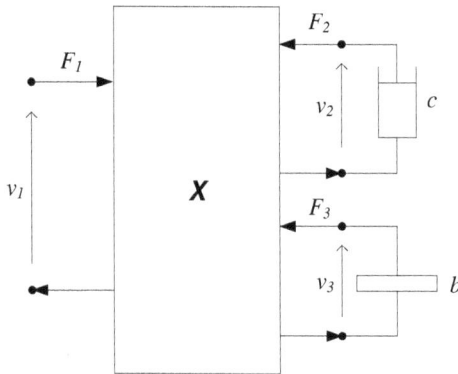

Fig. 5.1   One-port mechanical network N consisting of one damper, one inerter, and a finite number of springs [Chen and Smith (2009b)].

Letting $D := T^{-1}$, where T is the diagonal matrix in (5.4), the realizability condition of sL as X is equivalent to the paramountcy of DRD

by (5.3). Therefore, as shown in [Chen and Smith (2009b), Lemma 5], the following result is obtained.

Lemma 5.1. *[Chen and Smith (2009b)] Any positive-real function $Y(s)$ is realizable as the admittance of a one-port network N consisting of one damper, one inerter, and a finite number of springs in Fig. 5.1, where the impedance of X exists, if and only if $Y(s)$ can be written in the form of (5.5), where any R_i, $i = 1, 2, \ldots, 6$, is the entry of the third-order non-negative matrix R in (5.5) and there exists a diagonal matrix $D = diag\{1, x, y\}$ with $x > 0$ and $y > 0$ such that DRD is paramount.*

As introduced in Chapter 4, the number of elements in X can be no more than six.

Now, an equivalent realizability condition is needed that does not rely on the existence of the parameter matrix D, and one also needs to determine the configurations covering all the cases, which contain as few springs as possible.

After a series of derivations in [Chen and Smith (2009b)], the desired realizability condition has been established, which is restated as follows.

Lemma 5.2. *[Chen and Smith (2009b)] Consider a non-negative definite matrix R as defined in (5.3). If there exists any first- or second-order minor of R being zero, then there exists an invertible diagonal matrix $D = diag\{1, x, y\}$ with $x > 0$ and $y > 0$ such that DRD is paramount.*

Lemma 5.3. *[Chen and Smith (2009b)] Consider a non-negative definite matrix R as defined in (5.3) with all the first- and second-order minors being non-zero. Then, there exists an invertible diagonal matrix $D = diag\{1, x, y\}$ with $x > 0$ and $y > 0$ such that DRD is a paramount matrix if and only if one of the following conditions holds:*

1. $R_4 R_5 R_6 < 0$;
2. $R_4 R_5 R_6 > 0$, $R_1 > (R_4 R_5 / R_6)$, $R_2 > (R_4 R_6 / R_5)$ and $R_3 > (R_5 R_6 / R_4)$;
3. $R_4 R_5 R_6 > 0$, $R_3 < (R_5 R_6 / R_4)$ and $R_1 R_2 R_3 + R_4 R_5 R_6 - R_1 R_6^2 - R_2 R_5^2 \geq 0$;
4. $R_4 R_5 R_6 > 0$, $R_2 < (R_4 R_6 / R_5)$ and $R_1 R_2 R_3 + R_4 R_5 R_6 - R_1 R_6^2 - R_3 R_4^2 \geq 0$;
5. $R_4 R_5 R_6 > 0$, $R_1 < (R_4 R_5 / R_6)$ and $R_1 R_2 R_3 + R_4 R_5 R_6 - R_3 R_4^2 - R_2 R_5^2 \geq 0$.

Theorem 5.1. *[Chen and Smith (2009b)] A positive-real function $Y(s)$ is realizable as the admittance of a one-port mechanical network N consisting of one damper, one inerter, and an arbitrary number of springs in Fig. 5.1, where the impedance of the three-port network X exists, if and only if $Y(s)$ can be written in the form of (5.5) where R_i, $i = 1, 2, \ldots, 6$, is the entry of the third-order non-negative matrix R in (5.5), such that the condition of either Lemma 5.2 or Lemma 5.3 holds.*

Using the construction of Tellegen (see Fig. 4.2), it is clear that any positive-real admittance satisfying the condition of Theorem 5.1 is realizable with one damper, one inerter, and no more than six springs. Furthermore, Chen and Smith [Chen and Smith (2009b)] determined covering configurations containing no more than four springs, stated as follows.

Theorem 5.2. *Consider a positive-real function $Y(s)$ in the form of (5.5) where any R_i, $i = 1, 2, \ldots, 6$, is the entry of the non-negative definite matrix R as defined in (5.3). If the condition of Lemma 5.2 holds, then $Y(s)$ is realizable as the admittance of a series-parallel mechanical network consisting of one damper, one inerter, and no more than three springs by the Foster preamble. If one of the five conditions in Lemma 5.3 holds, then $Y(s)$ is realizable as the admittance of one of the five configurations in Fig. 5.2 containing one damper, one inerter, and four springs, which correspond to these five conditions, respectively.*

Moreover, the mechanical element extraction approach utilized in [Chen and Smith (2009b)] has been successfully applied to solve the synthesis problem of one-port damper-spring-inerer (resp., RLC) networks (see [Chen and Smith (2008); Chen *et al.* (2015c); Hughes and Smith (2012)]). In [Chen and Smith (2008)], extracting the spring and inerter has been made to solve the realization of any positive-real function as the admittance of a one-port network containing one spring, one inerter, and a finite number of dampers. In [Chen *et al.* (2015c)], together with the minimal synthesis results of three-port resistive networks, the methodology in [Chen and Smith (2009b)] has been further applied to solve the case where there are no more than three springs.

In summary, the mechanical element extraction approach utilized to solve the synthesis problem of any one-port mechanical network containing m dampers, l inerters, and a finite number of springs, for $m, k = 1, 2, \ldots$, is stated as an algorithm below.

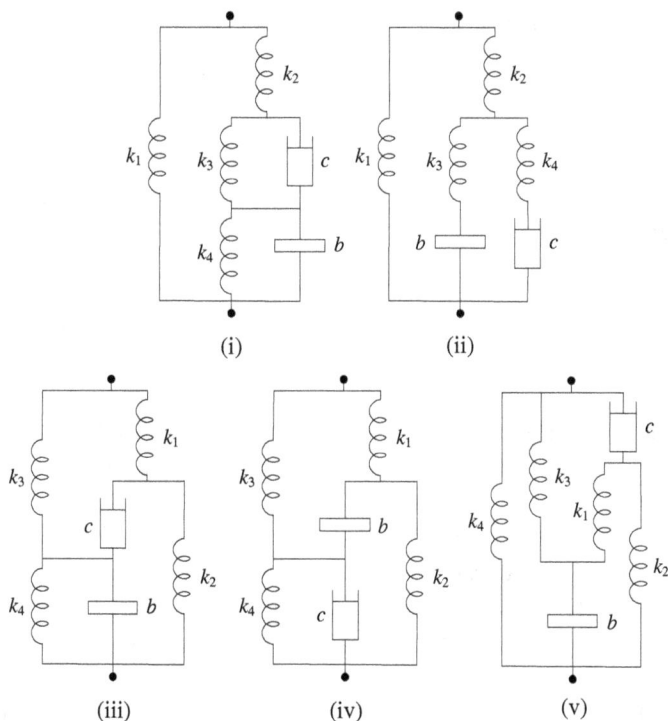

Fig. 5.2 Five configurations realizing the network containing one damper, one inerter and a finite number of springs [Chen and Smith (2009b)].

Algorithm 5.1

Step 1. *Extract m dampers and l inerters to form an $(m+l+1)$-port network X consisting of only springs. The impedance matrix (or admittance matrix) of X is assumed to exist, which is denoted as sL with $L \in \mathbb{S}^{m+l+1}$ being non-negative definite.*

Step 2. *The admittance $Y(s)$ of the one-port mechanical network is determined as $Y(s) = \alpha_0(s)/\beta_0(s)$, where the coefficients of $\alpha_0(s)$ and $\beta_0(s)$ are in terms of the entries of L, and the element values of dampers and inerters $(c_i, b_j, i = 1, 2, \ldots, m, j = 1, 2, \ldots, l)$.*

Step 3. *By utilizing the transformation $\mathfrak{T}(L) = R$, where $\mathfrak{T}(\cdot)$ is related to c_i and b_j, the expression of $Y(s)$ is correspondingly equivalent to $Y(s) = \alpha(s)/\beta(s)$, where the coefficients of $\alpha(s)$ and $\beta(s)$ are only in terms of the entries of R.*

Step 4. *Based on the realization of the non-negative definite matrix $L \in$*

$\mathbb{S}^{m+l+1}$ *as the impedance matrix of an* $(m + l + 1)$*-port resistive network, derive an explicit realizability condition of any positive-real admittance as a one-port mechanical network containing* m *dampers,* l *inerters, and a finite number of springs.*

Step 5. *Determine a set of configurations that can realize all the cases of the realizability condition, where realization configurations contain as few springs as possible.*

5.2   Realization of a Special Class of Admittances with One Damper, One Inerter, and Finite Springs

This section investigates the mechanical realization problem of a special class of positive-real admittances, which is common in vehicle suspension designs.

Smith [Smith (2002)] first discussed such a specific class of admittances, where the positive-realness condition is established and it is shown that any such a positive-real admittance is realizable as a one-port mechanical network containing no more than five elements through the Foster preamble. As discussed above, it is meaningful to consider the class of realizations in which the number of dampers and inerters is restricted to one in each case.

A necessary and sufficient condition is derived for the above special class of positive-real admittances to be realizable by employing one damper, one inerter, and a finite number of springs, under the assumption that the impedance of the three-port network after extracting the damper and the inerter exists. Correspondingly, realization configurations with explicit element values are presented, where one employs four springs and the other two. Based on the relationship between the topological properties of the n-port network and the fact that its impedance does not exist, the realization condition and the corresponding configurations are finally obtained, after removing the constraint on the existence of the impedance of the three-port spring network.

The mechanical admittance $Y(s)$ to be discussed is in the form of

$$Y(s) = k \frac{a_0 s^2 + a_1 s + 1}{s(d_0 s^2 + d_1 s + 1)}, \tag{5.6}$$

where $a_i > 0$, $i = 0, 1$, $d_j > 0$, $j = 0, 1$, and $k > 0$. Letting $p(s) := a_0 s^2 + a_1 s + 1$ and $q(s) := d_0 s^2 + d_1 s + 1$, the resultant of $p(s)$ and $q(s)$ is obtained as

$$R_0(p, q, s) = \begin{vmatrix} a_0 & a_1 & 1 & 0 \\ 0 & a_0 & a_1 & 1 \\ d_0 & d_1 & 1 & 0 \\ 0 & d_0 & d_1 & 1 \end{vmatrix} = (a_0 - d_0)^2 - (a_0 d_1 - a_1 d_0)(a_1 - d_1).$$

A positive-realness criterion for such a function was shown in [Smith (2002); Chen and Smith (2009a)].

Lemma 5.4. *[Smith (2002); Chen and Smith (2009a)] A given function* $Y(s)$ *in the form of* (5.6) *with* $a_i > 0$, $i = 0, 1$, $d_j > 0$, $j = 0, 1$, *and* $k > 0$, *is positive-real, if and only if* $a_0 d_1 - a_1 d_0 \geq 0$, $a_0 - d_0 \geq 0$, *and* $a_1 - d_1 \geq 0$.

In [Chen and Smith (2009b)], it is shown through a counterexample that not all positive-real admittances (5.6) can be realized employing one damper and one inerter. This section aims to solve the following question: what is the necessary and sufficient condition for a positive-real admittance $Y(s)$ in the form of (5.6) to be realizable with one damper, one inerter, and a finite number of springs, and what about the covering configurations?

5.2.1 Realizability Conditions when the Impedance of Spring Network Exists

As discussed in Section 5.1, any network consisting of one damper, one inerter, and a finite number of springs can be formulated as shown in Fig. 5.1, where the three-port network X contains only springs. In this subsection, the realization problem of $Y(s)$ in the form of (5.6) will be discussed under the assumption that the impedance matrix of X exists. It will be shown that the results can be obtained following the results in Theorems 5.1 and 5.2.

To make it easier to check the realizability condition for admittance $Y(s)$ in the form of (5.6), it seems natural to convert the admittance $Y(s)$ from the form of (5.5) to the following form

$$Y(s) = \frac{\alpha_3 s^3 + \alpha_2 s^2 + \alpha_1 s + 1}{\beta_4 s^4 + \beta_3 s^3 + \beta_2 s^2 + \beta_1 s}, \tag{5.7}$$

where the realizability conditions are only in terms of the coefficients α_i, $i = 1, 2, 3$, and β_j, $j = 1, 2, 3, 4$. Therefore, the following equations hold:

$$\alpha_3 = R_2 R_3 - R_6^2, \quad \alpha_2 = R_3, \quad \alpha_1 = R_2,$$
$$\beta_4 = \det(R), \quad \beta_3 = R_1 R_3 - R_5^2, \quad \beta_2 = R_1 R_2 - R_4^2, \quad \beta_1 = R_1. \tag{5.8}$$

In order to derive Theorem 5.20, some lemmas are established, where

$$W_1 := \alpha_1\alpha_2 - \alpha_3,$$
$$W_2 := \alpha_2\beta_1 - \beta_3,$$
$$W_3 := \alpha_1\beta_1 - \beta_2,$$
$$W := 2\alpha_1\alpha_2\beta_1 + \beta_4 - \alpha_1\beta_3 - \alpha_2\beta_2 - \alpha_3\beta_1$$

are used to simplify the expressions.

Lemma 5.5. *For a function $Y(s)$ in the form of (5.7), where $\alpha_i \geq 0$, $i = 1, 2, 3$, and $\beta_j \geq 0$, $j = 1, 2, 3, 4$, $Y(s)$ can be expressed as (5.5), that is,*

$$Y(s) = \frac{(R_2 R_3 - R_6^2)s^3 + R_3 s^2 + R_2 s + 1}{s(\det R s^3 + (R_1 R_3 - R_5^2)s^2 + (R_1 R_2 - R_4^2)s + R_1)}.$$

where R_i, $i = 1, 2, \ldots, 6$, is the entry of a third-order non-negative definite matrix R defined in (5.3), that is,

$$R := \begin{bmatrix} R_1 & R_4 & R_5 \\ R_4 & R_2 & R_6 \\ R_5 & R_6 & R_3 \end{bmatrix},$$

which satisfies (5.8), if and only if $W_1 \geq 0$, $W_2 \geq 0$, $W_3 \geq 0$, and $W^2 = 4W_1 W_2 W_3$.

Proof. *Sufficiency.* Supposing that $W_1 \geq 0$, $W_2 \geq 0$, $W_3 \geq 0$, and $W^2 = 4W_1 W_2 W_3$, it suffices to show that one can determine a non-negative definite matrix R in (5.3) whose entries satisfy (5.8).

Let $R_2 = \alpha_1$, $R_3 = \alpha_2$, and $R_1 = \beta_1$. Since $W_1, W_2, W_3 \geq 0$, one can obtain $\alpha_3 = R_2 R_3 - R_6^2$, $\beta_3 = R_1 R_3 - R_5^2$, and $\beta_2 = R_1 R_2 - R_4^2$, by properly choosing the values of R_4^2, R_5^2, and R_6^2. Furthermore, one can show that $R_4^2 R_5^2 R_6^2 = W_1 W_2 W_3$. Since $W^2 = 4W_1 W_2 W_3$, properly assigning the signs of R_4, R_5, and R_6 can always guarantee that $W = 2R_4 R_5 R_6$, which gives the following relationship: $\beta_4 = 2R_4 R_5 R_6 - 2\alpha_1\alpha_2\beta_1 + \alpha_3\beta_1 + \alpha_1\beta_3 + \alpha_2\beta_2 = 2R_4 R_5 R_6 - 2R_1 R_2 R_3 + R_1(R_2 R_3 - R_6^2) + R_2(R_1 R_3 - R_5^2) + R_3(R_1 R_2 - R_4^2) = R_1 R_2 R_3 - R_1 R_6^2 - R_2 R_5^2 - R_3 R_4^2 + 2R_4 R_5 R_6 = \det(R)$.

Now, one has obtained all the equations in (5.8). Therefore, one can express (5.7) as (5.5). Furthermore, it can be verified that R as defined in (5.3) is non-negative definite.

Necessity. From (5.8), it follows that $W_1 = R_6^2 \geq 0$, $W_2 = R_5^2 \geq 0$, $W_3 = R_4^2 \geq 0$, and $W^2 = 4W_1 W_2 W_3 = 4R_4^2 R_5^2 R_6^2$.

Thus, the lemma is proved. $\qquad\square$

Lemma 5.6. *Consider a non-negative definite matrix R in the form of (5.3), and α_1, α_2, α_3, β_1, β_2, β_3, and β_4 satisfying (5.8). Then, there exists at least one of the first- or second-order minors of R being zero if and only if at least one of α_1, α_2, α_3, β_1, β_2, β_3, W_1, W_2, W_3, $(\beta_1 - W/(2W_1))$, $(\alpha_1 - W/(2W_2))$ and $(\alpha_2 - W/(2W_3))$ is zero.*

Proof. Since the variables α_1, α_2, α_3, β_1, β_2, β_3, and β_4 satisfy (5.8), from $R_1 = \beta_1$, $R_2 = \alpha_1$, and $R_3 = \alpha_2$ one has $R_6^2 = W_1$, $R_5^2 = W_2$, and $R_4^2 = W_3$. In addition, from $\beta_4 = \det(R)$, one obtains

$$
\begin{aligned}
2R_4R_5R_6 &= \beta_4 - R_1R_2R_3 + R_1R_6^2 + R_2R_5^2 + R_3R_4^2 \\
&= 2\alpha_1\alpha_2\beta_1 + \beta_4 - \alpha_1\beta_3 - \alpha_2\beta_2 - \alpha_3\beta_1 = W.
\end{aligned}
\tag{5.9}
$$

Consequently, the following equations are obtained:

$$
R_1 - \frac{R_4R_5}{R_6} = R_1 - \frac{2R_4R_5R_6}{2R_6^2} = \beta_1 - \frac{W}{2W_1},
\tag{5.10}
$$

$$
R_2 - \frac{R_4R_6}{R_5} = R_2 - \frac{2R_4R_5R_6}{2R_5^2} = \alpha_1 - \frac{W}{2W_2},
\tag{5.11}
$$

$$
R_3 - \frac{R_5R_6}{R_4} = R_3 - \frac{2R_4R_5R_6}{2R_4^2} = \alpha_2 - \frac{W}{2W_3}.
\tag{5.12}
$$

It is known that there exists at least one of the first- or second-order minors of R being zero if and only if one of the following twelve equations is satisfied: $R_1 = 0$, $R_2 = 0$, $R_3 = 0$, $R_4 = 0$, $R_5 = 0$, $R_6 = 0$, $R_1R_2 - R_4^2 = 0$, $R_1R_3 - R_5^2 = 0$, $R_2R_3 - R_6^2 = 0$, $R_1R_6 - R_4R_5 = 0$, $R_4R_6 - R_2R_5 = 0$, and $R_3R_4 - R_5R_6 = 0$. Therefore, one has $R_1 = 0 \Leftrightarrow \beta_1 = 0$, $R_2 = 0 \Leftrightarrow \alpha_1 = 0$, $R_3 = 0 \Leftrightarrow \alpha_2 = 0$, $R_4 = 0 \Leftrightarrow W_3 = 0$, $R_5 = 0 \Leftrightarrow W_2 = 0$, $R_6 = 0 \Leftrightarrow W_1 = 0$, $R_1R_2 - R_4^2 = 0 \Leftrightarrow \beta_2 = 0$, $R_1R_3 - R_5^2 = 0 \Leftrightarrow \beta_3 = 0$, and $R_2R_3 - R_6^2 = 0 \Leftrightarrow \alpha_3 = 0$.

When $R_4R_5R_6 \neq 0$, the following conditions are satisfied

$$
R_1R_6 - R_4R_5 = 0 \Leftrightarrow R_1 - \frac{R_4R_5}{R_6} = 0 \Leftrightarrow \beta_1 - \frac{W}{2W_1} = 0,
$$

$$
R_4R_6 - R_2R_5 = 0 \Leftrightarrow R_2 - \frac{R_4R_6}{R_5} = 0 \Leftrightarrow \alpha_1 - \frac{W}{2W_2} = 0,
$$

$$
R_3R_4 - R_5R_6 = 0 \Leftrightarrow R_3 - \frac{R_5R_6}{R_4} = 0 \Leftrightarrow \alpha_2 - \frac{W}{2W_3} = 0.
$$

Combining the conditions shown in the above equations, one obtains the condition of the lemma. □

By Lemma 5.2, for any matrix R satisfying Lemma 5.6, there always exists an invertible matrix $D = \text{diag}\{1, x, y\}$ such that DRD is paramount.

Lemma 5.7. *Consider a non-negative definite matrix R in the form of (5.3), whose first- and second-order minors are all non-zero, and the variables α_1, α_2, α_3, β_1, β_2, β_3, and β_4 satisfying (5.8). Then, there exists an invertible matrix $D = \text{diag}\{1, x, y\}$ with $x > 0$ and $y > 0$ such that DRD is paramount, if and only if one of the following conditions holds:*

1. *$W < 0$;*
2. *$W > 0$, $\beta_1 > (W/(2W_1))$, $\alpha_1 > (W/(2W_2))$, $\alpha_2 > (W/(2W_3))$;*
3. *$W > 0$, $\alpha_2 < (W/(2W_3))$ and $\beta_4 + \alpha_1\beta_3 + \alpha_3\beta_1 - \alpha_2\beta_2 \geq 0$;*
4. *$W > 0$, $\alpha_1 < (W/(2W_2))$ and $\beta_4 + \alpha_2\beta_2 + \alpha_3\beta_1 - \alpha_1\beta_3 \geq 0$;*
5. *$W > 0$, $\beta_1 < (W/(2W_1))$ and $\beta_4 + \alpha_1\beta_3 + \alpha_2\beta_2 - \alpha_3\beta_1 \geq 0$.*

Proof. Since an invertible matrix $D = \text{diag}\{1, x, y\}$ with $x > 0$ and $y > 0$ exists such that DRD is paramount if and only if the conditions of Lemma 5.3 hold, one can prove the equivalence between the conditions of Lemma 5.3 and the conditions of this lemma. Since the variables α_1, α_2, α_3, β_1, β_2, β_3, and β_4 are defined in (5.8), one has (5.9), (5.10), (5.11), and (5.12) from the above discussion.

Furthermore, one can obtain

$$
\begin{aligned}
R_1R_2R_3 + R_4R_5R_6 - R_1R_6^2 - R_2R_5^2 &= \beta_1\alpha_1\alpha_2 + \frac{W}{2} + \beta_1W_1 + \alpha_1W_2 \\
&= \frac{\beta_4 + \alpha_1\beta_3 + \alpha_3\beta_1 - \alpha_2\beta_2}{2},
\end{aligned}
$$

$$
\begin{aligned}
R_1R_2R_3 + R_4R_5R_6 - R_1R_6^2 - R_3R_4^2 &= \beta_1\alpha_1\alpha_2 + \frac{W}{2} + \beta_1W_1 + \alpha_2W_3 \\
&= \frac{\beta_4 + \alpha_2\beta_2 + \alpha_3\beta_1 - \alpha_1\beta_3}{2},
\end{aligned}
$$

$$
\begin{aligned}
R_1R_2R_3 + R_4R_5R_6 - R_3R_4^2 - R_2R_5^2 &= \beta_1\alpha_1\alpha_2 + \frac{W}{2} + \alpha_2W_3 + \alpha_1W_2 \\
&= \frac{\beta_4 + \alpha_1\beta_3 + \alpha_2\beta_2 - \alpha_3\beta_1}{2}.
\end{aligned}
$$

Since all the first- and second-order minors of R are non-zero, it follows that W_1, W_2, W_3, and W in the above equations never go to zero. Thus, the lemma is proved. $\square$

Consequently, the following theorem is obtained, which is equivalent to Theorem 5.1.

Theorem 5.3. *A positive-real function $Y(s)$ is realizable as the admittance of a one-port mechanical network consisting of one damper, one inerter, and a finite number of springs in Fig. 5.1, where the impedance of the three-port network X exists, if and only if $Y(s)$ can be written in the form of (5.7), where the coefficients satisfy $\alpha_i \geq 0$, $i = 1, 2, 3$, and $\beta_j \geq 0$, $j = 1, 2, 3, 4$, and further satisfy the conditions of Lemma 5.5, as well as the conditions of either Lemma 5.6 or Lemma 5.7.*

Proof. *Sufficiency.* It follows from Lemma 5.5 that $Y(s)$ can also be expressed as (5.5), where R is defined in (5.3) and non-negative definite, and the non-negative coefficients $\alpha_i \geq 0$, $i = 1, 2, 3$, and $\beta_j \geq 0$, $j = 1, 2, 3, 4$, satisfy (5.8). Furthermore, if the conditions of Lemma 5.6 hold, then there must exist at least one minor of R being zero, which by Lemma 5.2 implies that there must exist an invertible $D = \text{diag}\{1, x, y\}$ such that DRD is paramount; if the conditions of Lemma 5.7 hold, then there also exists such an invertible matrix. Finally, by Lemma 5.1, the sufficiency part is proved.

Necessity. It follows from Lemma 5.1 that $Y(s)$ can be written in the form of (5.5) where R as defined in (5.3) is non-negative definite. Then, it is clear that $Y(s)$ can also be expressed as (5.7) with (5.8), which implies that $\alpha_i \geq 0$, $i = 1, 2, 3$, $\beta_j \geq 0$, $j = 1, 2, 3, 4$, and the conditions of Lemma 5.5 hold. Since the conditions of either Lemma 5.2 or Lemma 5.3 hold, one concludes that the conditions of either Lemma 5.6 or Lemma 5.7 are satisfied.   □

Now consider the realization of admittance (5.6), that is,

$$Y(s) = k\frac{a_0 s^2 + a_1 s + 1}{s(d_0 s^2 + d_1 s + 1)},$$

where $a_i > 0$, $i = 0, 1$, $d_j > 0$, $j = 0, 1$, and $k > 0$.

It is known that the resultant satisfies $R_0(p, q, s) = 0$ if and only if a positive-real $Y(s)$ in the form of (5.6) can be written in the following form:

$$Y(s) = k\frac{as + 1}{s(ds + 1)}, \tag{5.13}$$

where $a \geq 0$, $d \geq 0$, and $a - d \geq 0$ [Gohberg *et al.* (1982)]. Therefore, one has the following result.

Theorem 5.4. *Consider a positive-real function $Y(s)$ in the form of (5.6), where $a_i > 0$, $i = 0, 1$, $d_j > 0$, $j = 0, 1$, and $k > 0$. If $R_0(p, q, s) :=$*

$(a_0 - d_0)^2 - (a_0 d_1 - a_1 d_0)(a_1 - d_1) = 0$, *then it can be realized as the admittance of a network containing at most one damper and two springs.*

Proof. From the discussion above, it is known that if $R_0(p, q, s) = 0$, then $Y(s)$ can be written as (5.13), where $a \geq 0$, $d \geq 0$, and $a - d \geq 0$. Furthermore, it is obvious that $Y(s) = k/s + k(a - d)/(ds + 1)$. Therefore, $Y(s)$ is realizable as a configuration containing at most one damper and two springs, by the Foster preamble. □

Based on Theorem 5.4, in order to investigate the realizability conditions of (5.6), it suffices to consider the case of $R_0(p, q, s) \neq 0$. The next theorem presents the realizability condition for admittance (5.6) to be realizable as a network in Fig. 5.1.

Theorem 5.5. *Consider a positive-real function $Y(s)$ in the form of (5.6), where $a_i > 0$, $i = 0, 1$, $d_j > 0$, $j = 0, 1$, $k > 0$, and $R_0(p, q, s) := (a_0 - d_0)^2 - (a_0 d_1 - a_1 d_0)(a_1 - d_1) \neq 0$. Then, $Y(s)$ is realizable as the admittance of a one-port mechanical network consisting of one damper, one inerter, and a finite number of springs in Fig. 5.1, where the impedance of the three-port network X exists, if and only if*

$$\frac{d_0^2}{(a_0 d_1 - a_1 d_0)(a_1 - d_1)} \geq 1 \tag{5.14}$$

or

$$a_0 d_1 - a_1 d_0 = 0. \tag{5.15}$$

Proof. *Necessity.* By Theorem 5.20, $Y(s)$ in the form of (5.6) with $R_0(p, q, s) \neq 0$ can be expressed as (5.7). Thus, there are two possible cases.

For the first case, the coefficients in (5.7) satisfy

$$\alpha_3 = 0, \ \alpha_2 = a_0, \ \alpha_1 = a_1, \ \beta_4 = 0, \ \beta_3 = \frac{d_0}{k}, \ \beta_2 = \frac{d_1}{k}, \ \beta_1 = \frac{1}{k}, \tag{5.16}$$

which are all non-negative. Furthermore, $W^2 = 4W_1 W_2 W_3$ implies (5.15).

For the second case, multiplying a common factor $(Ts + 1)$ with $T > 0$, the coefficients in (5.7) can be expressed as

$$\alpha_3 = a_0 T, \ \alpha_2 = a_0 + a_1 T, \ \alpha_1 = a_1 + T, \ \beta_4 = d_0 T/k,$$
$$\beta_3 = (d_0 + d_1 T)/k, \ \beta_2 = (d_1 + T)/k, \ \beta_1 = 1/k, \ T > 0. \tag{5.17}$$

Moreover, $W^2 = 4W_1W_2W_3$ implies $(a_1 - d_1)T^2 - (a_0d_1 - a_1d_0) = 0$. If $a_1 = d_1$, then $a_0d_1 = a_1d_0$, which is the condition derived in the first case. If $a_1 \neq d_1$, then $a_0d_1 \neq a_1d_0$, by which the positive-realness condition (see Lemma 5.4) is equivalent to $a_0d_1 - a_1d_0 > 0$, $a_0 - d_0 \geq 0$, and $a_1 - d_1 > 0$. This implies that $a_0 > 0$, $a_1 > 0$, $d_1 > 0$, and $d_0 \geq 0$. Therefore, $\alpha_1 > 0$, $\alpha_2 > 0$, $\alpha_3 > 0$, $\beta_1 > 0$, $\beta_2 > 0$, $\beta_3 > 0$, and $\beta_4 \geq 0$. Moreover, T can be expressed as

$$T = \sqrt{\frac{a_0d_1 - a_1d_0}{a_1 - d_1}}. \tag{5.18}$$

It can be verified that $W_1 = a_1T^2 + a_1^2T + a_0a_1 > 0$, $W_2 = ((a_1-d_1)T + (a_0-d_0))/k > 0$, $W_3 = (a_1-d_1)/k > 0$, $\beta_1 - W/(2W_1) = ((a_1+d_1)T^2 + 2a_1d_1T + (a_0d_1 + a_1d_0))/(2ka_1(T^2 + a_1T + a_0)) > 0$, $\alpha_1 - W/(2W_2) = ((a_1-d_1)T^2 + 2(a_0-d_0)T + (a_0d_1 - a_1d_0))/(2((a_1-d_1)T + (a_0-d_0))) > 0$, $\alpha_2 - W/(2W_3) = (a_1d_0 - a_0d_1)/(a_1 - d_1) < 0$. Thus, the condition of Lemma 5.6 is not satisfied, which implies that the condition of Lemma 5.7 must hold. It is noted that $W = 2a_1(\sqrt{(a_0d_1 - a_1d_0)(a_1 - d_1)} + (a_0 - d_0))/k > 0$. Consequently, one concludes that only Condition 3 of Lemma 5.7 holds. Thus, it follows that $\beta_4 + \alpha_1\beta_3 + \alpha_3\beta_1 - \alpha_2\beta_2 \geq 0$. Since the equation $\beta_4 + \alpha_1\beta_3 + \alpha_3\beta_1 - \alpha_2\beta_2 = 2\sqrt{a_0d_1 - a_1d_0}(d_0 - \sqrt{(a_0d_1 - a_1d_0)(a_1 - d_1)})/(k\sqrt{a_1 - d_1})$ holds, one obtains (5.14).

Sufficiency. When (5.15) holds, one can express $Y(s)$ in the form of (5.7) with the coefficients α_i, $i = 1, 2, 3$, and β_j, $j = 1, 2, 3, 4$, satisfying (5.16), implying that all the coefficients are non-negative and the condition of Lemma 5.6 must hold. The positive-realness of $Y(s)$ guarantees that $W_1 \geq 0$, $W_2 \geq 0$, $W_3 \geq 0$, and $W^2 = 4W_1W_2W_3$, because of (5.15).

When (5.14) holds, together with the positive-realness of $Y(s)$ (see Lemma 5.4), one concludes that $a_i > 0$, $i = 0, 1$ and $d_j > 0$, $j = 0, 1$. Simultaneously multiplying the numerator and the denominator of (5.6) by a common factor $(Ts + 1)$, where T is in (5.18), $Y(s)$ can be expressed as (5.7) with the coefficients satisfying (5.17), which are all positive. From the necessity part of the proof above, both the condition of Lemma 5.5 and Condition 3 of Lemma 5.7 are satisfied.

Thus, by Theorem 5.20, $Y(s)$ is realizable as the required network. $\square$

In order to make the realization be available, it is necessary to provide explicit network configurations that will cover the realizability conditions. The following two theorems address the two conditions (5.14) and (5.15) of Theorem 5.5, respectively.

Theorem 5.6. *Consider a positive-real function $Y(s)$ in the form of (5.6) where $a_i > 0$, $i = 0, 1$, $d_j > 0$, $j = 0, 1$, $k > 0$, and $R_0(p, q, s) \neq 0$. If Condition (5.14) holds, then $Y(s)$ is realizable as the configuration in Fig. 5.3 with the element values satisfying*

$$k_1 = \frac{k a_0 d_1 (a_0 d_1 - a_1 d_0)[(a_1 - d_1)T + (a_0 - d_0)]}{d_0(a_1 - d_1)(d_1 T + d_0)[(a_0 - d_0)T + (a_0 d_1 - a_1 d_0)]},$$

$$k_2 = \frac{k a_0 a_1 d_1 T[(a_1 - d_1)T + (a_0 - d_0)]^2}{(a_1 - d_1)(d_1 T + d_0)[(a_0 - d_0)T + (a_0 d_1 - a_1 d_0)]^2},$$

$$k_3 = \frac{k a_0 T[(a_1 - d_1)T + (a_0 - d_0)]}{(d_1 T + d_0)[(a_0 - d_0)T + (a_0 d_1 - a_1 d_0)]},$$

$$k_4 = \frac{k a_0 T[d_0 T + (a_1 d_0 - a_0 d_1)][(a_1 - d_1)T + (a_0 - d_0)]}{(d_1 T + d_0)[(a_0 - d_0)T + (a_0 d_1 - a_1 d_0)]^2},$$

$$b = \frac{k a_0^2 (a_0 d_1 - a_1 d_0)[(a_1 - d_1)T + (a_0 - d_0)]}{(a_1 - d_1)[(a_0 - d_0)T + (a_0 d_1 - a_1 d_0)]^2},$$

$$c = \frac{k a_0^2 d_1^2 (a_0 d_1 - a_1 d_0)[(a_1 - d_1)T + (a_0 - d_0)]^2}{(a_1 - d_1)^2 (d_1 T + d_0)^2[(a_0 - d_0)T + (a_0 d_1 - a_1 d_0)]^2},$$

(5.19)

where T is defined in (5.18).

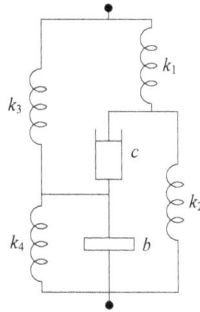

Fig. 5.3   The configuration in Theorem 5.6, where $b > 0$, $c > 0$, $k_i > 0$, $i = 1, 2, 3$, and $k_4 \geq 0$.

Proof. Suppose that condition (5.14) holds, from the proof of the sufficiency part of Theorem 5.5, one has that $a_i > 0$, $i = 0, 1$ and $d_j > 0$, $j = 0, 1$. Multiply the numerator and the denominator of (5.6) by $(Ts + 1)$ simultaneously, where T is defined as (5.18). From that proof, it is known that $Y(s)$ can be expressed as (5.7) with the coefficients α_1, α_2, α_3, β_1, β_2, and β_3 satisfying (5.17), which are all positive, and the conditions of Lemma 5.5 and the third condition of Lemma 5.7 are satisfied, which lead to the fact that $Y(s)$ can be expressed as (5.5), where non-negative definite R is defined as (5.3) and satisfies (5.8), whose elements also satisfy the third condition of Lemma 5.3. Then, it is shown in [Chen and Smith (2009b)] that $Y(s)$ is realizable as the configuration in Fig. 5.3, with $k_1 = (R_2 R_3 - R_6^2)(R_1 - R_4 R_5/R_6)(R_3 - R_5 R_6/R_4)/(\det R(R_1 R_3 - R_5^2)(R_2 - R_4 R_6/R_5)(-R_5/(R_4 R_6)))$, $k_2 = R_6^2(R_2 R_3 - R_6^2)(R_1 - R_4 R_5/R_6)/((R_4 R_6 - R_2 R_5)^2(R_1 R_3 - R_5^2))$, $k_3 = (R_2 R_3 - R_6^2)/((R_1 R_3 - R_5^2)(R_2 - R_4 R_6/R_5))$, $k_4 = (R_1 R_2 R_3 + R_4 R_5 R_6 - R_1 R_6^2 - R_2 R_5^2)(R_2 R_3 - R_6^2)/((R_4 R_6 - R_2 R_5)^2(R_1 R_3 - R_5^2))$, $b = (R_2 R_3 - R_6^2)^2/(R_2 R_5 - R_4 R_6)^2$, and $c = (R_2 R_3 - R_6^2)^2(R_1 R_6 - R_4 R_5)^2/((R_1 R_3 - R_5^2)^2(R_2 R_5 - R_4 R_6)^2)$. It follows from (5.8) that (5.9)–(5.12) hold. Therefore, the above element value expressions are equivalent to

$$k_1 = \frac{\alpha_3(\beta_1 - \frac{W}{2W_1})(\alpha_2 - \frac{W}{2W_3})}{\beta_4 \beta_3(\alpha_1 - \frac{W}{2W_2})(-\frac{2W_2}{W})},$$

$$k_2 = \frac{\alpha_3(\alpha_1 \alpha_2 - \alpha_3)(\beta_1 - \frac{W}{2W_1})}{\beta_3(\alpha_2 \beta_1 - \beta_3)(\alpha_1 - \frac{W}{2W_2})^2},$$

$$k_3 = \frac{\alpha_3}{\beta_3(\alpha_1 - \frac{W}{2W_2})},$$

$$k_4 = \frac{\alpha_3(\beta_4 + \alpha_1 \beta_3 + \alpha_3 \beta_1 - \alpha_2 \beta_2)}{2\beta_3(\alpha_2 \beta_1 - \beta_3)(\alpha_1 - \frac{W}{2W_2})^2},$$

$$b = \frac{\alpha_3^2}{(\alpha_2 \beta_1 - \beta_3)(\alpha_1 - \frac{W}{2W_2})^2},$$

$$c = \frac{\alpha_3^2(\alpha_1 \alpha_2 - \alpha_3)(\beta_1 - \frac{W}{2W_1})^2}{\beta_3^2(\alpha_2 \beta_1 - \beta_3)(\alpha_1 - \frac{W}{2W_2})^2}.$$

Since (5.17) holds, where T is defined as (5.18), the values can be further expressed as in (5.19). It can be seen that k_1, k_2, k_3, b, $c > 0$, and $k_4 \geq 0$. $\square$

Theorem 5.7. *Consider a positive-real function $Y(s)$ in the form of (5.6) where $a_i > 0$, $i = 0, 1$, $d_j > 0$, $j = 0, 1$, $k > 0$, and $R_0(p, q, s) \neq 0$. If Condition (5.15) holds, then $Y(s)$ is realizable with no more than one inerter, one damper, and two springs.*

Proof. The positive-realness condition of $Y(s)$ implies that $a_0 \geq d_0$. Moreover, it implies from $a_0 d_1 = a_1 d_0$ and $R_0(p, q, s) \neq 0$ that $a_0 > d_0$. Therefore, the following two cases are discussed, respectively.

If $a_0 d_1 = a_1 d_0$ and $a_0 > d_0 = 0$, then $d_1 = 0$. Therefore,

$$Y(s) = k a_0 s + k a_1 + \frac{k}{s},$$

which is realizable with one spring, one inerter and no more than one damper with the element values being $k_1 = k > 0$, $b = k a_0 > 0$, and $c = k a_1 \geq 0$, respectively.

If $a_0 d_1 = a_1 d_0$ and $a_0 > d_0 > 0$, then

$$Y(s) = \frac{k}{s} + k \frac{(a_0 - d_0)s + (a_1 - d_1)}{d_0 s^2 + d_1 s + 1}$$

$$= \frac{k}{s} + \left(\frac{d_0}{k(a_0 - d_0)} s + \frac{1}{k(a_0 - d_0)s + k(a_1 - d_1)} \right)^{-1},$$

which is realizable with two springs, one inerter, and no more than one damper with the element values being $k_1 = k > 0$, $k_2 = k(a_0 - d_0)/d_0 > 0$, $b = k(a_0 - d_0) > 0$, and $c = k(a_1 - d_1) \geq 0$, respectively.   □

5.2.2 Final Realization Results

The discussion in the previous subsection is under the assumption that the impedance of the three-port network X in Fig. 5.1 exists. This subsection will further investigate the realization problem of $Y(s)$ when this constraint is removed in order to completely solve the problem.

As discussed in Chapter 2, similarly to RLC electrical networks, any damper-spring-inerter mechanical network can be described by an augmented graph $\mathcal{G}$ with each element or each port corresponding to an edge, and each velocity node corresponding to a vertex. The subgraph consisting of all the edges corresponding to ports (called port edges) is port graph $\mathcal{G}_p$, and the subgraph consisting of all the edges corresponding to elements (called network edges) is network graph $\mathcal{G}_e$. The augmented $\mathcal{G}$ can always be regarded as being connected, for otherwise one can obtain the connected case by letting any velocity node of a component be common with that of another. When the number of ports is one, the augmented graph $\mathcal{G}$ can also be regarded as being nonseparable, since obviously any one-port damper-spring-inerter network can be equivalent to one whose augmented graph is nonseparable.

Lemma 5.8. *Consider a positive-real function $Y(s)$ in the form of (5.6), where $a_i > 0$, $i = 0, 1$, $d_j > 0$, $j = 0, 1$, $k > 0$, and $R_0(p, q, s) \neq 0$. Then, $Y(s)$ is realizable as the admittance of a one-port mechanical network consisting of one damper, one inerter, and a finite number of springs in*

Fig. 5.1, with its augmented graph $\mathcal{G}$ non-separable and the impedance of X not existing, if and only if $Y(s)$ is realizable as the admittance of the network as shown in Fig. 5.4, where its augmented graph is non-separable, the two-port network L consists of only springs, with the impedance of L existing.

Proof. *Sufficiency.* It is obvious that the network in Fig. 5.4 is a special case of Fig. 5.1. Let $\mathcal{G}^{\mathcal{X}}$ denote the augmented graph of X. Then, $\mathcal{G}^{\mathcal{X}}$ must be non-separable because of the non-separability of $\mathcal{G}$. Since the two edges corresponding to the damper and the inerter respectively constitute a cut-set of $\mathcal{G}$, it is clear that the two port edges corresponding to ports terminated with the damper and the inerter form a cut-set of $\mathcal{G}^{\mathcal{X}}$. By Theorem 2.18, the impedance of X does not exist.

Necessity. It is noted that $Z(s) = Y^{-1}(s)$ has a zero at $s = 0$. According to [Seshu (1959), Theorem 2], there must be a path $\mathcal{P}(a, a')$ consisting of only springs, where a and a' are two terminals of N.

Next step is to show that the two port edges corresponding to the ports terminated with the inerter and the damper constitute a cut-set of the nonseparable augmented graph $\mathcal{G}^{\mathcal{X}}$ of X. By Theorem 2.18, there must exist a cut-set among the three port edges of $\mathcal{G}^{\mathcal{X}}$. It is shown in [Seshu and Reed (1961), Theorem 3-3] that any nonseparable graph with at least two edges must be cyclically connected, which means that any two vertices can be placed in a circuit. Thus, it implies that the number of edges of any cut-set of $\mathcal{G}^{\mathcal{X}}$ is at least two. Therefore, it suffices to show that any such cut-set can never contain the edge that corresponds to the only port of N whose two terminals are a and a'. Assume that there exists such a cut-set $\mathcal{C}$ among those three port edges containing the edge e_p. Then, by the property of the cut-set, $\mathcal{C}$ separates $\mathcal{G}$ into two parts, which results in terminals a and a' located in each part. Thus, all the paths $\mathcal{P}(a, a')$ of the network in Fig. 5.1 must contain a damper, or an inerter, or both, which contradicts the conclusion stated at the beginning of the necessity part.

Consequently, by the property of cut-set and the analysis of all the possible cases, the network in Fig. 5.1 can always be equivalent to the one in Fig. 5.4, where L contains only springs. Since it has been shown that any cut-set never contains the port edge, one can prove that L must have a well-defined impedance because no cut-set is contained in its port graph. □

Next, based on Lemma 5.9, a necessary and sufficient condition for any positive-real function to be the admittance of the network in Fig. 5.4 is derived in Lemma 5.10. The realization is then given in Lemma 5.11.

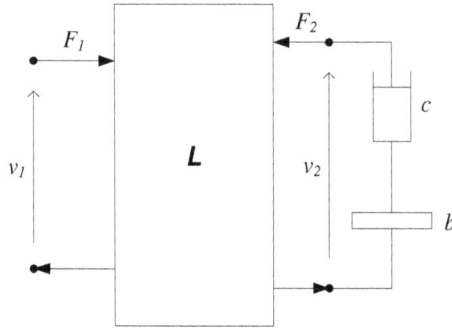

Fig. 5.4   The general network discussed in Lemma 5.8, where $b > 0$, $c > 0$, L consists of only springs, and the impedance of L exists.

Lemma 5.9. *Consider any 2×2 symmetric matrix $Q \in \mathbb{S}^2$. If Q is non-negative definite, then there must exist an invertible matrix $D = \text{diag}\{1, x\}$ with $x > 0$ such that DQD is paramount.*

Proof. This lemma can be easily proved by the definition of paramountcy and the property of non-negative definiteness.   □

Lemma 5.10. *A positive-real function $Y(s)$ is realizable as the admittance of a one-port mechanical network consisting of one damper, one inerter, and a finite number of springs in Fig. 5.4, where L consists of only springs and the impedance of L exists, if and only if $Y(s)$ can be written in the form of*

$$Y(s) = \frac{\alpha_2 s^2 + \alpha_1 s + 1}{\beta_3 s^3 + \beta_2 s^2 + \beta_1 s}, \tag{5.20}$$

where $\alpha_2 \geq 0$, $\beta_2 \geq 0$, $\beta_3 \geq 0$, $\alpha_1 > 0$, $\beta_1 > 0$, and the coefficients satisfy $\alpha_1 \beta_1 - \beta_2 = 0$.

Proof. *Necessity.* It can be verified that the admittance of the network in Fig. 5.4 satisfies

$$\frac{\hat{F}_1}{\hat{v}_1} = \frac{(b/c)cP_2 s^2 + (b/c)s + 1}{(b/c)(cP_1 P_2 - cP_3^2)s^3 + (b/c)P_1 s^2 + P_1 s}, \tag{5.21}$$

where $b > 0$, $c > 0$ and

$$P := \begin{bmatrix} P_1 & P_3 \\ P_3 & P_2 \end{bmatrix} \tag{5.22}$$

is the impedance of L, which is paramount by Theorem 4.1. Furthermore, (5.21) can be expressed as

$$Y(s) = \frac{mQ_2s^2 + ms + 1}{m(Q_1Q_2 - Q_3^2)s^3 + mQ_1s^2 + Q_1s}, \tag{5.23}$$

where $m = b/c$, and Q is obtained through

$$Q := \begin{bmatrix} Q_1 & Q_3 \\ Q_3 & Q_2 \end{bmatrix} = T \begin{bmatrix} P_1 & P_3 \\ P_3 & P_2 \end{bmatrix} T, \tag{5.24}$$

where $T = \mathrm{diag}\{1, \sqrt{c}\}$. Therefore, Q must be non-negative definite. Letting $\alpha_1 = m$, $\alpha_2 = mQ_2$, $\beta_1 = Q_1$, $\beta_2 = mQ_1$, and $\beta_3 = m(Q_1Q_2 - Q_3^2)$, one obtains $\alpha_2 \geq 0$, $\beta_1 \geq 0$, $\beta_2 \geq 0$, $\beta_3 \geq 0$, $\alpha_1 > 0$, and $\alpha_1\beta_1 = \beta_2 = mQ_1$. By the positive-realness of Q, $\beta_1 = 0$ yields $\beta_2 = \beta_3 = 0$, which implies the non-existence of $Y(s)$. Thus, one obtains $\beta_1 > 0$.

Sufficiency. It suffices to show that $Y(s)$ can be written in the form of (5.21), where $b > 0$, $c > 0$, and P as defined in (5.22) is paramount.

Since the positive-realness of $Y(s)$ guarantees that $\alpha_2\beta_1 - \beta_3 \geq 0$, let $m = \alpha_1$, $Q_1 = \beta_1$, $Q_2 = \alpha_2/\alpha_1$, and $Q_3^2 = (\alpha_2\beta_1 - \beta_3)/\alpha_1$. Consequently, it can be verified that Q as defined in (5.24) is non-negative definite. Since $\alpha_1\beta_1 - \beta_2 = 0$ holds, it can be verified that $\alpha_1 = m$, $\alpha_2 = mQ_2$, $\beta_1 = Q_1$, $\beta_2 = mQ_1$, and $\beta_3 = m(Q_1Q_2 - Q_3^2)$. This implies that $Y(s)$ can also be expressed as (5.23) with $m > 0$, and Q as defined in (5.24) being non-negative definite. By Lemma 5.9, there must exist an invertible matrix $D = \mathrm{diag}\{1, x\}$, where $x > 0$, such that

$$P := \begin{bmatrix} P_1 & P_3 \\ P_3 & P_2 \end{bmatrix} = \begin{bmatrix} 1 & 0 \\ 0 & x \end{bmatrix} \begin{bmatrix} Q_1 & Q_3 \\ Q_3 & Q_2 \end{bmatrix} \begin{bmatrix} 1 & 0 \\ 0 & x \end{bmatrix}$$

is paramount. Therefore, one obtains $Q_1 = P_1$, $Q_2 = P_2/x^2$, and $Q_3 = P_3/x$, converting (5.23) to

$$Y(s) = \frac{m(1/x^2)P_2s^2 + ms + 1}{m((1/x^2)P_1P_2 - (1/x^2)P_3^2)s^3 + mP_1s^2 + P_1s}.$$

Letting $c = 1/x^2$ and $b = mc$ gives (5.21), where $b > 0$, $c > 0$, and P is paramount. □

Lemma 5.11. *Consider a positive-real function $Y(s)$ in the form of (5.6), where $a_i > 0$, $i = 0, 1$, $d_j > 0$, $j = 0, 1$, $k > 0$, and $R_0(p, q, s) \neq 0$. Then, $Y(s)$ is realizable as the admittance of a one-port mechanical network consisting of one damper, one inerter, and a finite number of springs in Fig. 5.4, where L consists of only springs and the impedance of L exists,* *if and only if* $a_1 = d_1 > 0$. *Moreover, if the condition holds, then* $Y(s)$ *is realizable as the configuration in Fig. 5.5 with the element values satisfying*

$$k_1 = k, \ k_2 = \frac{(a_0 - d_0)k}{d_0}, \ b = (a_0 - d_0)k, \ c = \frac{(a_0 - d_0)k}{a_1}. \tag{5.25}$$

Proof. From Lemma 5.10, the condition can be easily derived. Furthermore, the admittance of the network shown in Fig. 5.5 is calculated to be $Y(s) = (bc(k_1 + k_2)s^2 + bk_1k_2s + ck_1k_2)/(s(bcs^2 + bk_2s + ck_2))$. Substituting (5.25), whose values are non-negative, into the above equation results in (5.6) because $a_1 = d_1 > 0$. Therefore, this lemma is proved.   □

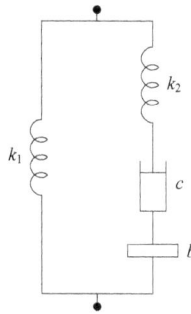

Fig. 5.5   The configuration in Lemma 5.11, where the impedance of X does not exist when the configuration is expressed in the form of Fig. 5.1.

Now, the final condition is presented.

Theorem 5.8. *Consider a positive-real function* $Y(s)$ *in the form of* (5.6), *where* $a_i > 0$, $i = 0, 1$, $d_j > 0$, $j = 0, 1$, $k > 0$, *and* $R_0(p, q, s) \neq 0$. *Then,* $Y(s)$ *is realizable as the admittance of a one-port mechanical network consisting of one inerter, one damper, and a finite number of springs, which is the network* N *shown in Fig. 5.1, if and only if*

$$(a_0d_1 - a_1d_0)(a_1 - d_1) - d_0^2 \leq 0. \tag{5.26}$$

Proof. *Necessity.* When the impedance of X exists, by Theorem 5.5, $Y(s)$ satisfies (5.14) or (5.15), that is, $d_0^2/((a_0d_1 - a_1d_0)(a_1 - d_1)) \geq 1$ or $a_0d_1 - a_1d_0 = 0$. When the impedance of X does not exist and the augmented graph $\mathcal{G}$ of N is non-separable, by Lemma 5.11 one obtains $a_1 = d_1 > 0$.

When the impedance of X does not exist and the augmented graph $\mathcal{G}$ of N is separable, one can always obtain an equivalent network N' in the form of Fig. 5.1, which belongs to one of the above two cases. Combining the above conditions, (5.26) is proved.

Sufficiency. Since condition (5.26) holds, it implies that at least one of the conditions of Theorem 5.5 and Lemma 5.11 must hold. Thus, admittance (5.6) is realizable as the required network by Theorem 5.6, Theorem 5.7, and Lemma 5.11. □

5.3   Realizations of a Special Class of Admittances with Strictly Lower Complexity than Canonical Configurations

This section continues to investigate the realization problem of the admittance $Y(s)$ in the form of (5.6), that is,

$$Y(s) = k\frac{a_0s^2 + a_1s + 1}{s(d_0s^2 + d_1s + 1)},$$

where $a_i > 0$, $i = 0, 1$, $d_j > 0$, $j = 0, 1$, and $k > 0$.

As shown in the previous section, the resultant of $p(s) := a_0s^2 + a_1s + 1$ and $q(s) := d_0s^2 + d_1s + 1$ in s is obtained as $R_0(p, q, s) := (a_0-d_0)^2-(a_0d_1 - a_1d_0)(a_1 - d_1)$. A necessary and sufficient condition for $Y(s)$ in the form (5.6) to be positive-real is presented in Lemma 5.4, that is, $a_0d_1 - a_1d_0 \geq 0$, $a_0 - d_0 \geq 0$, and $a_1 - d_1 \geq 0$ hold simultaneously.

For any positive-real admittance $Y(s)$ in the form of (5.6), with $a_i > 0$, $i = 0, 1$, $d_j > 0$, $j = 0, 1$, and $k > 0$, it is shown in [Smith (2002)] that $Y(s)$ is realizable as a five-element configuration as in Fig. 5.6, by using the Foster preamble (the elements can be fewer than five for some special cases). When $R_0(p, q, s) > 0$, the configuration is shown in Fig. 5.6(a); when $R_0(p, q, s) < 0$, the configuration is shown in Fig. 5.6(b). Since the canonical configurations for $R_0(p, q, s) = 0$ contain no more than three elements by the Foster preamble, which is fairly simple, they are not drawn. The configurations generally contain either two dampers, two springs, and one inerter, or two dampers and three springs.

In this section, the minimal realization problem of $Y(s)$ will be discussed under the following measure describing the complexity.

Definition 5.1. Given a set of indices to describe networks, if a network N_1 has at least one of the indices less than that of a network N_2, with its

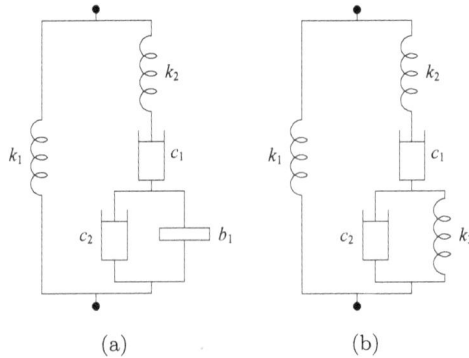

Fig. 5.6   The canonical realization configuration of $Y(s)$ in the form of (5.6). (a) for $R_0(p, q, s) > 0$; (b) for $R_0(p, q, s) < 0$.

other indices no greater, then the network N_1 is of strictly lower complexity than the network N_2.

Remark 5.1. Since this class of admittances is mainly used in mechanical systems, the following three indices are considered: (a) the total number of elements; (b) the number of dampers; (c) the number of inerters. Element values in this section are assumed to be positive and finite.

For the canonical realization configuration of $Y(s)$ in the form (5.6), the three indices are five elements, two dampers, and one inerter when $R_0(p, q, s) > 0$; five elements, two dampers, and zero inerter when $R_0(p, q, s) < 0$.

This section focuses on investigating the condition for an admittance $Y(s)$ in the form of (5.6) with $a_i > 0$, $i = 0, 1$, $d_j > 0$, $j = 0, 1$, $k > 0$, and $R_0(p, q, s) \neq 0$, to be realizable as a network of strictly lower complexity than the canonical configuration shown in Fig. 5.6. Since the complexity of realizations for $R_0(p, q, s) = 0$ is fairly low and it is only a special case, it is assumed that $R_0(p, q, s) \neq 0$.

5.3.1   Cases with Zero Coefficients

In this subsection, first consider the cases with zero coefficients since they are relatively simple.

Lemma 5.12. *Consider a function $Y(s)$ in the form (5.6), where $a_i > 0$, $i = 0, 1$, $d_j > 0$, $j = 0, 1$, and $k > 0$. Then, $Y(s)$ is positive-real with at* *least one of* a_0, a_1, d_0, *and* d_1 *being zero, if and only if at least one of the following two conditions holds:*

1. $d_0 = 0$ *and* $a_1 - d_1 \geq 0$;
2. $a_1 = 0$, $d_1 = 0$, *and* $a_0 - d_0 \geq 0$.

Proof. *Sufficiency.* It suffices to show that the three inequalities in Lemma 5.4 hold.

For Condition 1, $d_0 = 0$ makes the three inequalities be equivalent to $a_0 d_1 \geq 0$, $a_0 \geq 0$, and $a_1 - d_1 \geq 0$, which obviously hold because of $a_1 - d_1 \geq 0$ and the assumption that all the four coefficients be non-negative. For Condition 2, $a_1 = 0$ and $d_1 = 0$ make the three inequalities be equivalent to $a_0 - d_0 \geq 0$, which obviously holds.

Necessity. Suppose that $Y(s)$ is positive-real with at least one of the four coefficients a_0, a_1, d_0, and d_1 being zero.

If there is exactly one coefficient being zero, then it is only possible that $d_0 = 0$ to ensure the positive-realness of $Y(s)$ in Lemma 5.4. In this case, the positive-realness condition of $Y(s)$ in Lemma 5.4 becomes $a_1 - d_1 \geq 0$.

If there are exactly two coefficients being zero, then only the cases when (i) $a_0 = 0$ and $d_0 = 0$; (ii) $a_1 = 0$ and $d_1 = 0$; (iii) $d_0 = 0$ and $d_1 = 0$ are possible to guarantee the positive-realness of $Y(s)$ in Lemma 5.4. When $a_0 = 0$ and $d_0 = 0$, $Y(s)$ is positive-real if and only if $a_1 - d_1 \geq 0$. When $a_1 = 0$ and $d_1 = 0$, $Y(s)$ is positive-real if and only if $a_0 - d_0 \geq 0$. When $d_0 = 0$ and $d_1 = 0$, $Y(s)$ can always be positive-real, implying that $a_1 - d_1 \geq 0$ must hold.

If there are exactly three coefficients being zero, then only the cases when (1) $a_0 = 0$, $d_0 = 0$, and $d_1 = 0$; (2) $a_1 = 0$, $d_0 = 0$, and $d_1 = 0$ are possible to ensure the positive-realness of $Y(s)$ in Lemma 5.4. It is obvious that the two cases can always be positive-real, which implies $a_1 - d_1 \geq 0$.

If there are exactly four coefficients being zero, then $Y(s)$ is always positive-real, and $a_1 - d_1 \geq 0$ always holds.

Summarizing all the above discussions, the two cases in the theorem are verified. $\square$

Theorem 5.9. *Consider a positive-real function* $Y(s)$ *in the form* (5.6), *where* $a_i > 0$, $i = 0,1$, $d_j > 0$, $j = 0,1$, $k > 0$, *and* $R_0(p,q,s) \neq 0$. *If at least one of the four coefficients* a_0, a_1, d_0, *and* d_1 *is zero, then* $Y(s)$ *is realizable as the admittance of a one-port mechanical network containing no more than four elements. Furthermore, the realization is of strictly lower complexity than the canonical configuration.*

Proof. When Condition 1 of Lemma 5.12 holds, since $d_0 = 0$ and $a_1 - d_1 \geq 0$, $Y(s)$ can be written as

$$Y(s) = k\frac{a_0 s^2 + a_1 s + 1}{d_1 s^2 + s},$$

and $R_0(p, q, s) = a_0(a_0 + d_1^2 - a_1 d_1)$. If $R_0(p, q, s) > 0$, then $Y(s)$ can be written as

$$Y(s) = \frac{k}{s} + \left(\frac{d_1}{ka_0} + \left(\frac{ka_0^3 s}{R_0(p,q,s)} + \frac{ka_0^2(a_1 - d_1)}{R_0(p,q,s)}\right)^{-1}\right)^{-1}.$$

Therefore, $Y(s)$ is realizable as the admittance of a network consisting of one inerter, one spring, and no more than two dampers, which is of strictly lower complexity than the canonical configuration in Fig. 5.6(a). If $R_0(p, q, s) < 0$, then $Y(s)$ can be written as

$$Y(s) = \frac{k}{s} + \frac{ka_0}{d_1} + \left(\frac{a_0 d_1}{-kR_0(p,q,s)} + \frac{a_0 d_1^2 s}{-kR_0(p,q,s)}\right)^{-1}.$$

Therefore, $Y(s)$ is realizable as the admittance of a network consisting of two dampers and two springs, which is of strictly lower complexity than the canonical configuration in Fig. 5.6(b).

When Condition 2 of Lemma 5.12 holds, since $a_1 = 0$, $d_1 = 0$, and $a_0 - d_0 \geq 0$, $Y(s)$ can be written as

$$Y(s) = k\frac{a_0 s^2 + 1}{s(d_0 s^2 + 1)} = \frac{k}{s} + \frac{k(a_0 - d_0)s}{d_0 s^2 + 1}$$

and $R_0(p, q, s) = (a_0 - d_0)^2$. Since it is assumed that $R_k \neq 0$, it implies that $R_0(p, q, s) > 0$. $Y(s)$ is realizable as the admittance of a network consisting of one inerter and two springs, which is of strictly lower complexity than the canonical configuration in Fig. 5.6(a). □

In the remaining part of this section, to further discuss the realizability condition, it is assumed that $a_i > 0$, $i = 0, 1$, $d_j > 0$, $j = 0, 1$, $k > 0$, and $R_0(p, q, s) \neq 0$.

5.3.2 *Preliminary Lemmas*

Based on the principle of frequency-inverse duality, a function $Y(s)$ is realizable as the admittance of a network whose one-terminal-pair labeled graph is $\mathcal{N}$ (denoted as $N \in \mathcal{N}$) if and only if $Y^{-1}(s^{-1})$ is realizable as the admittance of a network whose one-terminal-pair labeled graph is $\mathrm{GDu}(\mathcal{N})$ (denoted as $N^{id} \in \mathrm{GDu}(\mathcal{N})$).

Consider a positive-real function $Y(s)$ in the form (5.6), where $a_i > 0$, $i = 0, 1$, $d_j > 0$, $j = 0, 1$, and $k > 0$. It is noted that

$$Y^{-1}(s^{-1}) = \frac{d_0}{a_0 k} \cdot \frac{s^2/d_0 + d_1 s/d_0 + 1}{s(s^2/a_0 + a_1 s/a_0 + 1)},$$

which actually belongs to the same class of functions as (5.6). Therefore, the realization of $Y(s)$ in the form of (5.6) as the admittance of any network satisfying $N^{id} \in \mathrm{GDu}(\mathcal{N})$ can be determined from that of $Y(s)$ as $N \in \mathcal{N}$ through the transformations $a_0 \to 1/d_0$, $a_1 \to d_1/d_0$, $d_0 \to 1/a_0$, $d_1 \to a_1/a_0$, and $k \to d_0/(a_0 k)$.

Lemma 5.13. *A positive-real function $Y(s)$ in the form of (5.6), with $a_i > 0$, $i = 0, 1$, $d_j > 0$, $j = 0, 1$, $k > 0$, and $R_0(p, q, s) \neq 0$, cannot be realized as the admittance of a one-port spring-inerter network.*

Proof. Assume that $Y(s)$ is realizable as the admittance of a spring-inerter network (lossless network). Then, according to [Baher (1984), Chapter 3] the even part of $Y^{-1}(s)$ is equal to zero, that is, Ev $(Y^{-1}(s)) = 0$. Therefore, it follows that

$$\text{Ev } (Y^{-1}(s)) = \frac{1}{2} \left(Y^{-1}(s) + Y^{-1}(-s) \right)$$
$$= \frac{2s^2((a_0 d_1 - a_1 d_0)s^2 + (d_1 - a_1))}{k(a_0 s^2 + a_1 s + 1)(a_0 s^2 - a_1 s + 1)} = 0,$$

which means that $2s^2((a_0 d_1 - a_1 d_0)s^2 + (d_1 - a_1)) = 0$ holds for all s. Therefore, one obtains $a_1 - d_1 = 0$ and $a_0 d_1 - a_1 d_0 = 0$, which contradicts the assumption that $R_0(p, q, s) := (a_0 - d_0)^2 - (a_0 d_1 - a_1 d_0)(a_1 - d_1) \neq 0$. $\square$

Lemma 5.14. *The network graph of a network N with two terminals, a and a', realizing an admittance $Y(s)$ in the form of (5.6) with $a_i > 0$, $i = 0, 1$, $d_j > 0$, $j = 0, 1$, and $k > 0$, must contain a path $\mathcal{P}(a, a')$ and a cut-set $\mathcal{C}(a, a')$ whose edges correspond to only springs.*

Proof. It is obvious that $Z(s) = Y^{-1}(s)$ has a pole at $s = \infty$ and a zero at $s = 0$. By [Seshu (1959), Theorem 2], the lemma can be easily proved. $\square$

Lemma 5.15. *A positive-real function $Y(s)$ in the form of (5.6), with $a_i > 0$, $i = 0, 1$, $d_j > 0$, $j = 0, 1$, and $k > 0$, cannot be realized as the admittance of any class of networks in Fig. 5.7.*

Proof. For any network in Fig. 5.7(b), its admittance must contain a pole at $s = j\omega_0$ with $\omega_0 > 0$, which contradicts the fact that all the coefficients are positive. Similarly, one can prove the case of Fig. 5.7(a). $\square$

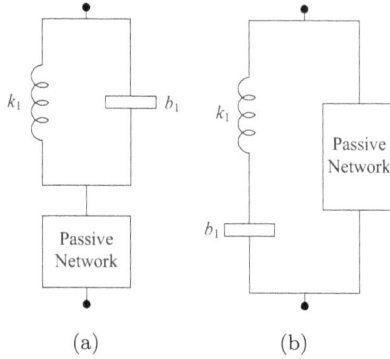

Fig. 5.7   The class of networks discussed in Lemma 5.15.

5.3.3   *Realizations with No More than Four Elements*

Lemma 5.16. *A positive-real function $Y(s)$ in the form of (5.6), with $a_i > 0$, $i = 0, 1$, $d_j > 0$, $j = 0, 1$, $k > 0$, and $R_0(p, q, s) \neq 0$, cannot be realized as the admittance of a one-port mechanical network containing no more than three elements.*

Proof. Assume that the realization is possible. By the method of enumeration, the network graph of any one-port network containing no more than three elements or its dual is shown in Fig. 5.8. It follows from Lemmas 3.7 and 5.13–5.15 that the realization configuration can be either a spring or a three-element configuration as shown in Fig. 5.9, which implies that $R_0(p, q, s) = 0$. Consequently, the lemma can be proved by contradiction. $\square$

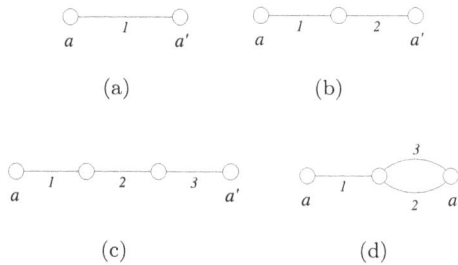

Fig. 5.8   One half of the network graphs of the one-port networks containing no more than three elements.

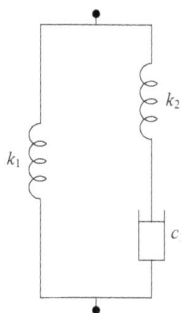

Fig. 5.9   The three-element configuration corresponding to the case when $R_0(p, q, s) \neq 0$.

Lemma 5.17. *A positive-real function $Y(s)$ in the form of (5.6), with $a_i > 0$, $i = 0, 1$, $d_j > 0$, $j = 0, 1$, $k > 0$, and $R_0(p, q, s) \neq 0$, cannot be realized as the admittance of the configuration in Fig. 5.10.*

Proof. This lemma can be proved by Lemmas 3.7 and 5.16.   □

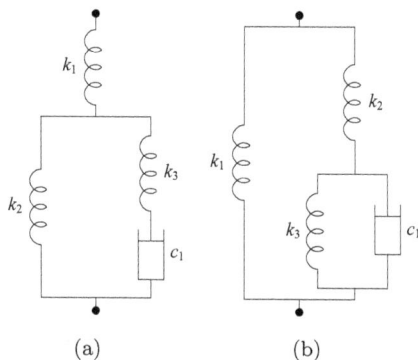

(a)   (b)

Fig. 5.10   The four-element configurations discussed in Lemma 5.17.

Theorem 5.10. *A positive-real function $Y(s)$ in the form of (5.6), with $a_i > 0$, $i = 0, 1$, $d_j > 0$, $j = 0, 1$, $k > 0$, and $R_0(p, q, s) \neq 0$, is realizable as the admittance of a one-port mechanical network consisting of no more than four elements, if and only if $Y(s)$ is realizable as the admittance of at least one of the configurations in Fig. 5.11.*

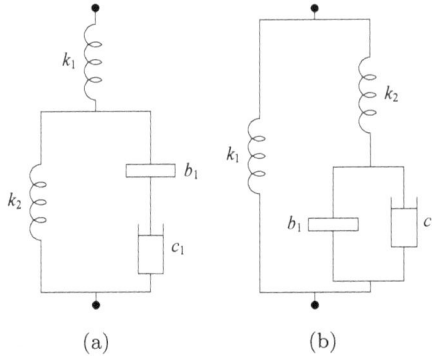

Fig. 5.11   The four-element configurations realizing $Y(s)$ in the form of (5.6), with $a_i > 0$, $i = 0, 1$, $d_j > 0$, $j = 0, 1$, $k > 0$, and $R_0(p, q, s) \neq 0$. The one-terminal-pair labeled graph of the configuration in (b) is the graph dual to that of the configuration in (a).

Proof. *Sufficiency.* The admittance of the configuration in Fig. 5.11(a) can be obtained as

$$Y(s) = \frac{b_1 k_2^{-1} s^2 + b_1 c_1^{-1} s + 1}{b_1 k_1^{-1} k_2^{-1} s^3 + b_1 c_1^{-1} (k_1^{-1} + k_2^{-1}) s^2 + (k_1^{-1} + k_2^{-1}) s},$$

which can be expressed as (5.6), where $a_0 = b_1 k_2^{-1} > 0$, $a_1 = b_1 c_1^{-1} > 0$, $d_0 = b_1 k_1^{-1} k_2^{-1} (k_1^{-1} + k_2^{-1})^{-1} > 0$, $d_1 = b_1 c_1^{-1} > 0$, and $k = (k_1^{-1} + k_2^{-1})^{-1} > 0$. It follows that $R_0(p, q) = b_1^2 k_2^{-4} (k_1^{-1} + k_2^{-1})^{-2} \neq 0$.

Since the one-terminal-pair labeled graph of the configuration in Fig. 5.11(b) is the dual to that of the configuration in Fig. 5.11(a), it directly implies that the configuration in Fig. 5.11(b) can also realize a positive-real admittance function $Y(s)$ in the form of (5.6), with $a_i > 0$, $i = 0, 1$, $d_j > 0$, $j = 0, 1$, $k > 0$, and $R_0(p, q, s) \neq 0$.

Necessity. Suppose that $Y(s)$ in the form of (5.6), with $a_i > 0$, $i = 0, 1$, $d_j > 0$, $j = 0, 1$, $k > 0$, and $R_0(p, q, s) \neq 0$, is realizable with no more than four elements. By Lemma 5.16, it suffices to consider the four-element configuration that may not always be equivalent to one containing fewer elements.

The method of enumeration is utilized for the proof. One half of the network graphs of the possible realizations are listed in Fig. 5.12, and other possible graphs are dual with them.

By Lemma 5.14, there must be a path $\mathcal{P}(a, a')$ and a cut-set $\mathcal{C}(a, a')$ consisting of only springs for possible realizations. It is noted that any network whose network graph is one of Figs. 5.12(a), 5.12(b), and 5.12(e), satisfying the above constraint, can always be equivalent to one containing no more than three elements. Therefore, these three network graphs can be eliminated.

For Fig. 5.12(c), Edge 1 and Edge 4 must correspond to springs by Lemma 5.14. Together with Lemma 5.15, either Edge 2 or Edge 3 must correspond to a damper. If the other edge correspond to a spring, then one obtains the configuration in Fig. 5.10(a), which cannot realize $Y(s)$ by Lemma 5.17. Therefore, only the configuration in Fig. 5.11(a) is possible for the network graph in Fig. 5.12(c). Similarly, the only configuration realizing $Y(s)$ is Fig. 5.12(d), which can be equivalent to the one in Fig. 5.11(b) by Lemma 3.7. Finally, any possible realization of $Y(s)$, whose network graph is the dual to a graph in Fig. 5.12, can be equivalent to one of the configurations in Fig. 5.11, based on the principle of frequency-inverse duality.   □

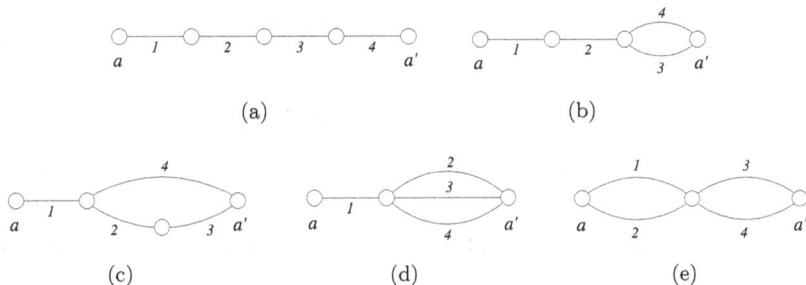

Fig. 5.12   One half of the network graphs of the four-element realizations of $Y(s)$ in the form of (5.6), with $a_i > 0$, $i = 0, 1$ $d_j > 0$, $j = 0, 1$, $k > 0$, and $R_0(p, q, s) \neq 0$.

Theorem 5.11. *A positive-real function $Y(s)$ in the form of (5.6), with $a_i > 0$, $i = 0, 1$, $d_j > 0$, $j = 0, 1$, $k > 0$, and $R_0(p, q, s) \neq 0$, is realizable as the admittance of the configuration in Fig. 5.11(a), if and only if*

$$a_0 - d_0 > 0, \qquad (5.27)$$
$$a_1 - d_1 = 0. \qquad (5.28)$$

Moreover, if the condition is satisfied, then the values of elements are expressed as

$$k_1 = \frac{ka_0}{d_0}, \tag{5.29}$$

$$k_2 = \frac{ka_0}{a_0 - d_0}, \tag{5.30}$$

$$b_1 = \frac{ka_0^2}{a_0 - d_0}, \tag{5.31}$$

$$c_1 = \frac{ka_0^2}{a_1(a_0 - d_0)}. \tag{5.32}$$

Proof. *Necessity.* The admittance of the configuration in Fig. 5.11(a) can be obtained as

$$Y(s) = \frac{b_1 k_2^{-1} s^2 + b_1 c_1^{-1} s + 1}{b_1 k_1^{-1} k_2^{-1} s^3 + b_1 c_1^{-1} (k_1^{-1} + k_2^{-1}) s^2 + (k_1^{-1} + k_2^{-1}) s}.$$

Since $Y(s)$ in the form of (5.6), with $a_i > 0$, $i = 0, 1$, $d_j > 0$, $j = 0, 1$, $k > 0$, and $R_0(p, q, s) \neq 0$, is realizable as the configuration in Fig. 5.11(a), it follows that

$$a_0 = b_1 k_2^{-1}, \tag{5.33}$$

$$a_1 = b_1 c_1^{-1}, \tag{5.34}$$

$$d_0 = b_1 k_1^{-1} k_2^{-1} (k_1^{-1} + k_2^{-1})^{-1}, \tag{5.35}$$

$$d_1 = b_1 c_1^{-1}, \tag{5.36}$$

$$k = (k_1^{-1} + k_2^{-1})^{-1}. \tag{5.37}$$

From (5.34) and (5.36), it is obvious that (5.28) holds. From (5.33) and (5.35), one obtains

$$\frac{a_0}{d_0} = \frac{k_1^{-1} + k_2^{-1}}{k_1^{-1}},$$

from which, together with (5.37), one can express k_1 as (5.29). Substituting k_1 into (5.37), k_1 is obtained in the form of (5.30), from which it implies (5.27). Substituting k_2 into (5.33), one obtains (5.31). Finally, from (5.34), one obtains (5.32).

Sufficiency. Consider a positive-real function $Y(s)$ in the form of (5.6), where $a_i > 0$, $i = 0, 1$, $d_j > 0$, $j = 0, 1$, $k > 0$, $R_0(p, q, s) \neq 0$, and Conditions (5.27) and (5.28) hold. Calculate the element values by (5.29)–(5.32). Since (5.28) holds, (5.33)–(5.37) must hold. Therefore, the admittance of the configuration in Fig. 5.11(a) is equivalent to (5.6), proving the sufficiency. □

Based on the principle of frequency-inverse duality, through the transformations $a_0 \rightarrow 1/d_0$, $a_1 \rightarrow d_1/d_0$, $d_0 \rightarrow 1/a_0$, $d_1 \rightarrow a_1/a_0$, and $k \rightarrow d_0/(a_0 k)$, one obtains the realizability condition of Fig. 5.11(b), which is $a_0 - d_0 > 0$ and $a_0 d_1 - a_1 d_0 = 0$.

In summary, the following conclusion is reached.

Theorem 5.12. *A positive-real function $Y(s)$ in the form of (5.6), with $a_i > 0$, $i = 0, 1$, $d_j > 0$, $j = 0, 1$, $k > 0$, and $R_0(p, q, s) \neq 0$, is realizable as the admittance of a one-port mechanical network consisting of no more than four elements, if and only if $R_0(p, q, s) > 0$ and $(a_0 d_1 - a_1 d_0)(a_1 - d_1) = 0$. Moreover, only the configurations in Fig. 5.11 are needed to fulfill the condition.*

Proof. *Sufficiency.* It is noted that the condition of this theorem is equivalent to $a_0 - d_0 > 0$, and either $a_1 - d_1 = 0$ or $a_0 d_1 - a_1 d_0 = 0$. When $a_0 - d_0 > 0$ and $a_1 - d_1 = 0$, $Y(s)$ is realizable as the configuration in Fig. 5.11(a) by Theorem 5.11. Based on the principle of frequency-inverse duality, the case when $a_0 - d_0 > 0$ and $a_0 d_1 - a_1 d_0 = 0$ can be proved.

Necessity. By Theorem 5.10, if a given positive-real function $Y(s)$ in the form of (5.6), with $a_i > 0$, $i = 0, 1$, $d_j > 0$, $j = 0, 1$, $k > 0$, and $R_0(p, q, s)$, is realizable as the admittance of a one-port mechanical network consisting of no more than four elements, then $Y(s)$ is realizable as the configuration in Fig. 5.11. By Theorem 5.11 and the principle of frequency-inverse duality, the condition of the theorem is satisfied. □

Corollary 5.1. *If a positive-real function $Y(s)$ in the form of (5.6), with $a_i > 0$, $i = 0, 1$, $d_j > 0$, $j = 0, 1$, $k > 0$, and $R_0(p, q, s) \neq 0$, is realizable as the admittance of a one-port mechanical network consisting of no more than four elements, then $Y(s)$ is realizable as the admittance of a four-element network of strictly lower complexity than its canonical realization configuration in Fig. 5.6.*

Proof. This corollary directly follows from Theorem 5.12. □

5.3.4 *Realizations of Five-Element Damper-Spring Networks*

In the previous subsection, the realization problem of the networks consisting of no more than four elements, which are of strictly lower complexity than the canonical configuration shown in Fig. 5.6, has been discussed. In order to further solve the problem formulated in this section, it is necessary to consider the irreducible five-element networks since the networks with more elements cannot satisfy the requirement.

It is known from Lemma 5.13 that any positive-real function $Y(s)$ in the form of (5.6), with $a_i > 0$, $i = 0, 1$, $d_j > 0$, $j = 0, 1$, $k > 0$, and $R_0(p, q, s) \neq 0$, cannot be realized as the admittance of a spring-inerter (lossless) network. To discuss the realizations consisting of two kinds of elements, it suffices to investigate the damper-spring (RL) networks, since springs must be contained according to Lemma 5.14. Considering the strictly lower complexity, there are no more than two dampers. Therefore, only the networks containing one damper and a finite number of springs or the networks containing two dampers and a finite number of springs are possible.

As discussed in [Smith (2002)], if $Y(s)$ in the form of (5.6) is realizable as the admittance of the former kind of networks, then $R_0(p, q, s) = 0$. This means that it can be eliminated. The next theorem gives the realizability condition for the latter case.

Theorem 5.13. *A positive-real function $Y(s)$ in the form of (5.6), with $a_i > 0$, $i = 0, 1$, $d_j > 0$, $j = 0, 1$, $k > 0$, and $R_0(p, q, s) \neq 0$, is realizable as the admittance of a damper-spring network consisting of two dampers and a finite number of springs, if and only if $R_0(p, q, s) < 0$. Moreover, if the condition is satisfied, then $Y(s)$ is realizable as the configuration in Fig. 5.13 with $k_1 = k$, $k_2 = k(A - B)(B - C)/(B(B - D))$, $k_3 = k(A - D)(C - D)/(D(B - D))$, $c_1 = k(A - B)(B - C)/(B - D)$, and $c_2 = k(A - D)(C - D)/(B - D)$, where $A, C = (a_1 \pm \sqrt{a_1^2 - 4a_0})/2$ and $B, D = (d_1 \pm \sqrt{d_1^2 - 4d_0})/2$.*

Proof. *Necessity.* Since $R_0(p, q, s) \neq 0$, the McMillan degree of $Y(s)$ is three. By [Smith (2002), Theorem 4], the coefficients must satisfy $R_k < 0$.

Sufficiency. Since $R_0(p, q, s) := (a_0 - d_0)^2 - (a_0 d_1 - a_1 d_0)(a_1 - d_1) < 0$, one obtains $a_0 d_1 - a_1 d_0 > 0$.

In addition, it is known from [Smith (2002), Theorem 4] that $Y(s)$ is of McMillan degree three. From [Smith (2002); Guillemin (1957)], $Y(s)$ can be expressed as

$$Y(s) = k \frac{(As + 1)(Cs + 1)}{s(Bs + 1)(Ds + 1)},$$

where $A > B > C > D > 0$ and $k > 0$. Therefore, it is the admittance of a network with two dampers and a finite number of springs. Furthermore,

$Y(s)$ can be expressed as

$$Y(s) = \frac{k}{s} + \frac{k(A-B)(B-C)}{(B-D)(Bs+1)} + \frac{k(A-D)(C-D)}{(B-D)(Ds+1)}.$$

Therefore, $Y(s)$ is realizable as the admittance of the configuration in Fig. 5.13, with the element values satisfying $k_1 = k$, $k_2 = k(A-B)(B-C)/(B(B-D))$, $k_3 = k(A-D)(C-D)/(D(B-D))$, $c_1 = k(A-B)(B-C)/(B-D)$, and $c_2 = k(A-D)(C-D)/(B-D)$, which are positive and finite. Since $a_0 = AC$, $a_1 = A+C$, $d_0 = BD$, and $d_1 = B+D$, by solving them one obtains the expressions of A, B, C, and D, as stated in the theorem.   □

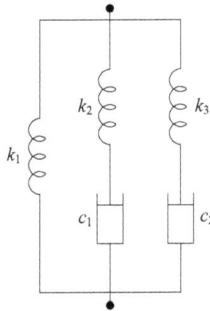

Fig. 5.13   The network that can fulfill the realizability conditions of Theorem 5.13.

Now, the next corollary follows easily.

Corollary 5.2. *If a positive-real function $Y(s)$ in the form of (5.6), with $a_i > 0$, $i = 0, 1$, $d_j > 0$, $j = 0, 1$, $k > 0$, and $R_0(p, q, s) \neq 0$, is realizable as the admittance of a damper-spring network, then $Y(s)$ is realizable as the admittance of a damper-spring network of strictly lower complexity than its canonical realization configuration in Fig. 5.6.*

Proof. As discussed above, it is only possible for the damper-spring network to contain one or two dampers, and the coefficients satisfy $R_0(p, q, s) = 0$, provided that the number of dampers is one.

Now, it suffices to discuss the network consisting of two dampers and a finite number of springs. According to Theorem 5.13, one has $R_0(p, q, s) < 0$, which does not satisfy the condition of Theorem 5.12. This means that the realization network can always be equivalent to the two-damper three-spring case, and the canonical realization must be the one in Fig. 5.6(b).   □

5.3.5 *Realizations of Five-Element Damper-Spring-Inerter Networks*

This subsection discusses the realization problem as a one-port mechanical network consisting of one damper, one inerter, and three springs, in order to guarantee a five-element damper-spring-inerter network of strictly lower complexity than the canonical configuration in Fig. 5.6.

Lemma 5.18. *A positive-real function $Y(s)$ in the form of (5.6), with $a_i > 0$, $i = 0, 1$, $d_j > 0$, $j = 0, 1$, $k > 0$, and $R_0(p, q, s) \neq 0$, cannot be realized as the admittance of the configurations in Fig. 5.14.*

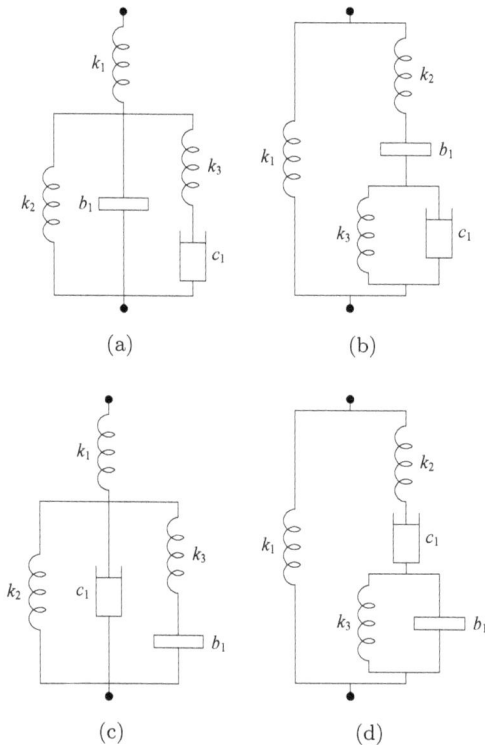

Fig. 5.14   The series-parallel configurations discussed in Lemma 5.18.

Proof. The admittance of the configuration in Fig. 5.14(a) is obtained as
$$Y(s) = (b_1 k_2^{-1} k_3^{-1} s^3 + b_1 c_1^{-1} k_2^{-1} s^2 + (k_2^{-1} + k_3^{-1}) s + c_1^{-1}) / (b_1 k_1^{-1} k_2^{-1} k_3^{-1} s^4 +$$

$b_1 c_1^{-1} k_1^{-1} k_2^{-1} s^3 + (k_1^{-1} k_2^{-1} + k_2^{-1} k_3^{-1} + k_1^{-1} k_3^{-1}) s^2 + c_1^{-1} (k_1^{-1} + k_2^{-1}) s)$. Since $R_0(p, q, s) \neq 0$, $Y(s)$ is realizable as the configuration in Fig. 5.14(a), if and only if there exists $T > 0$ such that the following equations hold:

$$a_0 T = b_1 c_1 k_2^{-1} k_3^{-1}, \quad a_0 + a_1 T = b_1 k_2^{-1}, \quad a_1 + T = c_1 (k_2^{-1} + k_3^{-1}), \quad (5.38)$$

$$d_0 T = \frac{b_1 c_1 k_1^{-1} k_2^{-1} k_3^{-1}}{k_1^{-1} + k_2^{-1}}, \quad d_0 + d_1 T = \frac{b_1 k_1^{-1} k_2^{-1}}{k_1^{-1} + k_2^{-1}}, \quad (5.39)$$

$$d_1 + T = \frac{k_1^{-1} k_2^{-1} + k_2^{-1} k_3^{-1} + k_1^{-1} k_3^{-1}}{c_1^{-1} (k_1^{-1} + k_2^{-1})}, \quad k = \frac{1}{k_1^{-1} + k_2^{-1}}. \quad (5.40)$$

To prove the lemma, it is necessary to show that $T > 0$ does not exist. After a series of calculations, it is verified that (5.38)–(5.40) are equivalent to

$$k_1 = \frac{k a_0}{d_0}, \quad k_2 = \frac{k a_1 (T^2 + a_1 T + a_0)}{(a_0 - d_0) T}, \quad k_3 = \frac{k a_0}{a_0 - d_0}, \quad (5.41)$$

$$b_1 = \frac{k a_0 (a_0 + a_1 T)}{a_0 - d_0} = \frac{k a_0^2 (d_0 + d_1 T)}{d_0 (a_0 - d_0)}, \quad (5.42)$$

$$
\begin{aligned}
c_1 &= \frac{k a_0 a_1 (T^2 + a_1 T + a_0)}{(a_0 + a_1 T)(a_0 - d_0)} \\
&= \frac{k a_0^2 a_1 (d_1 + T)(T^2 + a_1 T + a_0)}{(a_0 - d_0)(a_1 d_0 T^2 + (a_0^2 + a_1^2 d_0) T + a_0 a_1 d_0)}.
\end{aligned}
\quad (5.43)
$$

From (5.42), one obtains $a_0 - d_0 > 0$ and $a_1 d_0 - a_0 d_1 = 0$. The second equality of (5.43) is equivalent to $a_1 (a_0 - d_0) T^2 = 0$, which obviously cannot hold for any $T > 0$. Therefore, there does not exist any $T > 0$ such that (5.38)–(5.40) hold simultaneously.

Using a similar argument, the conclusion of the configuration in Fig. 5.14(c) can also be proved. Furthermore, since the one-terminal-pair labeled graph of the configuration in Fig. 5.14(b) (resp., Fig. 5.14(d)) is the dual to the configuration in Fig. 5.14(a) (resp., Fig. 5.14(c)), one can also prove the cases of Figs. 5.14(b) and 5.14(c). □

Lemma 5.19. *Consider a positive-real function $Y(s)$ in the form of (5.6), where $a_i > 0$, $i = 0, 1$, $d_j > 0$, $j = 0, 1$, $k > 0$, and $R_0(p, q, s) \neq 0$. If the condition of Theorem 5.12 does not hold, then $Y(s)$ cannot be realized as the admittance of any series-parallel network consisting of one damper, one inerter, and three springs.*

Proof. By the principle of frequency-inverse duality, it only needs to discuss one half of all the possible network graphs presented in Fig. 5.15, where other graphs are dual to them.

Since the condition of Theorem 5.12 does not hold, any five-element realization network of $Y(s)$ cannot be equivalent to one containing fewer elements. This means that there do not exist two elements of the same kind in series or in parallel. Therefore, the graphs in Figs. 5.15(a) and 5.15(e) are first eliminated.

The network graphs in Figs. 5.15(b), 5.15(c), and 5.15(f)–5.15(h) are then eliminated, since the existence of a path $\mathcal{P}(a, a')$ corresponding to only springs by Lemma 5.14 implies that at least two springs are in series for these graphs.

For Fig. 5.15(l), it can be derived that Edge 1, Edge 2, and Edge 4 (Edge 5) must correspond to springs due to the existence of a cut-set $\mathcal{C}(a, a')$ corresponding to only springs by Lemma 5.14. Furthermore, Edge 3 and Edge 5 (Edge 4) must correspond to an inerter and a damper. Otherwise, it must be of the structure shown in Fig. 5.7(b). However, by the equivalence in Lemma 3.7 (Fig. 3.2), the only possible configuration is still equivalent to the one shown in Fig. 5.7(b). Therefore, the graph in Fig. 5.15(l) is eliminated.

For Fig. 5.15(d), Edge 1, Edge 5, and one of Edges 2–4 must correspond to springs, which implies that the possible realization network can always be equivalent to a four-element network by the equivalence in Lemma 3.7 (Fig. 3.2).

For Fig. 5.15(i), by Lemma 5.14 it implies that Edge 1, Edge 2 (or Edge 3), and Edge 4 (or Edge 5) must correspond to springs, and Edge 3 (or Edge 2) and Edge 5 (or Edge 4) must correspond to a damper and an inerter, respectively. Therefore, the possible realization network must be of the structure in Fig. 5.7(a), which cannot realize $Y(s)$ by Lemma 5.15. The only possible realizations for Fig. 5.15(j) are configurations in Figs. 5.14(a) and 5.14(c).

Furthermore, it can be verified that all the possible networks whose network graphs are shown in Fig. 5.15(k) can be equivalent to four-element networks based on the equivalence in Lemma 3.7 (Fig. 3.2). As discussed in Lemma 5.18, the configurations in Fig. 5.14 cannot realize $Y(s)$.

Therefore, there is no series-parallel network satisfying the requirement.

$\square$

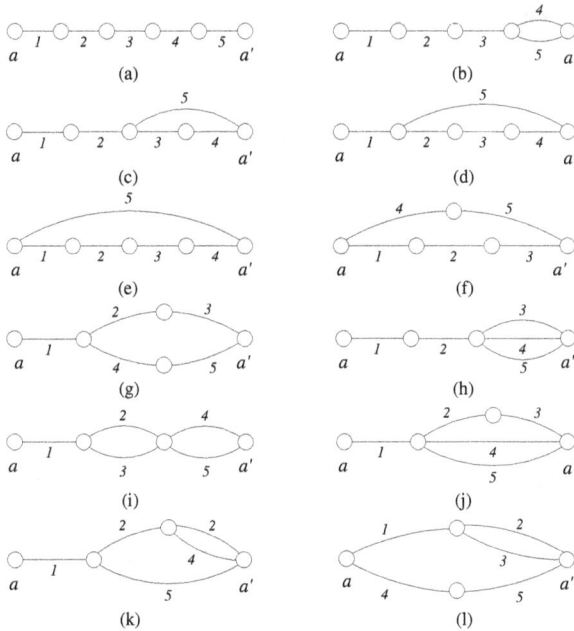

Fig. 5.15   One half of the network graphs for one-port five-element networks.

Finally, the bridge networks are considered, for which the following variables are defined: $W_1 := \alpha_1\alpha_2 - \alpha_3$, $W_2 := \alpha_2\beta_1 - \beta_3$, $W_3 := \alpha_1\beta_1 - \beta_2$, and $W := 2\alpha_1\alpha_2\beta_1 + \beta_4 - \alpha_1\beta_3 - \alpha_2\beta_2 - \alpha_3\beta_1$.

Lemma 5.20. *A positive-real function $Y(s)$ is realizable as the admittance of the configuration in Fig. 5.16 if and only if $Y(s)$ can be expressed as (5.7), that is,*

$$Y(s) = \frac{\alpha_3 s^3 + \alpha_2 s^2 + \alpha_1 s + 1}{\beta_4 s^4 + \beta_3 s^3 + \beta_2 s^2 + \beta_1 s},$$

where $\alpha_i > 0$, $i = 1, 2, 3$, $\beta_j > 0$, $j = 1, 2, 3, 4$, $W_k > 0$, $k = 1, 2, 3$, $W - 2\alpha_2 W_3 > 0$, and the following two equations hold:

$$W^2 - 4W_1 W_2 W_3 = 0, \tag{5.44}$$
$$\beta_4 + \alpha_1\beta_3 + \alpha_3\beta_1 - \alpha_2\beta_2 = 0. \tag{5.45}$$

Moreover, the element values are expressed as

$$k_1 = \frac{\alpha_1(\alpha_2\beta_1 - \beta_3)}{\alpha_3\beta_1^2}, \tag{5.46}$$

$$k_2 = \frac{\alpha_1(\alpha_2\beta_1 - \beta_3)}{(\alpha_1\alpha_2\beta_1 - \alpha_3\beta_1 - \alpha_1\beta_3)\beta_1}, \tag{5.47}$$

$$k_3 = \frac{\alpha_2\beta_1 - \beta_3}{\beta_1\beta_3}, \tag{5.48}$$

$$b_1 = \frac{\alpha_2\beta_1 - \beta_3}{\beta_1^2}, \tag{5.49}$$

$$c_1 = \frac{\alpha_1^2(\alpha_2\beta_1 - \beta_3)}{(\alpha_1\alpha_2 - \alpha_3)\beta_1^2}. \tag{5.50}$$

Proof. *Necessity.* First, the admittance of the configuration in Fig. 5.16 can be expressed as (5.7), with

$$\alpha_3 = b_1c_1k_1^{-1}(k_2^{-1} + k_3^{-1}), \alpha_2 = b_1(k_1^{-1} + k_2^{-1} + k_3^{-1}), \alpha_1 = c_1(k_2^{-1} + k_3^{-1}), \tag{5.51}$$

$$\beta_4 = b_1c_1k_1^{-1}k_2^{-1}k_3^{-1}, \quad \beta_3 = b_1k_3^{-1}(k_1^{-1} + k_2^{-1}), \tag{5.52}$$

$$\beta_2 = c_1(k_1^{-1}k_2^{-1} + k_2^{-1}k_3^{-1} + k_1^{-1}k_3^{-1}), \quad \beta_1 = k_1^{-1} + k_2^{-1}. \tag{5.53}$$

It is clear that $\alpha_i > 0$, $i = 1, 2, 3$ and $\beta_j > 0$, $j = 1, 2, 3, 4$. After a series of calculations, the above equations are verified to be equivalent to (5.50)–(5.49) and the following two equations:

$$\alpha_1\alpha_2^2\beta_1\beta_4 + \alpha_3\beta_3\beta_4 + \alpha_3^2\beta_1\beta_3 + \alpha_1\alpha_3\beta_3^2$$
$$= \alpha_2\alpha_3\beta_1\beta_4 + \alpha_1\alpha_2\beta_3\beta_4 + \alpha_1\alpha_2\alpha_3\beta_1\beta_3, \tag{5.54}$$

$$\alpha_1\alpha_2^2\beta_1\beta_2 + \alpha_3\beta_2\beta_3 + \alpha_3^2\beta_1^2 + \alpha_1\alpha_3\beta_1\beta_3 + \alpha_1^2\beta_3^2$$
$$= \alpha_2\alpha_3\beta_1\beta_2 + \alpha_1\alpha_2\beta_2\beta_3 + \alpha_1\alpha_2\alpha_3\beta_1^2 + \alpha_1^2\alpha_2\beta_1\beta_3. \tag{5.55}$$

Furthermore, (5.55) is equivalent to $\alpha_1\alpha_2\alpha_3\beta_1\beta_3 - \alpha_3^2\beta_1\beta_3 - \alpha_1\alpha_3\beta_3^2 = (\alpha_1\alpha_2^2\beta_1 + \alpha_3\beta_3 - \alpha_2\alpha_3\beta_1 - \alpha_1\alpha_2\beta_3)(\alpha_2\beta_2 - \alpha_1\beta_3 - \alpha_3\beta_1)$. Since all the values of the elements are positive and finite, one can calculate to obtain that $W_1 = \alpha_1\alpha_2 - \alpha_3 = b_1c_1(k_2^{-1} + k_3^{-1})^2 > 0$, $W_2 = \alpha_2\beta_1 - \beta_3 = b_1(k_1^{-1} + k_2^{-1})^2 > 0$, $W_3 = c_1k_2^{-2} > 0$, and $W - 2\alpha_2W_3 = 2b_1c_1k_1^{-1}k_2^{-1}k_3^{-1} > 0$. Since $W_1 > 0$ and $W_2 > 0$, one has

$$\alpha_1\alpha_2^2\beta_1 + \alpha_3\beta_3 - \alpha_2\alpha_3\beta_1 - \alpha_1\alpha_2\beta_3 = (\alpha_1\alpha_2 - \alpha_3)(\alpha_2\beta_1 - \beta_3)$$
$$= W_1W_2 > 0. \tag{5.56}$$

Therefore, it follows from (5.54) that

$$\beta_4 = \frac{\alpha_1\alpha_2\alpha_3\beta_1\beta_3 - \alpha_3^2\beta_1\beta_3 - \alpha_1\alpha_3\beta_3^2}{\alpha_1\alpha_2^2\beta_1 + \alpha_3\beta_3 - \alpha_2\alpha_3\beta_1 - \alpha_1\alpha_2\beta_3}$$

$$= \frac{(\alpha_1\alpha_2^2\beta_1 + \alpha_3\beta_3 - \alpha_2\alpha_3\beta_1 - \alpha_1\alpha_2\beta_3)(\alpha_2\beta_2 - \alpha_1\beta_3 - \alpha_3\beta_1)}{\alpha_1\alpha_2^2\beta_1 + \alpha_3\beta_3 - \alpha_2\alpha_3\beta_1 - \alpha_1\alpha_2\beta_3} \quad (5.57)$$

$$= \alpha_2\beta_2 - \alpha_1\beta_3 - \alpha_3\beta_1,$$

which implies (5.45). Substituting the above equation into W, one obtains $W = 2(\alpha_1\alpha_2\beta_1 - \alpha_1\beta_3 - \alpha_3\beta_1)$, implying

$$4(\alpha_1\alpha_3\beta_1\beta_3 - \alpha_1^2\alpha_2\beta_1\beta_3 - \alpha_1\alpha_2\alpha_3\beta_1^2 + \alpha_1^2\beta_3^2 + \alpha_3^2\beta_1^2 + \alpha_1\alpha_2^2\beta_1\beta_2$$

$$- \alpha_2\alpha_3\beta_1\beta_2 - \alpha_1\alpha_2\beta_2\beta_3 + \alpha_3\beta_2\beta_3)$$

$$= 4(\alpha_1\alpha_2\beta_1 - \alpha_3\beta_1 - \alpha_1\beta_3)^2 - 4(\alpha_1\alpha_2 - \alpha_3)(\alpha_2\beta_1 - \beta_3)(\alpha_1\beta_1 - \beta_2)$$

$$= W^2 - 4W_1W_2W_3.$$

$$(5.58)$$

Now, (5.55) and (5.58) yield (5.44).

Sufficiency. Let the element values satisfy (5.46)–(5.50). $W_1 > 0$ and $W_2 > 0$ indicates that $k_1 > 0$, $k_3 > 0$, $b_1 > 0$, and $c_1 > 0$. Then, substituting β_4 obtained from (5.45) into W, one obtains $W - 2\alpha_2W_3 = 2(\alpha_2\beta_2 - \alpha_1\beta_3 - \alpha_3\beta_1)$. Since $W - 2\alpha_2W_3 > 0$, it follows that $k_2 > 0$. Substituting β_4 obtained from (5.45) into (5.44), one obtains (5.55) immediately. Together with (5.56) and (5.57), one obtains (5.54). It is known from the necessity part that (5.51)–(5.53) must hold. □

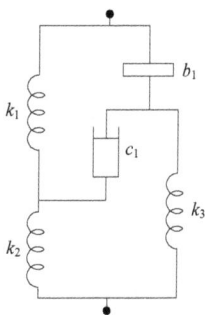

Fig. 5.16   The five-element bridge configuration, which can realize admittance (5.6).

The next lemma follows immediately.

Lemma 5.21. *A positive-real function $Y(s)$ in the form of* (5.6), *with $a_i > 0$, $i = 0, 1$, $d_j > 0$, $j = 0, 1$, $k > 0$, and $R_0(p, q, s) \neq 0$, is realizable as* the admittance of the configuration in Fig. 5.16, if and only if

$$(a_0 d_1 - a_1 d_0)(a_1 - d_1) - d_0^2 = 0. \tag{5.59}$$

Moreover, the element values are expressed as

$$k_1 = \frac{k(a_1 + T)\left((a_1 - d_1)T + (a_0 - d_0)\right)}{a_0 T}, \tag{5.60}$$

$$k_2 = \frac{k(a_1 + T)\left((a_1 - d_1)T + (a_0 - d_0)\right)}{(a_1 - d_1)T^2 + (a_1^2 - a_1 d_1 - d_0)T + a_1(a_0 - d_0)}, \tag{5.61}$$

$$k_3 = \frac{k\left((a_1 - d_1)T + (a_0 - d_0)\right)}{d_1 T + d_0}, \tag{5.62}$$

$$b_1 = k\left((a_1 - d_1)T + (a_0 - d_0)\right), \tag{5.63}$$

$$c_1 = \frac{k(a_1 + T)^2\left((a_1 - d_1)T + (a_0 - d_0)\right)}{a_1(T^2 + a_1 T + a_0)}, \tag{5.64}$$

where

$$T = \sqrt{\frac{a_0 d_1 - a_1 d_0}{a_1 - d_1}}. \tag{5.65}$$

Proof. *Necessity.* Suppose that a given positive-real function $Y(s)$ in the form of (5.6), with $a_i > 0$, $i = 0, 1$, $d_j > 0$, $j = 0, 1$, $k > 0$, and $R_0(p, q, s) \neq 0$, is realizable as the configuration in Fig. 5.16. Then, it follows from Lemma 5.20 that $Y(s)$ can be expressed as (5.7), with all the coefficients being positive and satisfying the condition of Lemma 5.20. Since $R_0(p, q, s) \neq 0$, the only possibility to express (5.6) as (5.7) with $\alpha_i > 0$, $i = 1, 2, 3$ and $\beta_j > 0$, $j = 1, 2, 3, 4$, is to multiply both the numerator and the denominator by $(Ts + 1)$ with $T > 0$. Consequently, it follows that

$$\alpha_3 = a_0 T, \ \alpha_2 = a_0 + a_1 T, \ \alpha_1 = a_1 + T,$$

$$\beta_4 = \frac{d_0 T}{k}, \ \beta_3 = \frac{d_0 + d_1 T}{k}, \ \beta_2 = \frac{d_1 + T}{k}, \ \beta_1 = \frac{1}{k}. \tag{5.66}$$

It is clear that $\alpha_i > 0$, $i = 1, 2, 3$ and $\beta_j > 0$, $j = 1, 2, 3, 4$. Furthermore, other conditions can be presented as follows:

$$W_1 = \alpha_1 \alpha_2 - \alpha_3 = a_1 T^2 + a_1^2 T + a_0 a_1 > 0, \tag{5.67}$$

$$W_2 = \alpha_2 \beta_1 - \beta_3 = \frac{(a_1 - d_1)T + (a_0 - d_0)}{k} > 0, \tag{5.68}$$

$$W_3 = \alpha_1 \beta_1 - \beta_2 = \frac{a_1 - d_1}{k} > 0, \tag{5.69}$$

$$W - 2\alpha_2 W_3 = \frac{(a_1 - d_1)T^2 + (a_0 d_1 - a_1 d_0)}{k} > 0, \tag{5.70}$$

$$W^2 - 4W_1 W_2 W_3 = \frac{\left((a_1 - d_1)T^2 - (a_0 d_1 - a_1 d_0)\right)^2}{k^2} = 0, \tag{5.71}$$

$$\beta_4 + \alpha_1\beta_3 + \alpha_3\beta_1 - \alpha_2\beta_2 = \frac{-(a_1 - d_1)T^2 + 2d_0T - (a_0d_1 - a_1d_0)}{k} = 0.$$

$$(5.72)$$

Thus, (5.69) implies $a_1 - d_1 > 0$. Then, T can be solved from (5.71) as (5.65). The constraint that $T > 0$ yields $a_0d_1 - a_1d_0 > 0$. Substituting the solved T into (5.72), one obtains (5.59).

Sufficiency. Suppose that $(a_0d_1 - a_1d_0)(a_1 - d_1) - d_0^2 = 0$ holds. Then, $d_0 > 0$ and the positive-realness of $Y(s)$ implies $a_0d_1 - a_1d_0 > 0$ and $a_1 - d_1 > 0$. Thus, one may choose

$$T = \sqrt{\frac{a_0d_1 - a_1d_0}{a_1 - d_1}} > 0.$$

Furthermore, multiplying both the numerator and the denominator of $Y(s)$ by the factor $(Ts + 1)$, one can express $Y(s)$ in the form of (5.7), with the coefficients satisfying (5.66).

It is clear that $\alpha_i > 0$, $i = 1, 2, 3$, $\beta_j > 0$, $j = 1, 2, 3, 4$, and (5.67)–(5.71) hold. Since $(a_0d_1 - a_1d_0)(a_1 - d_1) - d_0^2 = 0$, (5.72) is also satisfied. It is concluded that the condition of Lemma 5.20 holds, therefore $Y(s)$ is realizable as the configuration in Fig. 5.16.

The expressions of the element values presented in (5.60)–(5.64) are derived from (5.46)–(5.50) with the relation (5.66). □

Lemma 5.22. *A positive-real function $Y(s)$ in the form of (5.6), with $a_i > 0$, $i = 0, 1$, $d_j > 0$, $j = 0, 1$, $k > 0$, and $R_0(p, q, s) \neq 0$, cannot be realized as the admittance of the configurations in Fig. 5.17.*

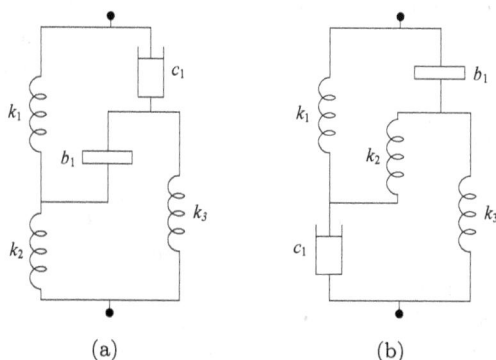

(a)　　　　　　(b)

Fig. 5.17　The five-element bridge configurations discussed in Lemma 5.22.

Proof. The admittance of the configuration in Fig. 5.17(a) can be expressed as (5.7), with $\alpha_3 = b_1 c_1 k_1^{-1}(k_2^{-1} + k_3^{-1})$, $\alpha_2 = b_1(k_2^{-1} + k_3^{-1})$, $\alpha_1 = c_1(k_1^{-1} + k_2^{-1} + k_3^{-1})$, $\beta_4 = b_1 c_1 k_1^{-1} k_2^{-1} k_3^{-1}$, $\beta_3 = b_1(k_1^{-1} k_2^{-1} + k_2^{-1} k_3^{-1} + k_1^{-1} k_3^{-1})$, $\beta_2 = c_1 k_3^{-1}(k_1^{-1} + k_2^{-1})$, and $\beta_1 = k_1^{-1} + k_2^{-1}$. It can be verified that the coefficients must satisfy $W - 2\alpha_2 W_3 = -2b_1 c_1 k_1^{-1}(k_1^{-1} + k_2^{-1})(k_2^{-1} + k_3^{-1}) < 0$. Similarly, the admittance of the configuration in Fig. 5.17(b) can also be expressed as (5.7), with the coefficients being positive and satisfying $W - 2\alpha_2 W_3 < 0$.

Assume that a given positive-real function $Y(s)$ in the form of (5.6), with $a_i > 0$, $i = 0, 1$, $d_j > 0$, $j = 0, 1$, $k > 0$, and $R_0(p, q, s) \neq 0$, is realizable as the configuration in Fig. 5.17(a) or Fig. 5.17(b). Then, $Y(s)$ can be expressed in the form of (5.7) with (5.66) being satisfied. As discussed in the proof of Lemma 5.21, one has $W - 2\alpha_2 W_3 = ((a_1 - d_1)T^2 + (a_0 d_1 - a_1 d_0))/k \geq 0$, which contradicts the hypothesis. ☐

Theorem 5.14. *Consider a positive-real function $Y(s)$ in the form of (5.6), where $a_i > 0$, $i = 0, 1$, $d_j > 0$, $j = 0, 1$, $k > 0$, $R_0(p, q, s) \neq 0$, and the condition of Theorem 5.12 does not hold. Then, $Y(s)$ is realizable as the admittance of a one-port mechanical network consisting of one damper, one inerter, and three springs, if and only if $Y(s)$ is realizable as the configuration in Fig. 5.16, that is, the coefficients satisfy $(a_0 d_1 - a_1 d_0)(a_1 - d_1) = d_0^2$.*

Proof. *Sufficiency.* The sufficiency obviously holds by Lemma 5.21.

Necessity. Since the condition of Theorem 5.12 does not hold, $Y(s)$ cannot be realized as a series-parallel network consisting of one damper, one inerter, and three springs by Lemma 5.19. By Lemma 5.14, the only possible bridge networks are shown in Figs. 5.16 and 5.17. Furthermore, by Lemmas 5.21 and 5.22, only the configuration in Fig. 5.16 can realize $Y(s)$ and the coefficients must satisfy $(a_0 d_1 - a_1 d_0)(a_1 - d_1) = d_0^2$. ☐

5.3.6 *Final Condition*

Theorem 5.15. *A positive-real function $Y(s)$ in the form of (5.6), where $a_i > 0$, $i = 0, 1$, $d_j > 0$, $j = 0, 1$, $k > 0$, $R_0(p, q, s) \neq 0$, is realizable as the admittance of a one-port mechanical network of strictly lower complexity than the canonical realization configuration in Fig. 5.6, if and only if $R_k > 0$ and $(a_1 - d_1)(a_0 d_1 - a_1 d_0)\left((a_1 - d_1)(a_0 d_1 - a_1 d_0) - d_0^2\right) = 0$.*

Proof. *Sufficiency.* The sufficiency can be proved by Theorem 5.12, Corollary 5.1, and Theorem 5.14.

Necessity. If $Y(s)$ is realizable with no more than four elements, then by Theorem 5.12 and Corollary 5.1, one has that $R_k > 0$ and $(a_0 d_1 - a_1 d_0)(a_1 - d_1) = 0$. If $Y(s)$ is realizable as an irreducible five-element network consisting of only two kinds of elements, then it could only be a damper-spring network. Then, by Corollary 5.2, any such a network cannot be of strictly lower complexity than the canonical configuration shown in Fig. 5.6.

For any five-element damper-spring-inerter network of strictly lower complexity than the corresponding canonical configuration, it could only contain one damper, one inerter, and three springs. By Theorem 5.14, it implies $R_k > 0$ and $(a_0 d_1 - a_1 d_0)(a_1 - d_1) = d_0^2$.

Combining all the discussions above, the necessity is proved.   □

5.4 Synthesis of a One-Damper One-Inerter Network Containing No More than Three Springs

This section investigates the realization problem of any positive-real function as the admittance of a one-port mechanical network consisting of one damper, one inerter, and no more than three springs, in order to further reduce the number of springs achieved in [Chen and Smith (2009b)].

Considering the limitation of space and weight for passive mechanical systems, it is essential to further reduce the number of springs if ever possible. Therefore, combining the results in this section, passive inerter-based mechanical control can become more effective for practical applications. In fact, the results contribute to the development of minimal realizations in general.

5.4.1 *Realizability Conditions under a Particular Assumption*

Consider a one-port mechanical network consisting of one damper, one inerter, and no more than three springs in Fig. 5.1, where X is a three-port network consisting of no more than three springs. In this subsection, it is to solve the synthesis problem under the assumption that the admittance of X exists. This assumption will be removed in the next subsection.

Suppose that the admittance of X exists. The Laplace transformed forces $\hat{F}$ and velocities $\hat{v}$ for the ports of the network X satisfy the following expression:

$$\begin{bmatrix} \hat{F}_1 \\ \hat{F}_2 \\ \hat{F}_3 \end{bmatrix} = \frac{1}{s} \begin{bmatrix} K_{11} & K_{12} & K_{13} \\ K_{12} & K_{22} & K_{23} \\ K_{13} & K_{23} & K_{33} \end{bmatrix} \begin{bmatrix} \hat{v}_1 \\ \hat{v}_2 \\ \hat{v}_3 \end{bmatrix} =: \frac{1}{s} K \begin{bmatrix} \hat{v}_1 \\ \hat{v}_2 \\ \hat{v}_3 \end{bmatrix}, \qquad (5.73)$$

where $(1/s)K$ is the admittance of X and $K \in \mathbb{S}^3$ is necessarily non-negative definite [Newcomb (1966)] due to the passivity of X. Together with the terminal constraints $\hat{F}_2 = -c\hat{v}_2$ and $\hat{F}_3 = -bs\hat{v}_3$, one obtains the admittance of the one-port network as

$$Y(s) = \frac{\hat{F}_1}{\hat{v}_1} = \frac{\alpha_3 s^3 + \alpha_2 s^2 + \alpha_1 s + \alpha_0}{s^4 + \beta_3 s^3 + \beta_2 s^2 + \beta_1 s}, \qquad (5.74)$$

where $\alpha_3 = K_{11}$, $\alpha_2 = (1/c)(K_{11}K_{22} - K_{12}^2)$, $\alpha_1 = (1/b)(K_{11}K_{33} - K_{13}^2)$, $\alpha_0 = (1/(bc))\det(K)$, $\beta_3 = (1/c)K_{22}$, $\beta_2 = (1/b)K_{33}$, $\beta_1 = (1/(bc))(K_{22}K_{33} - K_{23}^2)$, and $b, c > 0$.

Furthermore, according to the analogy to one-element-kind networks, it is obvious that $(1/s)K$ is realizable as the admittance of a network consisting of at most three springs if and only if K is the admittance of a three-port resistive network containing no more than three elements. Therefore, the following conclusion can be directly obtained from Theorem 4.11.

Theorem 5.16. *An admittance $(1/s)K$ with $K \in \mathbb{S}^3$ in the form of (5.73) is realizable as a three-port network consisting of no more than three springs, if and only if one of the following two conditions holds:*

1. *$K_{12}K_{13}K_{23} \leq 0$, $K_{11} - |K_{12}| - |K_{13}| \geq 0$, $K_{22} - |K_{12}| - |K_{23}| \geq 0$, $K_{33} - |K_{13}| - |K_{23}| \geq 0$, and at least three of K_{12}, K_{13}, K_{23}, $(K_{11} - |K_{12}| - |K_{13}|)$, $(K_{22} - |K_{12}| - |K_{23}|)$, and $(K_{33} - |K_{13}| - |K_{23}|)$ are zero.*

2. *$K_{12}K_{13}K_{23} \geq 0$ and at least one of the following three conditions holds with at least three of the six inequality signs being equality: a) $-|K_{13}| \leq 0$, $|K_{13}| \leq |K_{12}| \leq K_{11}$, $|K_{13}| \leq |K_{23}| \leq K_{33}$, and $|K_{12}| + |K_{23}| - |K_{13}| \leq K_{22}$; b) $-|K_{12}| \leq 0$, $|K_{12}| \leq |K_{13}| \leq K_{11}$, $|K_{12}| \leq |K_{23}| \leq K_{22}$, and $|K_{13}| + |K_{23}| - |K_{12}| \leq K_{33}$; c) $-|K_{23}| \leq 0$, $|K_{23}| \leq |K_{12}| \leq K_{22}$, $|K_{23}| \leq |K_{13}| \leq K_{33}$, and $|K_{12}| + |K_{13}| - |K_{23}| \leq K_{11}$.*

To reduce the number of parameters to six, the following transformation is used:

$$G := \begin{bmatrix} G_1 & G_4 & G_5 \\ G_4 & G_2 & G_6 \\ G_5 & G_6 & G_3 \end{bmatrix} = T \begin{bmatrix} K_{11} & K_{12} & K_{13} \\ K_{12} & K_{22} & K_{23} \\ K_{13} & K_{23} & K_{33} \end{bmatrix} T, \qquad (5.75)$$

where $T = \text{diag}\{1, 1/\sqrt{c}, 1/\sqrt{b}\}$. Thus, $Y(s)$ is equivalent to

$$Y(s) = \frac{G_1 s^3 + (G_1 G_2 - G_4^2)s^2 + (G_1 G_3 - G_5^2)s + \det(G)}{s\left(s^3 + G_2 s^2 + G_3 s + (G_2 G_3 - G_6^2)\right)}. \tag{5.76}$$

Thus, the following lemma can be obtained.

Lemma 5.23. *A positive-real function $Y(s)$ is realizable as the admittance of a one-port mechanical network consisting of one damper, one inerter, and no more than three springs in Fig. 5.1, where the admittance of X exists, if and only if $Y(s)$ can be written in the form of (5.76), where G as defined in (5.75) is non-negative definite, and there exists an invertible diagonal matrix $D = \text{diag}\{1, x, y\}$ with $x, y > 0$ such that the entries DGD satisfy the conditions of Theorem 5.16.*

Proof. *Necessity.* Let $(1/s)K$ be the admittance of the three-port spring network obtained by extracting one damper and one inerter. Let $x = \sqrt{c}$, $y = \sqrt{b}$, and $D = \text{diag}\{1, x, y\}$, where $b > 0$ and $c > 0$. Then, $Y(s)$ can be written in the form of (5.76), where $K = DGD$ in (5.75) satisfies the conditions of Theorem 5.16.

Sufficiency. Since DGD satisfies the conditions of Theorem 5.16, $(1/s)DGD$ is realizable as a three-port network consisting of at most three springs. Consequently, $Y(s)$ is the admittance of the network in Fig. 5.1, where $c = x^2$, $b = y^2$, and the admittance of X is $(1/s)DGD$. □

Lemma 5.24. *Consider a non-negative definite matrix G as defined in (5.75). If any first-order minor or second-order minor of G is zero, then there must exist an invertible diagonal matrix $D = \text{diag}\{1, x, y\}$ with $x > 0$ and $y > 0$ such that DGD satisfies the conditions of Theorem 5.16.*

Proof. First, it will be shown that if at least one of the first-order minors of G is zero, then there exists $D = \text{diag}\{1, x, y\}$ with $x > 0$ and $y > 0$ such that DGD satisfies Condition 1 of Theorem 5.16. If $G_4 = 0$ and other entries are nonzero, then Condition 1 of Theorem 5.16 is equivalent to $G_1 - y|G_5| \geq 0$, $x^2 G_2 - xy|G_6| \geq 0$, and $y^2 G_3 - y|G_5| - xy|G_6| \geq 0$ with at least two of the inequality signs being equality signs. By choosing $x = G_1|G_6|/(G_2|G_5|)$ and $y = G_1/|G_5|$, the first and second inequality signs become equality signs, and the third item always holds because $G_1 G_2 G_3 - G_1 G_6^2 - G_2 G_5^2 = \det(G) \geq 0$ when $G_4 = 0$. Similarly, all the other subcases can be proved.

Then, it will be shown that if at least one of the second-order minors is zero with all the first-order minors being nonzero, than there exists $D = \text{diag}\{1, x, y\}$ with $x > 0$ and $y > 0$ such that DGD satisfies Condition 2 of

Theorem 5.16. Indeed, if $G_1G_6 - G_4G_5 = 0$, then $G_1 = G_4G_5/G_6$, implying $G_4G_5G_6 > 0$ and $G_1/|G_4| = |G_5|/|G_6|$. Condition 2a of Theorem 5.16 becomes

$$\frac{|G_5|}{|G_6|} \le x \le \frac{G_1}{|G_4|}, \quad \frac{|G_6|}{G_3} \le \frac{y}{x} \le \frac{|G_4|}{|G_5|}, \tag{5.77}$$

$$(x|G_6| - |G_5|)(y/x) \le xG_2 - |G_4|, \tag{5.78}$$

with at least three inequality signs being equality signs. Letting $x = G_1/|G_4| = |G_5|/|G_6|$, the first and second inequality signs of (5.77) both become equality signs, which implies that (5.78) holds because $G_1G_2 - G_4^2 \ge 0$. Since $G_3 - G_5G_6/G_4 = G_3 - G_5^2/G_1 = (G_1G_3 - G_5^2)/G_1 \ge 0$, one can choose some $y > 0$ such that the second item of (5.77) holds with one equality sign. Similarly, the subcases of $G_4G_6 - G_2G_5 = 0$ and that of $G_3G_4 - G_5G_6 = 0$ can be proved.

It has been shown in [Chen (2007), pg. 46] that the following expressions hold: $(G_2G_3 - G_6^2)(G_1G_3 - G_5^2) - (G_3G_4 - G_5G_6)^2 = G_3\det G$, $(G_2G_3 - G_6^2)(G_1G_2 - G_4^2) - (G_4G_6 - G_2G_5)^2 = G_2\det G$, and $(G_1G_3 - G_5^2)(G_1G_2 - G_4^2) - (G_1G_6 - G_4G_5)^2 = G_1\det G$. Therefore, if a principal minor is zero, then the other two minors built from the same rows are both zero. □

Lemma 5.25. *Consider a non-negative definite matrix G in the form of (5.75) with all the first-order minors and all the second-order minors being non-zero. There exists an invertible diagonal matrix $D = \mathrm{diag}\{1, x, y\}$ with $x > 0$ and $y > 0$ such that DGD satisfies the conditions of Theorem 5.16, if and only if one of the following conditions holds: 1) $G_4G_5G_6 < 0$ and $\det(G) = 0$; 2) $G_4G_5G_6 > 0$ and $G_1G_2G_3 + G_4G_5G_6 - G_1G_6^2 - G_3G_4^2 = 0$; 3) $G_4G_5G_6 > 0$ and $G_1G_2G_3 + G_4G_5G_6 - G_1G_6^2 - G_2G_5^2 = 0$; 4) $G_4G_5G_6 > 0$ and $G_1G_2G_3 + G_4G_5G_6 - G_2G_5^2 - G_3G_4^2 = 0$.*

Proof. *Necessity.* First, consider the case of $G_4G_5G_6 < 0$. Condition 1 of Theorem 5.16 becomes $G_1 = x|G_4| + y|G_5|$, $x^2G_2 = x|G_4| + xy|G_6|$, and $y^2G_3 = y|G_5| + xy|G_6|$ with $x > 0$ and $y > 0$, from which one obtains

$$x = \frac{G_1|G_6| + |G_4||G_5|}{|G_4||G_6| + G_2|G_5|}, \quad y = \frac{G_1G_2 - G_4^2}{|G_4||G_6| + G_2|G_5|}, \tag{5.79}$$

and $\det(G) = 0$, which implies Condition 1 of the lemma.

Then, consider the case of $G_4G_5G_6 > 0$. Condition 2a of Theorem 5.16 becomes (5.77) and (5.78) with each of the three having one and only one equality sign, implying $G_1 > G_4G_5/G_6$ and $G_3 > G_5G_6/G_4$. Since $G_2 \ne G_4G_6/G_5$, it is only possible that $|G_5|/|G_6| < |x| = G_1/|G_4|$ and

$|G_6|/G_3 = |y|/|x| < |G_4|/|G_5|$. Thus, (5.78) with the equality sign yields $G_1G_2G_3 + G_4G_5G_6 - G_1G_6^2 - G_3G_4^2 = 0$, which implies Condition 2 of the lemma. Similarly, one can prove that Condition 2b of Theorem 5.16 implies Condition 3 of this lemma, and Condition 2c of Theorem 5.16 implies Condition 4 of this lemma.

Sufficiency. If Condition 1 of the lemma holds, let $x > 0$ and $y > 0$ satisfy (5.79). Then, it can be verified that Condition 1 of Theorem 5.16 holds, following the proof of the necessity part.

As shown in [Chen and Smith (2009b)], at most one of $G_1 - G_4G_5/G_6$, $G_2 - G_4G_6/G_5$, and $G_3 - G_5G_6/G_4$ can be negative because of $G_4G_5G_6 > 0$ and the non-negative definiteness of G. If Condition 2 of the lemma holds, let $x = G_1/|G_4|$ and $y/x = |G_6|/G_3$. Then, one obtains $\det(G) = G_5^2(G_4G_6/G_5 - G_2)$, implying $G_1 > G_4G_5/G_6$, $G_2 < G_4G_6/G_5$, and $G_3 > G_5G_6/G_4$. Thus, (5.77) holds with each of the two items having one equality sign. Since $G_1G_2G_3 + G_4G_5G_6 - G_1G_6^2 - G_3G_4^2 = 0$, (5.78) holds with an equality sign. Therefore, $D = \mathrm{diag}\{1, x, y\}$ with $x > 0$ and $y > 0$ exists such that DGD satisfies Condition 2 of Theorem 5.16. Similarly, one can prove that Condition 2 of Theorem 5.16 holds if Condition 3 or Condition 4 of this lemma holds. Since Condition 3 implies $G_1 > G_4G_5/G_6$, $G_2 > G_4G_6/G_5$, and $G_3 < G_5G_6/G_4$, and Condition 4 implies $G_1 < G_4G_5/G_6$, $G_2 > G_4G_6/G_5$, and $G_3 > G_5G_6/G_4$, the four conditions stated in the lemma have no overlap. □

Now, by Lemmas 5.23–5.25, the following theorem is obtained.

Theorem 5.17. *A positive-real function $Y(s)$ is realizable as the admittance of a one-port mechanical network consisting of one damper, one inerter, and no more than three springs in Fig. 5.1, where the admittance of X exists, if and only if $Y(s)$ can be written in the form of (5.76), where G as defined in (5.75) is non-negative definite and satisfies the conditions of Lemma 5.24 or Lemma 5.25.*

Proof. It can be proved by combining Lemmas 5.23–5.25. □

Theorem 5.18. *Consider a positive-real function $Y(s)$ in the form of (5.76), where G as defined in (5.75) is a non-negative definite matrix. If any first-order minor or second-order minor of G is zero, then $Y(s)$ is realizable as the admittance of a one-port series-parallel mechanical network consisting of one damper, one inerter, and no more than three springs, through the Foster preamble.*

Proof. *Case 1:* When $G_4 = 0$, one obtains

$$Y(s) = \frac{k_1}{s} + \left(\frac{s}{k_2} + \left(bs + \left(\frac{s}{k_3} + \frac{1}{c} \right)^{-1} \right)^{-1} \right)^{-1},$$

where $k_1 = \det(G)/(G_2G_3 - G_6^2)$, $k_2 = G_2G_5^2/(G_2G_3 - G_6^2)$, $k_3 = G_2G_5^2G_6^2/(G_2G_3 - G_6^2)^2$, $b = G_2^2G_5^2/(G_2G_3 - G_6^2)^2$, and $c = G_5^2G_6^2/(G_2G_3 - G_6^2)^2$. If $G_2G_3 - G_6^2 \neq 0$. Thus, $Y(s)$ is realizable as the admittance of the configuration in Fig. 5.18(a), with k_1, k_2, k_3, b, $c \geq 0$. Specially, if $G_2G_3 - G_6^2 = 0$, then $Y(s) = G_1/s$, which is realizable as the configuration in Fig. 5.18(a), with $k_1 = G_1 \geq 0$ and $k_2 = \infty$.

Case 2: When $G_5 = 0$, one obtains

$$Y(s) = \frac{k_1}{s} + \left(\frac{s}{k_2} + \left(c + \left(\frac{s}{k_3} + \frac{1}{bs} \right)^{-1} \right)^{-1} \right)^{-1},$$

where $k_1 = \det(G)/(G_2G_3 - G_6^2)$, $k_2 = G_3G_4^2/(G_2G_3 - G_6^2)$, $k_3 = G_3G_4^2G_6^2/(G_2G_3 - G_6^2)^2$, $b = G_4^2G_6^2/(G_2G_3 - G_6^2)^2$, and $c = G_3^2G_4^2/(G_2G_3 - G_6^2)^2$. If $G_2G_3 - G_6^2 \neq 0$, then $Y(s)$ is realizable as the admittance of the configuration in Fig. 5.18(b), with k_1, k_2, k_3, b, $c \geq 0$. Specially, if $G_2G_3 - G_6^2 = 0$, then $Y(s) = G_1/s$, which is realizable as the configuration in Fig. 5.18(b), with $k_1 = G_1 \geq 0$ and $k_2 = \infty$.

Case 3: When $G_6 = 0$, one obtains

$$Y(s) = \frac{k_1}{s} + \left(\frac{s}{k_2} + \frac{1}{bs} \right)^{-1} + \left(\frac{s}{k_3} + \frac{1}{c} \right)^{-1},$$

where $k_1 = \det(G)/(G_2G_3)$, $k_2 = G_5^2/G_3$, $k_3 = G_4^2/G_2$, $b = G_5^2/G_3^2$, and $c = G_4^2/G_2^2$. If $G_2G_3 \neq 0$, then $Y(s)$ is realizable as the admittance of the configuration in Fig. 5.18(c), with k_1, k_2, k_3, b, $c \geq 0$. Specially, if $G_2 = 0$ and $G_3 \neq 0$, then $Y(s) = (G_1s^2 + (G_1G_3 - G_5^2))/(s(s^2 + G_3)) = (G_1G_3 - G_5^2)/(G_3s) + 1/(G_3s/G_5^2 + G_3^2/(G_5^2s))$, which is realizable as the configuration in Fig. 5.18(c), with $k_1 = (G_1G_3 - G_5^2)/G_3 \geq 0$, $k_2 = G_5^2/G_3 \geq 0$, $b = G_5^2/G_3^2 \geq 0$, and $k_3 = \infty$; if $G_3 = 0$ and $G_2 \neq 0$, then $Y(s) = (G_1s + (G_1G_2 - G_4^2))/(s(s + G_2)) = (G_1G_2 - G_4^2)/(G_2s) + 1/(G_2s/G_4^2 + G_2^2/G_4^2)$, which is realizable as the configuration in Fig. 5.18(c), with $k_1 = (G_1G_2 - G_4^2)/(G_2s) \geq 0$, $k_3 = G_4^2/G_2 \geq 0$, $c = G_4^2/G_2^2 \geq 0$, and $k_2 = \infty$; if $G_2 = G_3 = 0$, then $Y(s) = G_1/s$, which is realizable as the configuration in Fig. 5.18(c), with $k_1 = G_1 \geq 0$, $k_2 = \infty$, and $k_3 = \infty$.

Case 4: When $G_1G_2G_3 = 0$, since G is non-negative definite, one obtains $G_4G_5G_6 = 0$, which can be referred to Cases 1–3.

Case 5: When all the entries are non-zero and $G_1G_6 - G_4G_5 = 0$, one obtains

$$Y(s) = \left(\frac{s}{k_1} + \left(\frac{k_2}{s} + bs \right)^{-1} + \left(\frac{k_3}{s} + c \right)^{-1} \right)^{-1},$$

where $k_1 = G_1$, $k_2 = G_1(G_1G_3 - G_5^2)/G_5^2$, $k_3 = G_1(G_1G_2 - G_4^2)/G_4^2$, $b = G_4^2/G_6^2$, and $c = G_1^2/G_4^2$. Thus, $Y(s)$ is realizable as the admittance of the configuration in Fig. 5.18(d), with k_1, k_2, k_3, b, $c \geq 0$.

Case 6: When all the entries are non-zero and $G_2G_5 - G_4G_6 = 0$, one obtains

$$Y(s) = \left(\frac{s}{k_1} + \left(\frac{k_2}{s} + \left(\frac{1}{c} + \left(\frac{k_3}{s} + bs \right)^{-1} \right)^{-1} \right)^{-1} \right)^{-1},$$

where $k_1 = G_1$, $k_2 = G_1(G_1G_2 - G_4^2)/G_4^2$, $k_3 = G_1^2(G_2G_3 - G_6^2)/(G_2G_5^2)$, $b = G_1^2/G_5^2$, and $c = G_1^2/G_4^2$. Thus, $Y(s)$ is realizable as the admittance of the configuration in Fig. 5.18(e), with k_1, k_2, k_3, b, $c \geq 0$.

Case 7: When all the entries are non-zero and $G_3G_4 - G_5G_6 = 0$, one obtains

$$Y(s) = \left(\frac{s}{k_1} + \left(\frac{k_2}{s} + \left(\frac{1}{bs} + \left(\frac{k_3}{s} + c \right)^{-1} \right)^{-1} \right)^{-1} \right)^{-1},$$

where $k_1 = G_1$, $k_2 = G_1(G_1G_3 - G_5^2)/G_5^2$, $k_3 = G_1^2(G_2G_3 - G_6^2)/(G_3G_4^2)$, $b = G_1^2/G_5^2$, and $c = G_1^2/G_4^2$. Thus, $Y(s)$ is realizable as the admittance of the configuration in Fig. 5.18(f), with k_1, k_2, k_3, b, $c \geq 0$.

Case 8: When all the entries are non-zero and at least one principal minor is zero, it follows from the discussion in the proof of Lemma 5.24 that all the minors built in the same rows (columns) are zero, which can be referred to Cases 5–7.

Finally, note that all the realization processes of the above cases belong to the method of the Foster preamble. Therefore, the proof is completed.

$\square$

Theorem 5.19. *Consider a positive-real function $Y(s)$ in the form of (5.76), where G as defined in (5.75) is a non-negative definite matrix with all the first-order minors and all the second-order minors being non-zero. If $Y(s)$ satisfies the conditions of Lemma 5.25, then it is realizable as the admittance of a one-port mechanical network consisting of one damper, one inerter and three springs, which is one of the configurations in Fig. 5.19.*

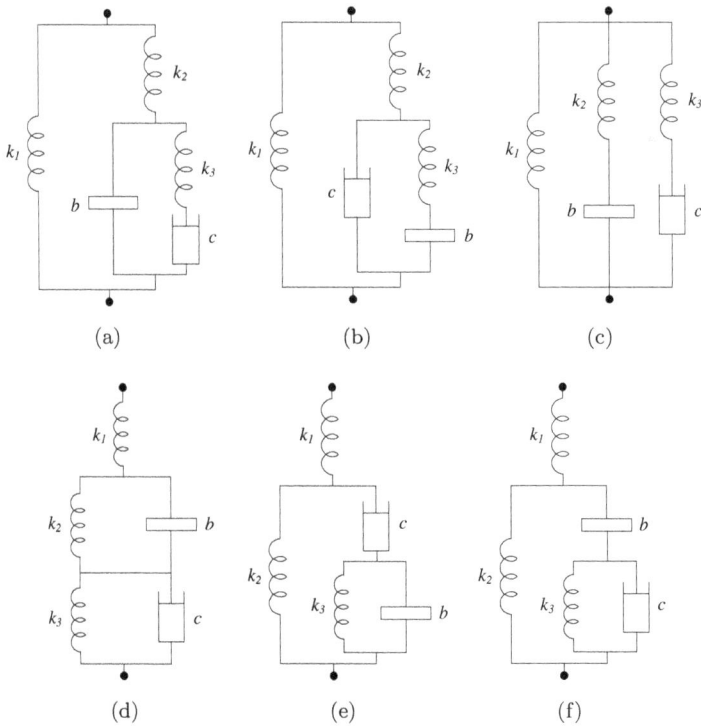

Fig. 5.18 The configurations covering the case when any first-order minor or second-order minor of G is zero, where the values of the elements are non-negative or infinity (corresponding to the element replaced by a short-circuit).

1. *If Condition 1 holds, then $Y(s)$ is realizable as the configuration in Fig. 5.19(a), with $k_1 = (G_1G_2 - G_4^2)/(G_2 - G_4G_6/G_5)$, $k_2 = G_6^2(G_1 - G_4G_5/G_6)(G_1G_2 - G_4^2)/(G_2G_5 - G_4G_6)^2$, $k_3 = G_4G_6(G_4G_5/G_6 - G_1)/(G_5(G_2 - G_4G_6/G_5))$, $b = (G_1G_2 - G_4^2)^2/(G_4G_6 - G_2G_5)^2$, and $c = (G_1G_6 - G_4G_5)^2/(G_4G_6 - G_2G_5)^2$.*

2. *If Condition 2 holds, then $Y(s)$ is realizable as the configuration in Fig. 5.19(b), with $k_1 = G_1G_6^2(G_1 - G_4G_5/G_6)/(G_3G_4^2)$, $k_2 = G_1G_5G_6/(G_3G_4)$, $k_3 = G_1(G_3 - G_5G_6/G_4)/G_3$, $b = G_1^2G_6^2/(G_3^2G_4^2)$, and $c = G_1^2/G_4^2$.*

3. *If Condition 3 holds, then $Y(s)$ is realizable as the configuration in Fig. 5.19(c), with $k_1 = G_1G_6^2(G_1 - G_4G_5/G_6)/(G_2G_5^2)$, $k_2 = G_1G_4G_6/(G_2G_5)$, $k_3 = G_1(G_2 - G_4G_6/G_5)/G_2$, $b = G_1^2/G_5^2$, and $c = G_1^2G_6^2/(G_2^2G_5^2)$.*

4. *If Condition 4 holds, then $Y(s)$ is realizable as the configuration in Fig. 5.19(d), with $k_1 = G_4^2(G_3 - G_5G_6/G_4)/(G_2G_3)$, $k_2 = (G_4G_5G_6)/(G_2G_3)$, $k_3 = G_5^2(G_2 - G_4G_6/G_5)/(G_2G_3)$, $b = G_5^2/G_3^2$, and $c = G_4^2/G_2^2$.*

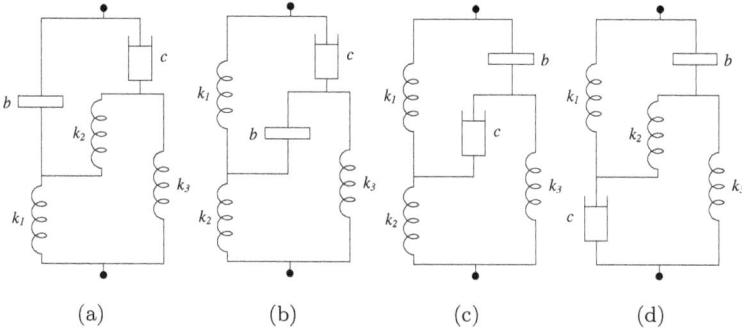

Fig. 5.19 The configurations covering all cases that satisfy the conditions of Lemma 5.25. In each case, $b > 0$, $c > 0$, $k_1 > 0$, $k_2 > 0$, and $k_3 > 0$.

Proof. *Condition 1:* Since $G_4G_5G_6 < 0$, $\det(G) = 0$, and G with first-order minors and second-order minors being non-zero is non-negative definite, it implies that $b > 0$, $c > 0$, $k_1 > 0$, $k_2 > 0$, $k_3 > 0$, and the admittance of the network in Fig. 5.19(a) is equivalent to (5.76).

Condition 2: Since $G_4G_5G_6 > 0$, $G_1G_2G_3 + G_4G_5G_6 - G_1G_6^2 - G_3G_4^2 = 0$, and G with first- and second-order minors non-zero is non-negative definite, it implies that $G_1 - G_4G_5/G_6 > 0$, $G_2 - G_4G_6/G_5 < 0$ and $G_3 - G_5G_6/G_4 > 0$ by the discussion in the proof of Lemma 5.25. Therefore, it follows that $b > 0$, $c > 0$, $k_1 > 0$, $k_2 > 0$, $k_3 > 0$, and the admittance of the network in Fig. 5.19(b) is equivalent to (5.76).

Condition 3: Since $G_4G_5G_6 > 0$, $G_1G_2G_3 + G_4G_5G_6 - G_1G_6^2 - G_2G_5^2 = 0$, and G with first- and second-order minors being non-zero is non-negative definite, it implies that $G_1 - G_4G_5/G_6 > 0$, $G_2 - G_4G_6/G_5 > 0$ and $G_3 - G_5G_6/G_4 < 0$ by the discussion in the proof of Lemma 5.25. Therefore, it follows that $b > 0$, $c > 0$, $k_1 > 0$, $k_2 > 0$, $k_3 > 0$, and the admittance of the network in Fig. 5.19(c) is equivalent to (5.76).

Condition 4: Since $G_4G_5G_6 > 0$, $G_1G_2G_3 + G_4G_5G_6 - G_2G_5^2 - G_3G_4^2 = 0$, and G with first- and second-order minors non-zero is non-negative definite, it implies that $G_1 - G_4G_5/G_6 < 0$, $G_2 - G_4G_6/G_5 > 0$ and

$G_3 - G_5G_6/G_4 > 0$ by the discussion in the proof of Lemma 5.25. Hence, it follows that $b > 0$, $c > 0$, $k_1 > 0$, $k_2 > 0$, $k_3 > 0$, and the admittance of the network in Fig. 5.19(d) is equivalent to (5.76).   □

To make the conditions concerning with admittance $Y(s)$ in the form of (5.76) become easier to check, the admittance will be further converted into the form of (5.74), with realizability conditions given in terms of α_3, α_2, α_1, α_0, β_3, β_2, and β_1. Thus, $\alpha_3 = G_1$, $\alpha_2 = G_1G_2 - G_4^2$, $\alpha_1 = G_1G_3 - G_5^2$, $\alpha_0 = \det(G)$, $\beta_3 = G_2$, $\beta_2 = G_3$, and $\beta_1 = G_2G_3 - G_6^2$. For simplicity, denote

$$W_1 := \alpha_3\beta_3 - \alpha_2, \ W_2 := \alpha_3\beta_2 - \alpha_1, \ W_3 := \beta_2\beta_3 - \beta_1,$$
$$W := \alpha_0 + 2\alpha_3\beta_2\beta_3 - \alpha_3\beta_1 - \alpha_2\beta_2 - \alpha_1\beta_3.$$

Then,

$$G_1 = \alpha_3, \ G_2 = \beta_3, \ G_3 = \beta_2, \ G_4^2 = W_1, \ G_5^2 = W_2,$$
$$G_6^2 = W_3, \ G_4G_5G_6 = \frac{W}{2}, G_1 - \frac{G_4G_5}{G_6} = \alpha_3 - \frac{W}{2W_3}, \quad (5.80)$$
$$G_2 - \frac{G_4G_6}{G_5} = \beta_3 - \frac{W}{2W_2}, \ G_3 - \frac{G_5G_6}{G_4} = \beta_2 - \frac{W}{2W_1},$$

and $W^2 = 4W_1W_2W_3$.

Theorem 5.20. *A positive-real function $Y(s)$ is realizable as the admittance of a one-port network consisting of one damper, one inerter, and no more than three springs in Fig. 5.1, where the admittance of X exists, if and only if $Y(s)$ can be written in the form of (5.74), where the coefficients satisfy α_0, α_1, α_2, α_3, β_1, β_2, $\beta_3 \geq 0$, W_1, W_2, $W_3 \geq 0$, $W^2 = 4W_1W_2W_3$, and also satisfy either 1) at least one of α_1, α_2, α_3, β_1, β_2, β_3, W_1, W_2, W_3, $(\beta_2 - W/(2W_1))$, $(\beta_3 - W/(2W_2))$, and $(\alpha_3 - W/(2W_3))$ is zero; or 2) one of the following holds with Condition 1 not being satisfied: a) $W < 0$ and $\alpha_0 = 0$; b) $W > 0$ and $\alpha_0 + \alpha_3\beta_1 + \alpha_2\beta_2 - \alpha_1\beta_3 = 0$; c) $W > 0$ and $\alpha_0 + \alpha_3\beta_1 + \alpha_1\beta_3 - \alpha_2\beta_2 = 0$; d) $W > 0$ and $\alpha_0 + \alpha_1\beta_3 + \alpha_2\beta_2 - \alpha_3\beta_1 = 0$.*

Proof. The proof is similar to the arguments used in Section 5.2, which is omitted here for brevity.   □

Moreover, Figs. 5.18 and 5.19(a)–5.19(d) are the realization configurations achieving Conditions 1 and 2a–2d of Theorem 5.20, respectively, whose expressions of element values can be obtained from those in terms of G_1 to G_6 through (5.80).

5.4.2 Final Realization Results

To complete the present study, consider the case when the impedance of X in Fig. 5.1 does not exist. The following lemma will be needed to derive Theorem 5.21 later.

Lemma 5.26. *If the admittance of a two-port network consisting of only springs exists, then it is realizable as the configuration in Fig. 5.20, with $k_1 \geq 0$, $k_2 \geq 0$, and $k_3 \geq 0$, or a configuration by switching the polarities of the two ports.*

Proof. For a two-port network consisting of only springs, its admittance Y_L is in the form of

$$Y_L = \frac{1}{s} \begin{bmatrix} K_{11} & K_{12} \\ K_{12} & K_{22} \end{bmatrix} := \frac{1}{s}K. \tag{5.81}$$

According to the analogy to one-element-kind networks, K as defined in (5.81) is necessarily paramount, that is, $K_{11} \geq |K_{12}|$ and $K_{22} \geq |K_{12}|$.

If $K_{12} \geq 0$, then Y_L in the form of (5.81) is realizable as shown in Fig. 5.20, where $k_1 = K_{11} - K_{12} \geq 0$, $k_2 = K_{12} \geq 0$, and $k_3 = K_{22} - K_{12} \geq 0$.

If $K_{12} < 0$, then Y_L in the form of (5.81) is realizable as shown in Fig. 5.20 by switching the polarities of the two ports, where $k_1 = K_{11} - |K_{12}| \geq 0$, $k_2 = |K_{12}| > 0$, and $k_3 = K_{22} - |K_{12}| \geq 0$. $\square$

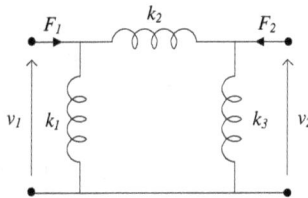

Fig. 5.20 The two-port configuration consisting of only springs.

Theorem 5.21. *A positive-real function $Y(s)$ is realizable as the admittance of a one-port mechanical network consisting of one damper, one inerter, and no more than three springs in Fig. 5.1, where the impedance of X does not exist, if and only if $Y(s)$ can be written in the form of*

$$Y(s) = \frac{\alpha_3 s^3 + \alpha_2 s^2 + \alpha_1 s + \alpha_0}{\beta_3 s^3 + \beta_2 s^2 + \beta_1 s}, \tag{5.82}$$

where $\alpha_i \geq 0$, $i = 0, 1, 2, 3$, $\beta_j \geq 0$, $j = 1, 2, 3$, and one of the following five conditions holds:

1. $\alpha_3 = \beta_2 = \beta_3 = 0$, $\alpha_1 > 0$, $\alpha_2 > 0$, and $\beta_1 > 0$;
2. $\alpha_3 = 0$, $\beta_2 > 0$, $\beta_3 > 0$, $\alpha_1\beta_1 - \alpha_0\beta_2 \geq 0$, $\alpha_1^2 + \alpha_0\beta_2^2 \geq \alpha_1\beta_1\beta_2$, $\alpha_0\beta_2 + \beta_2\beta_1^2 \geq \alpha_1\beta_1$, and $\alpha_1\beta_3 = \alpha_2\beta_2$;
3. $\beta_2 = 0$, $\alpha_3 > 0$, $\beta_3 > 0$, $\alpha_2\beta_1 - \alpha_0\beta_3 \geq 0$, $\alpha_2^2 + \alpha_0\beta_3^2 \geq \alpha_2\beta_1\beta_3$, $\alpha_0\beta_3 + \beta_3\beta_1^2 \geq \alpha_2\beta_1$, and $\alpha_1\beta_3 = \alpha_3\beta_1$;
4. $\beta_3 = 0$, $\alpha_3 > 0$, $\beta_2 > 0$, $\alpha_1\beta_1 - \alpha_0\beta_2 \geq 0$, $\alpha_1^2 + \alpha_0\beta_2^2 \geq \alpha_1\beta_1\beta_2$, $\alpha_0\beta_2 + \beta_2\beta_1^2 \geq \alpha_1\beta_1$, and $\alpha_3\beta_1 = \alpha_2\beta_2$;
5. $\alpha_3 > 0$, $\beta_2 > 0$, $\beta_3 > 0$, $\alpha_1\beta_3 + \alpha_2\beta_2 \geq \alpha_3\beta_1$, $\alpha_2\beta_2 + \alpha_3\beta_1 \geq \alpha_1\beta_3$, $\alpha_1\beta_3 + \alpha_3\beta_1 \geq \alpha_2\beta_2$, $\alpha_3 = \beta_2\beta_3$, and $\alpha_1^2\beta_3^2 + \alpha_2^2\beta_2^2 + \alpha_3^2\beta_1^2 + 4\alpha_0\alpha_3^2 = 2(\alpha_1\beta_3\alpha_2\beta_2 + \alpha_2\beta_2\alpha_3\beta_1 + \alpha_3\beta_1\alpha_1\beta_3)$.

Furthermore, the configurations in Fig. 5.21 can realize each of the five conditions above, respectively, where element values satisfy $b > 0$, $c > 0$, $k_1 \geq 0$, $k_2 \geq 0$, and $k_3 \geq 0$.

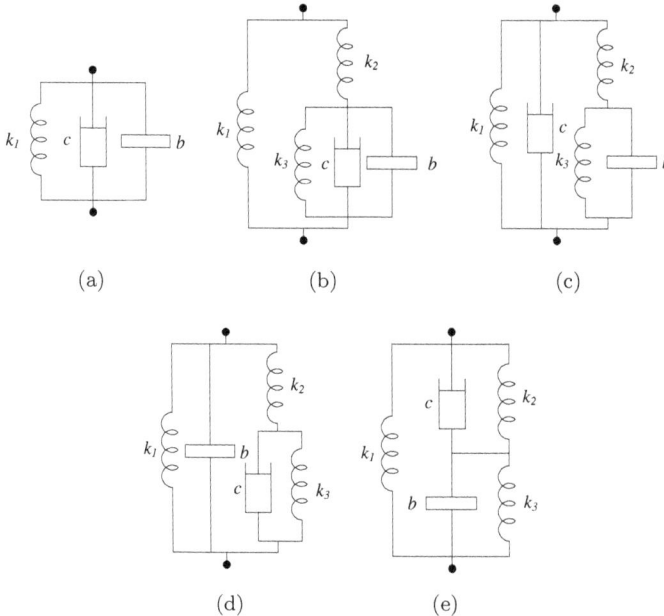

Fig. 5.21 The networks used to fulfill the conditions of Theorem 5.21. (a) for Condition 1; (b) for Condition 2; (c) for Condition 3; (d) for Condition 4; (e) for Condition 5. In each case, $b > 0$, $c > 0$, $k_1 \geq 0$, $k_2 \geq 0$, and $k_3 \geq 0$.

Proof. In this proof, call the network edges corresponding to the damper and the inerter as the damper edge and the inerter edge, respectively. Let $\mathcal{G}$ denote the augmented graph of the one-port mechanical network to be realized. When the admittance of X does not exist, it follows from Definition 2.9 and Theorem 2.18 that there must exist at least one circuit among the port edge, the damper edge, and the inerter edge of $\mathcal{G}$, respectively.

If each pair of the three edges constitute a circuit, then the possible realization can be equivalent to Fig. 5.21(a), where $b > 0$, $c > 0$, and $k_1 \geq 0$, because any other nodes can be eliminated by the generalized star-mesh transformation [Versfeld (1970)]. If there is only one circuit constituted by two edges of e_i, e_d, and e_p, then the other part of the network can be regarded as a two-port network, which consists of only springs and whose admittance exists. Therefore, based on Lemma 5.26, the realization can be one of the configurations in Figs. 5.21(b)–5.21(d), where $b > 0$, $c > 0$, $k_1 \geq 0$, $k_2 \geq 0$, and $k_3 \geq 0$.

If the three edges together form one circuit, then the realization can be equivalent to Fig. 5.21(e), where $b > 0$, $c > 0$, $k_1 \geq 0$, $k_2 \geq 0$, $k_3 \geq 0$, by using the generalized star-mesh transformation [Versfeld (1970)].

Now, it suffices to derive the realizability conditions for the networks in Fig. 5.21, each of which corresponds to one of the five conditions of this theorem.

The realizability condition of Fig. 5.21(a): The admittance of the configuration in Fig. 5.21(a) can be expressed as

$$Y(s) = \frac{bs^2 + cs + k_1}{s},\tag{5.83}$$

where $b > 0$, $c > 0$, and $k_1 \geq 0$. If $Y(s)$ is realizable as the configuration in Fig. 5.21(a), then it can be expressed as (5.82) with $\alpha_3 = 0$, $\alpha_2 = b$, $\alpha_1 = c$, $\alpha_0 = k_1$, $\beta_3 = 0$, $\beta_2 = 0$, and $\beta_1 = 1$. It is obvious that the coefficients are all non-negative and satisfy Condition 1. Conversely, if $\alpha_0 \geq 0$, $\alpha_1 \geq 0$, $\alpha_2 \geq 0$, $\alpha_3 \geq 0$, $\beta_1 \geq 0$, $\beta_2 \geq 0$, $\beta_3 \geq 0$ and they satisfy Condition 1, then letting $b = \alpha_2/\beta_1$, $c = \alpha_1/\beta_1$, and $k_1 = \alpha_0/\beta_1$ yields that $b > 0$, $c > 0$, $k_1 \geq 0$, and (5.83) is equivalent to (5.82).

The realizability condition of Fig. 5.21(b): The admittance of the configuration in Fig. 5.21(b) is obtained as

$$Y(s) = \frac{b(k_1 + k_2)s^2 + c(k_1 + k_2)s + (k_1 k_2 + k_2 k_3 + k_1 k_3)}{s\left(bs^2 + cs + (k_2 + k_3)\right)},\tag{5.84}$$

where $b > 0$, $c > 0$, $k_1 \geq 0$, $k_2 \geq 0$, and $k_3 \geq 0$. If $Y(s)$ is realizable as the configuration in Fig. 5.21(b), then it can be expressed as (5.82) with $\alpha_3 = 0$,

$\alpha_2 = b(k_1 + k_2)$, $\alpha_1 = c(k_1 + k_2)$, $\alpha_0 = k_1 k_2 + k_2 k_3 + k_1 k_3$, $\beta_3 = b$, $\beta_2 = c$, and $\beta_1 = k_2 + k_3$. It is obvious that the coefficients are all non-negative and satisfy Condition 2. Conversely, if $\alpha_0 \geq 0$, $\alpha_1 \geq 0$, $\alpha_2 \geq 0$, $\alpha_3 \geq 0$, $\beta_1 \geq 0$, $\beta_2 \geq 0$, $\beta_3 \geq 0$ and they satisfy Condition 2, then letting $b = \beta_3$, $c = \beta_2$, $k_1 = \alpha_1/\beta_2 - k_2$, $k_3 = \beta_1 - k_2$, and $k_2 = \sqrt{(\alpha_1\beta_1 - \alpha_0\beta_2)/\beta_2}$ gives that $b > 0$, $c > 0$, $k_1 \geq 0$, $k_2 \geq 0$, $k_3 \geq 0$, and (5.84) is equivalent to (5.82).

The realizability condition of Fig. 5.21(c): The admittance of the configuration in Fig. 5.21(c) is calculated to obtain

$$Y(s) = \frac{bcs^3 + b(k_1 + k_2)s^2 + c(k_2 + k_3)s + (k_1 k_2 + k_2 k_3 + k_1 k_3)}{s\left(bs^2 + (k_2 + k_3)\right)}, \quad (5.85)$$

where $b > 0$, $c > 0$, $k_1 \geq 0$, $k_2 \geq 0$, and $k_3 \geq 0$. If $Y(s)$ is realizable as the configuration in Fig. 5.21(c), then it can be expressed as (5.82) with $\alpha_3 = bc$, $\alpha_2 = b(k_1 + k_2)$, $\alpha_1 = c(k_2 + k_3)$, $\alpha_0 = k_1 k_2 + k_2 k_3 + k_1 k_3$, $\beta_3 = b$, $\beta_2 = 0$, and $\beta_1 = k_2 + k_3$. It is obvious that the coefficients are all non-negative and satisfy Condition 3. Conversely, if $\alpha_0 \geq 0$, $\alpha_1 \geq 0$, $\alpha_2 \geq 0$, $\alpha_3 \geq 0$, $\beta_1 \geq 0$, $\beta_2 \geq 0$, $\beta_3 \geq 0$ and they satisfy Condition 3, then letting $b = \beta_3$, $c = \alpha_3/\beta_3$, $k_1 = \alpha_2/\beta_3 - k_2$, $k_3 = \beta_1 - k_2$, and $k_2 = \sqrt{(\alpha_2\beta_1 - \alpha_0\beta_3)/\beta_3}$ provides that $b > 0$, $c > 0$, $k_1 \geq 0$, $k_2 \geq 0$, $k_3 \geq 0$, and (5.85) is equivalent to (5.82).

The realizability condition of Fig. 5.21(d): The admittance of the configuration in Fig. 5.21(d) is given as

$$Y(s) = \frac{bcs^3 + b(k_2 + k_3)s^2 + c(k_1 + k_2)s + (k_1 k_2 + k_2 k_3 + k_1 k_3)}{s\left(cs + (k_2 + k_3)\right)}, \quad (5.86)$$

where $b > 0$, $c > 0$, $k_1 \geq 0$, $k_2 \geq 0$, and $k_3 \geq 0$. If $Y(s)$ is realizable as the configuration in Fig. 5.21(d), then it can be expressed as (5.82) with $\alpha_3 = bc$, $\alpha_2 = b(k_2 + k_3)$, $\alpha_1 = c(k_1 + k_2)$, $\alpha_0 = k_1 k_2 + k_2 k_3 + k_1 k_3$, $\beta_3 = 0$, $\beta_2 = c$, and $\beta_1 = k_2 + k_3$. It is obvious that the coefficients are all non-negative and satisfy Condition 4. Conversely, if $\alpha_0 \geq 0$, $\alpha_1 \geq 0$, $\alpha_2 \geq 0$, $\alpha_3 \geq 0$, $\beta_1 \geq 0$, $\beta_2 \geq 0$, $\beta_3 \geq 0$ and they satisfy Condition 4, then letting $b = \alpha_3/\beta_2$, $c = \beta_2$, $k_1 = \alpha_1/\beta_2 - k_2$, $k_3 = \beta_1 - k_2$, and $k_2 = \sqrt{(\alpha_1\beta_1 - \alpha_0\beta_2)/\beta_2}$ yields that $b > 0$, $c > 0$, $k_1 \geq 0$, $k_2 \geq 0$, $k_3 \geq 0$, and (5.86) is equivalent to (5.82).

The realizability condition of Fig. 5.21(e): The admittance of the configuration in Fig. 5.21(e) is obtained as

$$Y(s) = \frac{bcs^3 + b(k_1 + k_2)s^2 + c(k_1 + k_3)s + (k_1 k_2 + k_2 k_3 + k_1 k_3)}{s\left(bs^2 + cs + (k_2 + k_3)\right)}, \quad (5.87)$$

where $b > 0$, $c > 0$, $k_1 \geq 0$, $k_2 \geq 0$, and $k_3 \geq 0$. If $Y(s)$ is realizable as the configuration in Fig. 5.21(e), then it can be expressed as (5.82) with $\alpha_3 = bc$,

$\alpha_2 = b(k_1 + k_2)$, $\alpha_1 = c(k_1 + k_3)$, $\alpha_0 = k_1 k_2 + k_2 k_3 + k_1 k_3$, $\beta_3 = b$, $\beta_2 = c$, and $\beta_1 = k_2 + k_3$. It is obvious that the coefficients are all non-negative and satisfy Condition 5. Conversely, if $\alpha_0 \geq 0$, $\alpha_1 \geq 0$, $\alpha_2 \geq 0$, $\alpha_3 \geq 0$, $\beta_1 \geq 0$, $\beta_2 \geq 0$, $\beta_3 \geq 0$ and they satisfy Condition 5, then letting $b = \beta_3$, $c = \beta_2$, $k_1 = (\alpha_1 \beta_3 + \alpha_2 \beta_2 - \alpha_3 \beta_1)/(2\alpha_3)$, $k_2 = (\alpha_2 \beta_2 - \alpha_1 \beta_3 + \alpha_3 \beta_1)/(2\alpha_3)$, and $k_3 = (\alpha_1 \beta_3 - \alpha_2 \beta_2 + \alpha_3 \beta_1)/(2\alpha_3)$ gives that $b > 0$, $c > 0$, $k_1 \geq 0$, $k_2 \geq 0$, $k_3 \geq 0$, and (5.87) is equivalent to (5.82). $\qquad\square$

Now, the final result is summarized as follows.

Theorem 5.22. *A positive-real function $Y(s)$ is realizable as the admittance of a one-port mechanical network consisting of one inerter, one damper, and no more than three springs, if and only if $Y(s)$ can be written in the form of*

$$Y(s) = \frac{\alpha_3 s^3 + \alpha_2 s^2 + \alpha_1 s + \alpha_0}{\beta_4 s^4 + \beta_3 s^3 + \beta_2 s^2 + \beta_1 s}, \qquad (5.88)$$

where $\alpha_i \geq 0$, $i = 0, 1, 2, 3$, and $\beta_j \geq 0$, $j = 1, 2, 3$, which satisfy the conditions of Theorem 5.20 when $\beta_4 = 1$, or the conditions of Theorem 5.21 when $\beta_4 = 0$. Moreover, $Y(s)$ is realizable as one of the configurations in Figs. 5.18, 5.19, and 5.21.

Proof. The theorem can be directly proved by combining Theorems 5.20 and 5.21. $\qquad\square$

5.4.3 *Some Examples*

Example 5.1. For a positive-real function

$$Y(s) = \frac{45s^3 + 99s^2 + 54s + 22}{9s^4 + 36s^3 + 27s^2 + 107s},$$

one can check that $Y(s)$ can be written in the form of (5.88) with $\alpha_3 = 5$, $\alpha_2 = 11$, $\alpha_1 = 6$, $\alpha_0 = 22/9$, $\beta_4 = 1$, $\beta_3 = 4$, $\beta_2 = 3$, and $\beta_1 = 107/9$. Therefore, the coefficients satisfy the conditions of Theorem 5.22 (Condition 2d of Theorem 5.20). By the discussion in this section, $Y(s)$ is realizable as the configuration in Fig. 5.19(d) with $b = 1$ kg, $c = 9/16$ Ns/m, $k_1 = 2$ N/m, $k_2 = 1/4$ N/m, and $k_3 = 11/4$ N/m.

Example 5.2. Given a positive-real function

$$Y(s) = \frac{2s^2 + 4s + 1}{s^3 + 4s^2 + s},$$

then $Y(s)$ can be written in the form of (5.88) with $\alpha_3 = 0$, $\alpha_2 = 2$, $\alpha_1 = 4$, $\alpha_0 = 1$, $\beta_4 = 0$, $\beta_3 = 1$, $\beta_2 = 4$, and $\beta_1 = 1$. Therefore, the coefficients satisfy the conditions of Theorem 5.22 (Condition 1 of Theorem 5.20). Consequently, $Y(s)$ is realizable as the configuration in Fig. 5.18(e), with $b = 4$ kg, $c = 1$ Ns/m, $k_1 = 2$ N/m, $k_2 = 2$ N/m, and $k_3 = 0$ N/m (two springs).

Example 5.3. Given a positive-real function

$$Y(s) = \frac{20s^3 + 44s^2 + 24s + 19}{4s^4 + 16s^3 + 12s^2 + 47s},$$

then the conditions of Theorem 5.22 does not hold when expressing $Y(s)$ in the form of (5.88) with $\alpha_3 = 5$, $\alpha_2 = 11$, $\alpha_1 = 6$, $\alpha_0 = 19/4$, $\beta_4 = 1$, $\beta_3 = 4$, $\beta_2 = 3$, and $\beta_1 = 47/4$. Therefore, $Y(s)$ cannot be realized as the admittance of a network consisting of one damper, one inerter, and no more than three springs. However, $Y(s)$ is realizable as the admittance of the series connection of a spring k_4 with Fig. 5.19(d) with $b = 1$ kg, $c = 9/16$ Ns/m, $k_1 = 15/8$ N/m, $k_2 = 3/8$ N/m, $k_3 = 21/8$ N/m, and $k_4 = 1/8$ N/m. This example also illustrates that the number of springs may not always be reducable to three.

Example 5.4. For the suspension system in [Papageorgiou and Smith (2006), Fig. 3], $K_{ap}(s)$ is presented in [Papageorgiou and Smith (2006), Eq. (27)], which can guarantee the same value of J_3 (see [Papageorgiou and Smith (2006), Eq. (5)]) as the optimization function $K(s)$ of J_3 when $k_s = 50$ kN/m (intermediate static stiffness range) by the YALMIP method. As a result, the admittance of the suspension strut becomes

$$Y(s) = \frac{5000}{s} + K_{ap}(s) = \frac{28901000s^2 + 50550000s + 4876200000}{s^3 + 1011s^2 + 97524s}.$$

From [Papageorgiou and Smith (2006), Fig. 12], it is known that $Y(s)$ is realizable with one damper, one inerter, and no more than three springs. By expressing $Y(s)$ in the form of (5.88) with $\alpha_3 = 0$, $\alpha_2 = 28901000$, $\alpha_1 = 50550000$, $\alpha_0 = 4876200000$, $\beta_4 = 0$, $\beta_3 = 1$, $\beta_2 = 1011$, and $\beta_1 = 97524$, one can check that the conditions of Theorem 5.22 (Condition 1) hold, which further illustrates the validity and benefits of the results in this section.

Example 5.5. Consider the quarter car model shown in Fig. 5.22, where the parameters are chosen as $m_s = 250$ kg, $m_u = 35$ kg, and $k_t = 150$ kN/m. The simulation results for the case, when $Y(s)$ is the admittance of the suspension strut consisting of one damper, one inerter, and no more than three springs, are compared with the results for the case when $Y(s)$ is the admittance of the suspension strut consisting of one inerter, one damper, and a finite number of springs, in terms of ride comfort (J_1) and tyre grip performance (J_3). The performance measures J_1 and J_3 defined in [Chen et al. (2012); Smith and Wang (2004)] are as follows:

$$J_1 = 2\pi(V\kappa)^{1/2}||sT_{\hat{z}_r \to \hat{z}_s}||_2, \quad J_3 = 2\pi(V\kappa)^{1/2}\left\|\frac{k_t}{s}T_{\hat{z}_r \to (\hat{z}_u - \hat{z}_r)}\right\|_2,$$

where $T_{\hat{z}_r \to \hat{z}_s}$ denotes the transfer function from z_r to z_s, $T_{\hat{z}_r \to (\hat{z}_u - \hat{z}_r)}$ denotes the transfer function from z_r to $z_u - z_r$, $||\cdot||_2$ denotes the standard H_2 norm, and the values of V and κ are taken as $V = 25$ m/s and $\kappa = 5 \times 10^{-7}$ m^{3}cycle^{-1} (a typical British principal road).

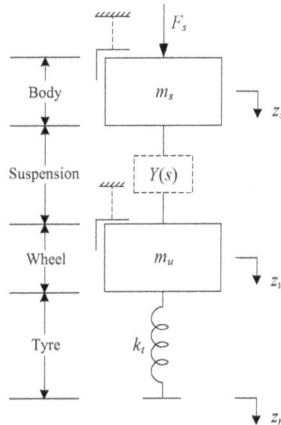

Fig. 5.22 A quarter-car vehicle suspension system model, where m_s denotes the sprung mass, m_u denotes the unsprung mass, k_t denotes the spring stiffness of the tyre, and $Y(s)$ denotes the mechanical admittance of the suspension strut consisting of dampers, springs, and inerters.

The comparison of ride comfort performance is shown in Fig. 5.23, where only slight degradation is observed when restricting the number of springs to three. For example, when $K = 60$ kN/m, the ride comfort performance measure J_1 for the suspension system with one inerter, one damper, and a finite number of springs is 1.5427, and the admittance corresponding to the optimal performance is realizable as the configuration in Fig. 5.2(i) with $c = 2.414$ kNs/m, $k_1 = 55.778$ kN/m, $k_2 = 5.2380 \times 10^5$ kN/m, $k_3 = 19.307$ kN/m, $k_4 = 5.404$ kN/m, and $b = 243.506$ kg. In comparison,

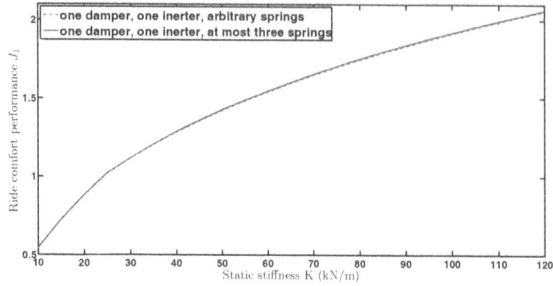

Fig. 5.23　Comparison of ride comfort performances.

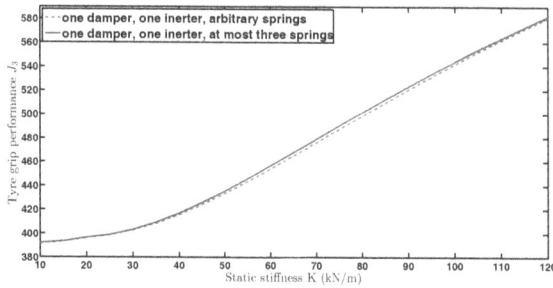

Fig. 5.24　Comparison of tyre grip performances.

the ride comfort performance measure J_1 for the suspension system with one inerter, one damper, and at most three springs is 1.5480 with only 0.341% reduction compared with the case with a finite number of springs, and the admittance corresponding to the optimal performance is realizable as the configuration in Fig. 5.19(c) with $c = 3.046$ kNs/m, $k_1 = 60$ kN/m, $k_2 = 1.132 \times 10^{12}$ kN/m, $k_3 = 22.093$ kN/m, and $b = 221.528$ kg. Since the value of k_2 is relatively large, the spring k_2 can be viewed as a rigid connection and so only two springs are effectively employed.

The comparison of dynamic tyre loads performance is shown in Fig. 5.24. Similar to the ride comfort performance, only slight degradation is observed when restricting the number of springs to three. For example, when $K = 60$ kN/m, the dynamic tyre loads performance measure J_3 for the suspension system with one inerter, one damper, and a finite number of springs is 454.15, and the admittance corresponding to the optimal performance is realizable as the configuration in Fig. 5.2(v), with $c = 2.446$ kNs/m, $k_1 = 343.71$ kN/m, $k_2 = 4.396$ kN/m, $k_3 = 17.566$ kN/m, $k_4 \doteq 56.520$ kN/m, and $b = 231.924$ kg. In comparison, J_3 for the suspension system with one inerter, one damper, and at most three springs is 456.805 with only 0.5845% reduction compared to the case with a finite number of springs, and the admittance corresponding to the optimal performance is realizable as the configuration in Fig. 5.19(c), with $c = 2.960$ kNs/m, $k_1 = 69.005$ kN/m, $k_2 = 459.77$ kN/m, $k_3 = 29.668$ kN/m, and $b = 212.058$ kg.

Therefore, it is concluded that both the comfort performances and the dynamic tyre loads performances are only slightly degraded by reducing the number of springs from four to three.

Chapter 6

Future Outlook

Recently, passive network synthesis has become much more practically essential due to its new motivation in inerter-based vibration control. Although a series of new results on this topic have appeared during the last decade, there are still many problems to be solved in order to make the inerter-based mechanical control approach more effective, especially for the economical synthesis problems of RLC (damper-spring-inerter) networks.

Some of the important problems to be further investigated include the following.

- The realizability problem of a biquadratic impedance as a one-port RLC network should be further investigated. As shown in Chapter 3, it is still unknown whether the least number of elements needed to realize any positive-real biquadratic impedance is seven or eight.
- In addition to biquadratic impedances, the electrical or mechanical realizability problems of other low-degree impedances (resp. admittances) should be further investigated, such as the bicubic impedance synthesis.
- It is essential to establish some alternative synthesis procedures suitable for any positive-real functions, which are more economical than the existing procedures, such as the classic Bott-Duffin synthesis procedure.
- The realizability of n-port resistive networks, when $n > 3$, should be investigated, which is a classical but still challenging problem.

Bibliography

Anderson, B. D. O. and Vongpanitlerd, S. (1973). *Network Analysis and Synthesis: A Modern Systems Theory Approach*, Prentice Hall, NJ.

Baher, H. (1984). *Synthesis of Electrical Networks*, Wiley, New York.

Balabanian, N. (1958). *Network Synthesis*, Prentice-Hall, NJ.

Bar-Lev, A. (1962). On the realization of nonminimum biquadratic driving point functions, *IEEE Transactions on Circuit Theory* **9**, 3, pp. 293–295.

Biorci, G. (1961). Sign matrices and realizability of conductance matrices, *Proceedings of the IEE-Part C: Monographs* **108**, 14, pp. 296–299.

Biorci, G. and Civalleri, P. P. (1961a). On the synthesis of resistive n-port networks, *IRE Transactions on Circuit Theory* **8**, 1, pp. 22–28.

Biorci, G. and Civalleri, P. P. (1961b). Conditions for the realizability of a conductance matrix, *IRE Transactions on Circuit Theory* **8**, 3, pp. 312–317.

Boesch, F. T. and Youla, D. C. (1965). Synthesis of $n + 1$ node resistor n-ports, *IEEE Transactions on Circuit Theory* **12**, 4, pp. 515–520.

Boesch, F. T. (1966). Some aspects of the analysis and synthesis of resistor networks, in *Text of Lectures Presented at NATO Advanced Study Inst. Network and Switching Theory*, pp. 188–204.

Bott, R. and Duffin, R. J. (1949). Impedance synthesis without use of transformers, *Journal of Applied Physics* **20**, pg. 816.

Brown, D. P. and Tokad, Y. (1961). On the synthesis of R networks, *IRE Transactions on Circuit Theory* **8**, 1, pp. 31–39.

Brown, D. P. and Reed, M. B. (1962a). On the sign pattern of network matrices, *Journal of the Franklin Institute* **273**, 3, pp. 179–186.

Brown, D. P. and Reed, M. B. (1962b). Necessary and sufficient conditions for R-graph synthesis, *Journal of the Franklin Institute* **273**, 6, pp. 472–481.

Brune, O. (1931). Synthesis of a finite two-terminal network whose drivingpoint impedance is a prescribed function of frequency, *Journal of Mathematical Physics* **10**, pp. 191–236.

Bruno, J. and Weinberg, L. (1971). Some results on paramount matrices, *IEEE Transactions on Circuit Theory* **18**, 5, pp. 560–562.

Cederbaum, I. (1958a). A generalization of the "non-amplification" property of resistive networks, *IRE Transactions on Circuit Theory* **5**, 3, pg. 224.

Cederbaum, I. (1958b). Condition for the impedance and admittance matrices of *n*-ports without ideal transformers, *Proceedings of the IEE-Part C: Monographs* **105**, 7, pp. 245–251.

Cederbaum, I. (1959). Application of matrix algebra to network theory, *IRE Transactions on Circuit Theory* **6**, 5, pp. 127–137.

Cederbaum, I. (1961). Topological considerations in the realization of resistive *n*-port networks, *IRE Transactions on Circuit Theory* **8**, 3, pp. 324–329.

Cederbaum, I. (1963). Paramount matrices and realization of resistive 3-port networks, *Proceedings of the IEE* **110**, 11, pp. 1960–1964.

Cederbaum, I. (1965). On equivalence of resistive *n*-port networks, *IEEE Transactions on Circuit Theory* **12**, 3, pp. 338–344.

Cederbaum, I. (1984). Some applications of graph theory to network analysis and synthesis, *IEEE Transactions on Circuits and Systems* **31**, 1, pp. 64–68.

Chang, S. (1969). On biquadratic functions with real noninterlaced poles and zeros, *IEEE Transactions on Circuit Theory* **16**, 2, pp. 250–252.

Chen, M. Z. Q. (2007). *Passive Network Synthesis of Restricted Complexity*. Ph.D. thesis, University of Cambridge, Cambridge, UK.

Chen, M. Z. Q. and Smith, M. C. (2008). Electrical and mechanical passive network synthesis, V. D. Blondel, S. P. Boyd and H. Kimura (Eds.), in *Recent Advances in Learning and Control*, Springer-Verlag, New York, **371**, pp. 35–50.

Chen, M. Z. Q., Papageorgiou, C., Scheibe, F., Wang, F.-C. and Smith, M. C. (2009). The missing mechanical circuit element, *IEEE Circuits and Systems Magazine* **9**, 1, pp. 10–26.

Chen, M. Z. Q. and Smith, M. C. (2009a). A note on tests for positive-real functions, *IEEE Transactions on Automatic Control* **54**, 2, pp. 390–393.

Chen, M. Z. Q. and Smith, M. C. (2009b). Restricted complexity network realizations for passive mechanical control, *IEEE Transactions on Automatic Control* **54**, 10, pp. 2290–2301.

Chen, M. Z. Q., Hu, Y. and Du, B. (2012). Suspension performance with one damper and one inerter, in *Proceedings of 24th Chinese Control and Decision Conference* (Taiyuan, China), pp. 3534–3539.

Chen, M. Z. Q., Wang, K., Zou, Y. and Lam, J. (2013a). Realizations of a special class of admittances with one damper and one inerter for mechanical control, *IEEE Transactions on Automatic Control* **58**, 7, pp. 1841–1846, 2013.

Chen, M. Z. Q., Wang, K., Shu, Z. and Li, C. (2013b). Realizations of a special class of admittances with strictly lower complexity than canonical forms, *IEEE Transactions on Circuits and Systems I: Regular Papers* **60**, 9, pp. 2465–2473.

Chen, M. Z. Q., Hu, Y., Huang, L. and Chen, G. (2014a). Influence of inerter on natural frequencies of vibration systems, *Journal of Sound and Vibration* **333**, 7, pp. 1874–1887.

Chen, M. Z. Q., Wang, K. and Chen, G. (2014b). A survey on synthesis of resistive *n*-port networks, in *Proceedings of 21st International Symposium on Mathematical Theory of Networks and Systems (MTNS2014)* (Groningen, The Netherlands), pp. 1562–1569.

Chen, M. Z. Q., Hu, Y., Li, C. and Chen, G. (2014c). Semi-active suspension with semi-active inerter and semi-active damper, in *Proceedings of the 19th IFAC World Congress* (Cape Town, South Africa), pp. 11225–11230.

Chen, M. Z. Q., Hu, Y., Li, C. and Chen, G. (2015a). Performance benefits of using inerter in semiactive suspensions, *IEEE Transactions on Control Systems Technology* **23**, 4, pp. 1571–1577.

Chen, M. Z. Q., Wang, K., Yin, M., Li, C., Zuo, Z. and Chen, G. (2015b). Synthesis of n-port resistive networks containing 2n terminals, *International Journal of Circuit Theory and Applications* **43**, pp. 427–437.

Chen, M. Z. Q., Wang, K., Zou, Y. and Chen, G. (2015c). Realization of three-port spring networks with inerter for effective mechanical control, *IEEE Transactions on Automatic Control* **60**, 10, pp. 2722–2727.

Chen, M. Z. Q., Hu, Y., Li, C. and Chen, G. (2016a). Application of semi-active inerter in semi-active suspensions via force tracking, *Journal of Vibration and Acoustics* **138**, 4, pp. 041014.

Chen, M. Z. Q., Wang, K. and Zou, Y. (2016b). A generalized theorem of Reichert for biquadratic minimum functions, *International Journal of Circuit Theory and Applications* **44**, pp. 1840–1858.

Chen, M. Z. Q., Wang, K., Li, C. and Chen, G. (2017). Realization of biquadratic impedances as five-element bridge networks, *IEEE Transactions on Circuits and Systems I: Regular Papers* **64**, 6, pp. 1599–1611.

Chen, M. Z. Q. and Hu, Y. (2019). *Inerter and Its Application in Vibration Control Systems*, Springer, Singapore.

Darlington, S. (1939). Synthesis of reactance 4-poles which produce prescribed insertion loss characteristics, *Journal of Mathematical Physics* **18**, pp. 257–353.

Dong, X., Liu, Y. and Chen, M. Z. Q. (2015). Application of inerter to aircraft landing gear suspension, in *Proceedings of the 34th Chinese Control Conference* (Hangzhou, China), pp. 2066–2071.

Eswaran, C. and Murti, V. G. K. (1973). Transformerless realization of biquadratic driving-point impedance functions, *IEEE Transactions on Circuit Theory* **20**, 3, pp. 314–316.

Foster, R. M. (1961). An open question, *IRE Transactions on Circuit Theory* **8**, 2, pg. 126.

Foster, R. M. (1962). Academic and theoretical aspects of circuit theory, *Proceedings of the IRE* **50**, 5, pp. 866–871.

Foster, R. M. (1963). Minimum biquadratic impedances, *IEEE Transactions on Circuit Theory* **10**, 4, pg. 527.

Gantmacher, F. R. (1980). *The Theory of Matrices*, Vol. II, Chelsea, New York.

Gould, R. (1958). Graphs and vector spaces, *Journal of Mathematical Physics* **37**, 3, pp. 193–214.

Gohberg, I., Lancaster, P. and Rodman, L. (1982). *Matrix Polynomials*, Academic Press.

Guillemin, E. A. (1957). *Synthesis of Passive Networks*, John Wiley & Sons.

Guillemin, E. A. (1959). How to grow your own trees from cut-set or tie-set matrices, *IRE Transactions on Circuit Theory* **6**, 5, pp. 110–126.

Guillemin, E. A. (1960). On the analysis and synthesis of single-element-kind network, *IRE Transactions on Circuit Theory* **7**, 3, pp. 303–312.

Guillemin, E. A. (1961). On the realization of an nth-order G matrix, *IRE Transactions on Circuit Theory* **8**, 3, pp. 318–323.

Halkias, C. C., Cederbaum, I. and Kim, W. H. (1962). Synthesis of resistive n-port networks with $n+1$ nodes, *IRE Transactions on Circuit Theory* **9**, 1, pp. 69–73.

Harary, F. (1969). *Graph Theory*, Addison-Wesley, MA.

Hu, Y., Chen, M. Z. Q. and Shu, Z. (2014). Passive vehicle suspensions employing inerters with multiple performance requirements, *Journal of Sound and Vibration* **333**, 8, pp. 2212–2225.

Hu, Y. and Chen, M. Z. Q. (2015). Performance evaluation for inerter-based dynamic vibration absorbers, *International Journal of Mechanical Sciences* **99**, pp. 297–307.

Hu, Y., Chen, M. Z. Q., Shu, Z. and Huang, L. (2015). Analysis and optimization for inerter-based isolators via fixed-point theory and algebraic solution, *Journal of Sound and Vibration* **346**, pp. 17–36.

Hu, Y. and Chen, M. Z. Q. (2017). Passive structural control with inerters for a floating offshore wind turbine, in *Proceedings of the 36th Chinese Control Conference* (Dalian, China), pp. 9266–9271.

Hu, Y., Chen, M. Z. Q., Xu, S. and Liu, Y. (2017a). Semiactive inerter and its application in adaptive tuned vibration absorbers, *IEEE Transactions on Control Systems Technology* **25**, 1, pp. 294–300.

Hu, Y., Chen, M. Z. Q. and Sun, Y. (2017b). Comfort-oriented vehicle suspension design with skyhook inerter configuration, *Journal of Sound and Vibration* **405**, pp. 34–47.

Hu, Y., Chen, M. Z. Q. and Smith, M. C. (2018a). Natural frequency assignment for mass-chain systems with inerters, *Mechanical Systems and Signal Processing* **108**, pp. 126–139.

Hu, Y., Wang, J., Chen, M. Z. Q., Sun, Y. and Li, Z. (2018b). Load mitigation for a barge-type floating offshore wind turbine via inerter-based passive structural control, *Engineering Structures* **177**, pp. 198–209.

Hughes, T. H. and Smith, M. C. (2012). Algebraic criteria for circuit realisations, K. Hüper and J. Trumpf (Eds.), *Mathematical System Theory*, CreateSpace, Charlotte, pp. 211–228.

Hughes, T. H. (2014). *On the Synthesis of Passive Networks Without Transfromers*. Ph.D. thesis, University of Cambridge, Cambridge, UK.

Hughes, T. H. and Smith, M. C. (2014). On the minimality and uniqueness of the Bott-Duffin realization procedure, *IEEE Transactions on Automatic Control* **59**, 7, pp. 1858–1873.

Hughes, T. H. (2017). Why RLC realizations of certain impedances need many more energy storage elements than expected, *IEEE Transactions on Automatic Control* **62**, 9, pp. 4333–4346.

Jiang, J. Z. (2010). *Passive Electrical and Mechanical Network Synthesis*. Ph.D. thesis, University of Cambridge, Cambridge, UK.

Jiang, J. Z. and Smith, M. C. (2011). Regular positive-real functions and five-element network synthesis for electrical and mechanical networks, *IEEE Transactions on Automatic Control* **56**, 6, pp. 1275–1290.

Jiang, J. Z. and Smith, M. C. (2012). Series-parallel six-element synthesis of biquadratic impedances, *IEEE Transactions on Circuits and Systems I: Regular Papers* **59**, 11, pp. 2543–2554.

Jiang, J. Z. and Zhang, S. Y. (2014). Synthesis of biquadratic impedances with a specific seven-element network, in *Proceedings of 2014 UKACC International Conference on Control* (Loughborough, U.K.), pp. 139–144.

Kalman, R. (2010). Old and new directions of research in system theory, in *Perspectives in Mathematical System Theory, Control, and Signal Processing*, J. C. Willems, S. Hara, Y. Ohta and H. Fujioka (Eds.), Springer-Verlag, New York, **398**, pp. 3–13.

Kalman, R. (2014). How old mathematics and old engineering became new engineering and new system theory, *Lecture in the ECE Distinguished Lecture Series*, Faculty of Applied Science and Engineering, University of Toronto.

Ladenheim, E. L. (1948). *A Synthesis of Biquadratic Impedances*, Master's thesis, Polytechnic Inst. of Brooklyn, N.Y.

Ladenheim, E. L. (1964). Three-reactive five-element biquadratic structures, *IEEE Transactions on Circuit Theory* **11**, 1, pp. 88–97.

Lavaei, J., Babakhani, A., Hajimiri, A. and Doyle, J. C. (2011). Solving largescale hybrid circuit-antenna problems, *IEEE Transactions on Circuits and Systems I: Regular Papers* **58**, 2, pp. 374–387.

Lempel, A. and Cederbaum, I. (1967). Parallel interconnection of n-port networks, *IEEE Transactions on Circuit Theory* **14**, 3, pp. 274–279.

Lin, P. M. (1965). A theorem on equivalent one-port networks, *IEEE Transactions on Circuit Theory* **12**, 4, pp. 619–621.

Lin, S., Oeding, L. and Sturmfels, B. (2011). *Electric Network Synthesis*, UC Berkeley. http://www.acritch.com/media/bass/electric-network-synthesis-v2.pdf. (Last accessed on February 28, 2019).

Liu, Y., Chen, M. Z. Q. and Tian, Y. (2015). Nonlinearities in landing gear model incorporating inerter, in *Proceeding of the 2015 IEEE International Conference on Information and Automation* (Lijiang, China), pp. 696–701.

Loughlin, T. A. and Slepian, P. (1979). A class of nonrealizable paramount matrices, *International Journal of Circuit Theory and Applications* **7**, 1, pp. 1–8.

Lupo, F. J. and Halkias, C. C. (1965). Synthesis of n-port networks on two-tree port-structures, *IEEE Transactions on Circuit Theory* **12**, 4, pp. 571–577.

Lupo, F. J. (1968). The synthesis of transformerless n-port networks on multitree port structures, *IEEE Transactions on Circuit Theory* **15**, 3, pp. 211–220.

Mukhtar, F., Yordanov, H. and Russer, P. (2011). Network model of on-chip antennas, *Advances in Radio Science* **9**, 17, pp. 237–239.

Murti, V. G. K. and Thulasiraman, K. (1967). Parallel connection of n-port networks, *Proceedings of the IEEE* **55**, 7, pp. 1216–1217.

Naidu, M. G. G., Reddy, P. S. and Thulasiraman, K. (1976). $(n + 2)$-node resistive n-port realizability of Y-matrices, *IEEE Transactions on Circuit and Systems* **23**, 5, pp. 254–260.

Nambiar, K. K. (1963). Answer to Foster's open question, *IEEE Transactions on Circuit Theory* **10**, 1, pp. 126–127.

Newcomb, R. W. (1966). *Linear Multiport Synthesis*, McGraw-Hill, New York.

Pantell, R. H. (1954). A new method of driving-point impedance synthesis, *Proceedings of the IRE* **42**, 5, p. 861.

Papageorgiou, C. and Smith, M. C. (2006). Positive real synthesis using matrix inequalities for mechanical networks: Application to vehicle suspension, *IEEE Transactions on Control Systems Technology* **14**, 3, pp. 423–435.

Reddy, P. S., Murti, V. G. K. and Thulasiraman, K. (1970). Realization of modified cut-set matrix and applications, *IEEE Transactions on Circuit Theory* **17**, 4, pp. 475–486.

Reddy, P. S. and Thulasiraman, K. (1972). Synthesis of $(n + 2)$-node resistive n-port networks, *IEEE Transactions on Circuit Theory* **19**, 1, pp. 20–25.

Reichert, M. (1969). Die kanonisch und übertragerfrei realisierbaren Zweipolfunktionen zweiten Grades (Transformerless and canonic realization of biquadratic immittance functions), *Arch. Elek. Übertragung* **23**, pp. 201–208.

Reis, T. and Stykel, T. (2011). Lyapunov balancing for passivity-preserving model reduction of RC circuits, *SIAM Journal on Applied Dynamical Systems* **10**, 1, pp. 1–34.

Reza, F. M. (1954). Synthesis without ideal transformers, *Journal of Applied Physics* **25**, pp. 807–808.

Richards, P. I. (1947). A special class of functions with positive real parts in a half-plane, *Duke Mathematical Journal* **14**, pp. 777–786.

Saeed, K. (2014). Carathéodory-Toeplitz based mathematical methods and their algorithmic applications in biometric image processing, *Applied Numerical Mathematics* **75**, pp. 2–21.

Seshu, S. (1959). Minimal realizations of the biquadratic minimum function, *IRE Transactions on Circuit Theory* **6**, 4, pp. 345–350.

Seshu, S. and Reed, M. B. (1961). *Linear Graphs and Electrical Networks*, Addison-Wesley, Boston.

Slepian, P. and Weinberg, L. (1958). Synthesis applications of paramount and dominant matrices, in *Proceedings of National Electronics Conference* **14**, pp. 611–630.

Smith, M. C. (2002). Synthesis of mechanical networks: the inerter, *IEEE Transactions on Automatic Control* **47**, 10, pp. 1648–1662.

Smith, M. C. and Wang, F.-C. (2004). Performance benefits in passive vehicle suspensions employing inerters, *Vehicle System Dynamics* **42**, 4, pp. 235–257.

Smith, M. C. (2008). Force-controlling mechanical device, Patent 7 316 303.

Steiglitz, K. and Zemanian, A. H. (1962). Sufficient conditions on pole and zero locations for rational positive-real functions, *IRE Transactions on Circuit Theory* **9**, 3, pp. 267–277.

Swaminathan, K. R. and Frisch, I. T. (1965). Necessary conditions for the realizability of n-port resistive networks with more than $n + 1$ nodes, *IEEE Transactions on Circuit Theory* **12**, 4, pp. 520–527.